STILL NO MIRACLES NEEDED

What if we don't need "mirac... problem? What if the technolog... what if we can use those existing tecimorogies to tricity, heat supplies, and energy security?

In a revised and updated edition of his award-winning climate bestseller, *No Miracles Needed*, the world's premier thinker on energy futures and one of the world's 100 most impactful people in the world in 2023, Mark Z. Jacobson reveals how nations, communities, and individuals can solve the climate crisis most effectively, while simultaneously eliminating air pollution and providing energy security. Mark explains how existing technologies can harness, store, and transmit energy from wind, water, and solar sources to ensure reliable electricity and heat supplies. It includes new, cutting-edge technologies, additional new real-life case studies about the solutions, and additional references. Written for everyone who cares about the future of our planet, this book advises individuals, policymakers, communities, and nations about what they can do to solve the problems identified, and the economic, health, and climate benefits of the solutions.

MARK Z. JACOBSON is a professor of Civil and Environmental Engineering and Director of the Atmosphere/Energy Program at Stanford University. He has published seven books and 190 papers. He received the 2018 Judi Friedman Lifetime Achievement Award. In 2022, he was ranked the world's most impactful scientist in the field of Meteorology and Atmospheric Sciences and the sixth-most impactful scientist in the field of Energy among those first publishing past 1985. In 2023, he was named one of *Worth Magazine*'s top 100 people globally "who have made an impact on the world this year." He was an expert witness in the first US climate case to have won at trial, *Held v. Montana*, and the first world climate case to have settled, *Navahine v. Hawaii*. He has appeared on the *Late Show with David Letterman* and cofounded The Solutions Project. His work is the scientific basis of the Green New Deal and 100 percent renewable energy laws worldwide.

PRAISE FOR *NO MIRACLES NEEDED*

"Pollution, climate catastrophe and energy security can all be addressed with [Mark's] simple plan ... This book is a godsend."

Mark Ruffalo

"... a compelling pitch: the world can rapidly get 100% of its energy from renewable sources with, as the title of his new book says, *no miracles needed*."

The Guardian

"Among the most important books you'll ever read ... This book should empower you – and with not a moment to spare!"

Bill McKibben

"... read energy systems expert Mark Jacobson's amazing new book ... and be informed and engaged to help tackle the defining challenge of our time."

Michael Mann, author of *The New Climate War*

"Many people believe or fear that we can't solve the climate crisis, because we just don't have the technologies in hand to do so. This book should lay that fear to rest, once and for all."

Naomi Oreskes, coauthor of *The Big Myth: How American Business Taught Us to Loathe Government and Love the Free Market*

"... a must-read for all who care about the future of our society and our planet, written by the world's premier thinker on energy futures."

Bob Howarth, Cornell University

"... sure to be one of the most important books that you will read this decade."

Peter Strachan, Robert Gordon University

"... everything you need to understand, and to join the fight against, the peril of our time."

Anthony R. Ingraffea, Cornell University

"A tireless and brilliant advocate for the environment, Professor Mark Jacobson's voice must be read, heard, and acted upon – now."

Heidi Hutner, Stony Brook University

STILL NO MIRACLES NEEDED

How Today's Technology Can Save Our Climate and Clean Our Air

Revised and Updated

Mark Z. Jacobson
Stanford University

CAMBRIDGE
UNIVERSITY PRESS

Shaftesbury Road, Cambridge CB2 8EA, United Kingdom

One Liberty Plaza, 20th Floor, New York, NY 10006, USA

477 Williamstown Road, Port Melbourne, VIC 3207, Australia

314–321, 3rd Floor, Plot 3, Splendor Forum, Jasola District Centre,
New Delhi – 110025, India

103 Penang Road, #05–06/07, Visioncrest Commercial, Singapore 238467

Cambridge University Press is part of Cambridge University Press & Assessment,
a department of the University of Cambridge.

We share the University's mission to contribute to society through the pursuit of
education, learning and research at the highest international levels of excellence.

www.cambridge.org
Information on this title: www.cambridge.org/9781009662475

DOI: 10.1017/9781009662482

© Mark Z. Jacobson 2023, 2026

This publication is in copyright. Subject to statutory exception and to the provisions
of relevant collective licensing agreements, no reproduction of any part may take
place without the written permission of Cambridge University Press & Assessment.

When citing this work, please include a reference to the DOI 10.1017/9781009662482

First published 2023 (version 2, March 2023)

Revised and updated edition published 2026

Cover image: Karl Hendon / Moment / Getty Images.

A catalogue record for this publication is available from the British Library

A Cataloging-in-Publication data record for this book is available from the Library of Congress

ISBN 978-1-009-66247-5 Paperback

Cambridge University Press & Assessment has no responsibility for the persistence
or accuracy of URLs for external or third-party internet websites referred to in this
publication and does not guarantee that any content on such websites is, or will
remain, accurate or appropriate.

For EU product safety concerns, contact us at Calle de José Abascal,
56, 1°, 28003 Madrid, Spain, or email eugpsr@cambridge.org

To those who dream of a world with both clean air and clean, renewable energy

Contents

List of Figures	*page* viii
Foreword	ix
Preface	xi
1 What Problems Are We Trying to Solve?	1
2 WWS Solutions for Electricity Generation	19
3 WWS Solutions for Electricity Storage	36
4 WWS Solutions for Transportation	51
5 WWS Solutions for Buildings	84
6 WWS Solutions for Industry	114
7 Solutions for Nonenergy Emissions	132
8 What Doesn't Help	138
9 Electricity Grids	205
10 Photovoltaics and Solar Radiation	225
11 Onshore and Offshore Wind Energy	246
12 Steps in Developing 100 Percent WWS Roadmaps	273
13 Keeping the Grid Stable with 100 Percent WWS	291
14 Timeline and Policies Needed to Transition	329
15 My Journey	351
References	393
Index	422

Figures

1.1	Estimated primary contributors to net observed global warming from 1750 to 2024.	page 6
2.1	Main generation, transmission, storage, and use components of a 100 percent WWS system to power the world for all purposes.	20
8.1	Low and high estimates of total 100-year-averaged CO_2-equivalent emissions from several electricity-producing technologies.	147
9.1	Production of single-phase alternating current (AC) by rotating a magnet in the presence of stationary wire coils.	211
9.2	Nikola Tesla in 1890 at age 34.	216
10.1	Diagram showing the substitution of a phosphorus (P) atom (left) and boron (B) atom (right) for a silicon (Si) atom in the lattice of a PV cell.	232
13.1	2022 percentages of total electricity generated by wind–water–solar (WWS) sources for the 15 countries or territories with the highest percentage of the total electricity they generated from WWS.	294

Foreword

More than any other single person, Mark Jacobson has been consistently and powerfully making the case for clean energy. For decades now, he's patiently assembled the data to show that the sun can replace combustion as the source of the power that humans require. We can catch its rays directly on photovoltaic arrays, and can take advantage of the fact that it differentially heats the Earth, producing the winds that turn those turbines. If you think about it, even hydropower relies on the sun, to evaporate water and move it uphill so it can fall as rain behind a dam.

When Jacobson began his work, we thought about these things as "alternative energy." But the last two or three years have made it clear that they're now ready for prime time – the vast majority of new generating capacity around the world comes from solar and wind power. I visited Jacobson at his solar-powered California home in 2024, mostly for the pleasure of listening to him gloat, just a little: His home state that year crossed some tipping point, and all of a sudden it was generating most of its power renewably for long stretches of the day; at night, batteries that had been soaking up excess sunshine all afternoon were often the biggest source of supply to the grid. It was what he had long predicted would happen – and not just in California. Texas, despite being the headquarters of hydrocarbons, was installing wind, sun, and batteries faster than any other state: The economics were just too good. Pakistanis put up solar panels that were the equivalent of a third of their national electric grid in 2024; China was fast becoming the world's first "electro-state."

All this, of course, has triggered the poor souls in Big Oil, who know that their product – energy – can now be delivered more cleanly and less expensively. That's why they bankrolled the Trump campaign – and in turn, that's why the new administration is doing everything it can think of to derail the clean energy revolution. That's a cruel and stupid decision, because in these same few years the planet's temperature is spiking dramatically; the best way to limit the damage is the rapid transition to the energy sources Jacobson so ably describes in this book. I do not mean to minimize

FOREWORD

Trump's power to cause trouble, but the economic case for clean energy is so strong that it will eventually prevail: Jacobson's vision is clearly the human future. The only question is how quickly we can get there.

I think this book is part of the answer to that question. It should inspire activists and ordinary citizens to do far more – to push for fast action at the state and local level, even if Washington, DC, is off limits for the moment. Above all, we need to get the essential message of this book out into the information bloodstream: There's nothing "alternative" about power from the sun and wind; we have the technology now, and we'd be fools not to use it.

Bill McKibben

Preface

During March, 2023, the first edition of this book, *No Miracles Needed*, was published. The premise of the book was we had 95 percent of the technologies needed to solve the air pollution, climate, and energy-security problems the world faces and knew how to create the rest. As such, we did not need *miracle* technologies to solve these three problems, contrary to what at least one well-known technologist and many pundits have claimed and what several industries have pushed for. Fast forward three years, and what has changed? The useful technologies we need – clean, renewable electricity and heat generators, energy storage equipment, and electric devices, machines, and vehicles – have been deployed in larger numbers, become more efficient, and decreased in cost. An economic (as opposed to political) path also exists for them to solve 80 percent of the three problems by 2030 and 100 percent by 2035 to 2050 at low cost while creating jobs and using similar or less land than current energy. Meanwhile, proposed *miracle* technologies, now largely promoted by fossil-fuel interests, agricultural interests, and nuclear interests, have done nothing but delay a solution and simultaneously increase air pollution, carbon dioxide emissions, fossil-fuel mining, fossil-fuel infrastructure, and/or costs. This new edition, which reexamines what we need to solve the three major problems stated, concludes that, fortunately, there are *Still No Miracles Needed* to reach a solution.

This book represents the culmination of my research since 1989 to understand and solve large-scale air-pollution, climate, and energy-security problems through clean, renewable energy systems. My trek started in 1978, at age 13, when I experienced dangerous smog for the first time and set a goal for myself to solve the air-pollution problem. Today, outdoor and indoor air pollution kills about 7.4 million people and causes billions more illnesses annually, making air-pollution the world's second-leading cause of death. In my later teens, in the early 1980s, I began learning about and hoping to help solve global warming as well. Yet decades later, in 2025, the globe continues to warm; glaciers and sea ice continue to melt; sea levels continue to rise; droughts and floods continue to increase; hurricanes and wildfires continue to intensify; air-pollution- and heat-related deaths continue to increase; malaria and dengue fever continue to spread; agricultural losses and famine continue to grow; and climate migration, species extinction, and coral reef damage continue to rise. Starting in the mid 2000s, I sought to address a third problem, energy insecurity. One type of energy insecurity arises because fossil fuels are limited resources. When they run out, economic, social, and political instability will ensue. Another type of insecurity arises because many countries rely on other countries for fuel, putting the receiving countries at risk of supply disruption. All three problems – air pollution, global warming, and energy insecurity – require an immediate and drastic solution.

The good news is that in recent years, the world has seen a fantastic acceleration in the growth of and international support for electrification and clean, renewable energy and storage. For example, the production of wind turbines; solar photovoltaics; batteries; battery-electric cars, trucks, buses, boats, ferries, and small planes; electric heat pumps; electric-induction cooktops; and of hydrogen from electricity has escalated rapidly. One reason is that the costs of these technologies have decreased as their penetration has increased. Another reason is that policies favoring these technologies have been implemented in more and more countries.

In equally good news, in 2023 and 2024, China, the country with the world's largest emissions of air pollutants and greenhouse gases, built by far the most wind turbines, solar photovoltaic systems, and battery-electric vehicles in the world. China's active participation in a solution is essential if nations hope to tackle the climate, air-pollution, and energy-security problems the world faces. Most other countries have also made progress, some incredibly. In fact, there are at least 10 countries that now provide 99.5 to 100 percent of their annually averaged electricity from just wind, solar, geothermal, and hydro sources. One US state, South Dakota, also produced 118 percent of the electricity it consumed, for the full year 2024, with just wind, hydro, and a tiny bit of solar. Ten other states supplied more than 50 percent of what they consumed from these technologies.

Yet, both population and per-person demand for energy has risen, and such increases are a challenge to a future solution. In addition, whereas significant progress has been made in the electricity sector, less progress has been made in the transportation, building, and industrial sectors. What is really holding us back from accelerating a transition across all sectors and countries? Among the barriers is the competition among solutions. The fossil-fuel, agricultural, and nuclear industries, in particular, are invested in the current energy infrastructure and want to keep their share of investment even though their technologies, going forward, are either unhelpful or damaging to the goals of zero air pollution, zero global warming, and zero energy insecurity.

Specifically, the fossil-fuel industry is trying to reinvent itself with four *miracle* technologies that cause more harm than good: carbon capture, synthetic direct air carbon capture, blue hydrogen, and nonhydrogen electro-fuels. These technologies all involve some type of carbon removal from the air or from an exhaust stream and require the creation and use of new, otherwise unneeded, electricity, equipment, and carbon dioxide pipelines. Even in the best case of using new renewable electricity to power them, these four processes always reduce less carbon dioxide than using the same renewable electricity to replace a fossil-fuel source. Further, replacing a fossil source with renewable electricity eliminates the air pollution, mining, and infrastructure from the source. Using the same renewable electricity for carbon capture, and so on, does none of that. Thus, these four technologies only increase carbon dioxide, air pollution, fossil-fuel mining, and fossil-fuel infrastructure because they tie up limited renewable electricity sources.

Similarly, the agriculture industry lobbies for bioenergy: solid biomass, liquid biofuels, and gaseous biogas. The nuclear industry lobbies for small and large nuclear reactors. These *miracle* technologies similarly increase carbon dioxide, air pollution, land requirements (in the case of bioenergy), water requirements, and energy-security problems (in the case of nuclear), relative to clean, renewable electricity.

Indeed, we still do not need such *miracle* technologies to solve air-pollution, climate, and energy-security problems. Instead, the solution is to transition combustion-based energy in all energy sectors to electricity and heat powered by 100 percent clean, renewable **wind, water, and solar (WWS)** and storage, while simultaneously eliminating nonenergy emissions. Given the limited time (until 2030 for an 80 percent transition and until 2035 to 2050 for a 100 percent transition) and limited funding available to implement a solution, it is essential to focus on known, effective solutions that can be deployed rapidly. Money spent on less-useful options causes more health, climate, and energy-insecurity damage.

The main idea behind the WWS solution comes from the fact that problems related to air pollution and climate arise from the same cause: combustion of fossil fuels, bioenergy fuels, and open biomass.

Fossil fuels include coal, oil, fossil gas (also called natural gas or thermogenic gas), and all their derivatives, such as gasoline, diesel, kerosene, jet fuel, and bunker fuel. Fossil fuels were all produced, over millions of years, by the breakdown of algae, phytoplankton, plants, and animals under high pressure and temperature, after they sunk deep into the Earth under layers of sand, silt, and rock. Whereas coal formed primarily from the compression of land-based plants and animals that died and fell into bogs and coastal swamps, oil and fossil gas formed from the compression of algae, phytoplankton, and sea life that fell to the ocean floor.

Bioenergy fuels include liquid biofuels, such as ethanol, biodiesel, and bio jet fuel, for transportation; solid biomass, such as wood, wood pellets, dung, and vegetation, for electricity generation and heat; and biogenic gas for electricity generation, heat, and transportation. Biofuels generally originate from crops, such as corn or soy. Biomass originates from any type of wood or plant material. Biogenic gas is produced by methanogenic bacteria that break down organic matter in places without oxygen, such as in marshes, bogs, and landfills. Biogenic gas is similar to fossil gas in that both contain substantial amounts of methane; however, biogenic and fossil gas are produced in very different ways.

Open biomass includes forests, woodland, grassland, savannah, agricultural crops, and agricultural residues. Burning any of these lands to clear them leads to pollution that affects health and climate. Open-biomass burning for land clearing is a nonenergy **anthropogenic** (human-caused) source of gases and particles that affect health and climate.

To solve air-pollution, global warming, and energy-security problems, it is necessary to move away from combustion by electrifying and providing direct heat without combustion. For the electricity and heat to remain clean and available for millennia to come while not creating other risks, they need to originate from clean, renewable and sustainable sources, namely, WWS.

WWS includes energy from **wind** (onshore and offshore wind electricity), **water** (hydroelectricity, tidal and ocean-current electricity, wave electricity, geothermal electricity, and geothermal heat), and **sunlight** (solar photovoltaic [PV] electricity, concentrated solar power [CSP] electricity and heat, and direct solar heat). WWS electricity and heat can power all current energy sectors, which include the electricity, transportation, building-heating and -cooling, industrial, agriculture/forestry/fishing, and military sectors. Worldwide, human-designed energy systems cause about 90 percent of anthropogenic air pollution and 75–80 percent of anthropogenic greenhouse gas emissions. This book discusses how to eliminate such emissions, but it also describes methods of eliminating nonenergy anthropogenic emissions, such as open-biomass burning for land clearing, that damage air quality and warm the planet.

PREFACE

Today, not only do we have 95–97 percent of the technologies we need to transition to WWS, we also know how to create the rest: primarily long-distance aircraft and ships, powered by hydrogen-fuel-cell electricity. As such, we do not need miracle technologies to solve these problems. We need the collective willpower of people around the world.

Why 100 percent clean, renewable energy and storage for everything? Why not 50 percent, 80 percent, or 99 percent? First, the health plus climate damage of every bit of pollution that we allow to remain in the air is so enormous that it is important both morally and economically to eliminate 100 percent of emissions. Second, 99 percent is not an ambitious goal to shoot for. Did Magellan aspire to circumnavigate 99 percent of the Earth? Did the Apollo 11 crew aspire to reach 99 percent of its way to the moon? No. The goal is 100 percent because that is the best society can do and will result in the healthiest air and most stable climate possible for future generations. Societies often strive for the best and safest.

Can we reach the goal of 100 percent WWS across all energy sectors and eliminate nonenergy emissions at that speed? This book examines this question and the scientific data and studies that say we can. It concludes that a transition among all energy and nonenergy sectors worldwide is economically possible with technology that almost all exists. The main obstacles are social and political.

This book is for lay readers concerned about the massive air-pollution, climate, and energy-security problems the world faces. To summarize, it discusses why no miracle technologies are needed to solve these problems in the short time we have left to do so. The solution is to use existing and known technologies to harness, store, and transmit energy extracted from the wind, the water, and the sun and to ensure reliable electricity and heat supplies worldwide. The book also discusses what technologies are not helpful or needed but are being pursued vigorously. The book gives information about what individuals, communities, and nations can do to solve the problems, as well as the cost, health, climate, and land benefits of the solution. Finally, the book discusses my personal journey into developing plans for transitioning countries, states, and cities to 100 percent WWS. Those plans, plus my fortunate meeting with several amazing, passionate people, led to the founding of a nonprofit organization, The Solutions Project. Armed with the plans, The Solutions Project then catalyzed a mass political movement, which led to the spread of information and, ultimately, the proposal and enactment of many 100 percent WWS policies and laws in countries, states, and cities and the commitment by over 400 international corporations to transition. One such proposed law arising from the plans was the Green New Deal.

1

WHAT PROBLEMS ARE WE TRYING TO SOLVE?

Why do we want to transition all of our energy to clean, renewable energy? Why don't we just continue burning fossil fuels until they run out, which may be in 50 to 150 years? For three major reasons. Namely, fossil fuels today cause massive air-pollution health damage, climate damage, and risks to the world's energy security. These three problems, which have the same root cause, require immediate and drastic solutions. The longer we wait to solve these problems, the more the damage accumulates. This chapter examines each problem, in turn.

1.1 THE AIR-POLLUTION TRAGEDY

Today, air pollution is the second-leading cause of human death and illness worldwide. It also kills and injures animals; impedes visibility; and harms plants, trees, crops, structures, tires, and art. Because air pollution causes such enormous loss and cost, controlling it is one of the greatest challenges of our time.

What is air pollution? **Air pollution** occurs when gases or aerosol particles in the air build up in concentration sufficiently high to cause direct or indirect damage to humans, plants, animals, other life forms, ecosystems, structures, or works of art.

What are gases and aerosol particles? A **gas** is a group of atoms or molecules that are not bonded to each other. Whereas a liquid occupies a fixed volume and a solid has a fixed shape, a gas is unconfined and freely expands with no fixed volume or shape.

An **aerosol particle** consists of 15 or more gas atoms or molecules, suspended in the air, that have bonded together and changed phase to become a liquid or solid. An aerosol particle can contain one chemical or a mixture of many different chemicals. An **aerosol** is an ensemble of aerosol particles.[1] Aerosol particles are distinguished from cloud drops, drizzle drops, raindrops, ice crystals, snowflakes, and hailstones in that the latter all start as an aerosol particle but accumulate far more water in them than the former.

1 WHAT PROBLEMS ARE WE TRYING TO SOLVE?

Gases and aerosol particles may be emitted into the air naturally or by humans (**anthropogenically**). They may also be produced chemically in the air from other gases or aerosol particles. Natural air-pollution problems on the Earth are as old as the planet itself. Volcanoes, natural fires, lightning, desert dust, sea spray, plant debris, pollen, spores, viruses, bacteria, and bacterial and animal metabolism have all contributed to natural air pollution.

Humans first emitted air pollutants when we burned wood for heating and cooking. Today, anthropogenic air pollution arises primarily from the burning of fossil fuels and bioenergy fuels used for energy, and from the burning of open biomass for land clearing or ritual, or due to arson or carelessness. Air pollutants also arise from the release of gases and/or particles to the air, such as from industrial processes or leaks, brake-pad wear, tire wear, and evaporation of solvents and pesticides, for example.

The main **fossil fuels** burned today are coal, fossil gas, and crude oil byproducts. Crude oil is refined into multiple products, including gasoline, diesel, kerosene, heating oil, naphtha, liquefied petroleum gas, jet fuel, and bunker fuel. **Bioenergy** fuels burned are either solid fuels, such as wood, vegetation, or dung (biomass); liquid fuels, such as methanol, ethanol, or biodiesel (biofuels); or gaseous fuels, such as landfill gas (biogas), produced by bacteria breaking down dead plant and animal material in the absence of oxygen. **Open biomass** includes forests, woodland, grassland, savannah, and agricultural residues. Anthropogenic emissions have contributed not only to indoor and outdoor air pollution, but also to acid rain, the Antarctic ozone hole, global stratospheric ozone loss, and global warming.

About 61 million people die each year from all causes worldwide. Air pollution causes about 7.4 million (12.1 percent) of the deaths, making it the second-leading cause of death after heart disease.[2,3] Air pollution also cause billions of illnesses each year. Of the annual air-pollution deaths, about 4.2 million are due to outdoor air pollution[3] and about 3.2 million are due to indoor air pollution.[2] Indoor air pollution arises because 2.6 billion people burn solid fuels (wood, dung, crop waste, coal) and kerosene indoors for cooking and heating.

The deaths and illnesses arise when air-pollution particles (mostly) and gases trigger or exacerbate heart disease, stroke, chronic obstruction pulmonary disease (chronic bronchitis and emphysema), lower respiratory tract infection (flu, bronchitis, and pneumonia), lung cancer, and asthma.

Almost half of all pneumonia deaths worldwide among children aged five and younger are due to air pollution.[2] Many children who die live in homes in which solid fuels or kerosene is burned for home heating and cooking. Their little lungs absorb a high concentration of aerosol particles in the air that result from fuel burning. They die of pneumonia because their immune systems weaken due to the assault of air pollutants on their respiratory systems. Most of the casualties are in developing countries, where indoor burning often still occurs on a large scale. These deaths and illnesses not only devastate families but also incur tremendous cost. The worldwide cost of all air-pollution death and illness may exceed $30 trillion per year today.[4]

Transition highlight
Of the 7.4 million deaths from air pollution worldwide each year, China and India absorb the brunt, with a combined total of 3.6 million deaths per year (49 percent of the total). Nigeria, Pakistan, Indonesia, Bangladesh, the Philippines, and Russia each suffer more than 100,000 air-pollution deaths per year. The highest per capita air-pollution death rates are in North Korea, Georgia, Chad, Nigeria, Bosnia and Herzegovina, and Somalia, respectively.

Of all air-pollution deaths, about 90 percent are due to air-pollution particles; the rest are due to air-pollution gases, primarily ozone. Around half the mass of particles in the air worldwide is in natural particles. However, natural particles are mostly large and thus don't penetrate deep into people's lungs or cause many air-pollution deaths. Conversely, combustion particles, which are almost all anthropogenic, are usually small and penetrate deep into people's lungs. Most combustion particles are also emitted in densely populated areas, so people breathe in these particles. As a result, about 90 percent of air-pollution deaths are due to energy, including all indoor air-pollution deaths.[5] Most of the remaining 10 percent of deaths are due to anthropogenic open biomass burning to clear land,[6] road dust, construction dust, natural wildfires, soil dust, sea spray, volcanic gases and particles, pollen, spores, bacteria, and viruses.[5]

Because combustion during energy production is the world's major source of air pollution, changing the world's energy infrastructure to eliminate combustion will largely eliminate air-pollution death and illness worldwide. This goal can be accomplished by transitioning to 100 percent clean, renewable energy and storage for everything.

1.2 GLOBAL WARMING

1.2.1 THE NATURAL GREENHOUSE EFFECT

Global warming is the human-caused increase in the average temperature of the Earth's lower atmosphere since the Industrial Revolution above and beyond the average temperature caused by the natural greenhouse effect.

The **natural greenhouse effect** is the Earth's average temperature with a pre–Industrial-Revolution atmosphere containing natural greenhouse gases, minus its temperature with no atmosphere. The natural greenhouse effect arose due to the buildup of natural greenhouse gases in the atmosphere since the formation of the Earth. **Greenhouse gases** are gases that are mostly transparent to sunlight but that absorb some of the heat emitted by the surface of the Earth. All objects in the universe, including the Earth, emit heat.

The Earth has three main sources of heat. The first and, by far, the most important, is sunlight, also called **solar radiation**. The Earth absorbs sunlight and converts it to heat, also called **infrared radiation**. About 99.97 percent of the heat emitted by the surface of the Earth originates from sunlight. The remaining 0.03 percent of heat originates from the interior of the Earth, from two sources, each in relatively equal proportions. One is heat left over from the formation of the Earth, called **primordial heat**. Due to gravitational compression of the Earth's interior during its formation, and despite heat loss over time, the temperature at the center of the Earth is still about 4,300 degrees Celsius. This heat transfers slowly to the surface of the Earth by **conduction**, which is the process by which molecules transfer energy to each other when they collide. Primordial heat also gets to the surface by volcanic activity. The other source of interior heat is heat released during the decay of radioactive elements in the Earth's interior. The main elements that decay include uranium, thorium, and a small fraction of potassium. The decay products of these elements decay further as well. The resulting heat transfers slowly to the surface, also by conduction.

Greenhouse gases in the Earth's atmosphere are transparent to sunlight, allowing the sunlight to penetrate to the Earth's surface. However, the same gases trap a portion of the Earth's outgoing heat, warming the ground and air near the ground. The more greenhouse gases present, the greater the trapping of heat and warming of the near-surface air. When the greenhouse gases are natural, the resulting warming is called the natural greenhouse effect.

The primary natural greenhouse gases in the Earth's atmosphere are water vapor, carbon dioxide (CO_2), ozone, nitrous oxide, methane, and oxygen gas. Oxygen gas is a weak greenhouse gas, but it is so abundant in the air (20.95 percent of all air molecules) that it has a nontrivial natural warming impact. Nitrogen gas, which comprises 78.08 percent of the molecules in the Earth's atmosphere, is not a greenhouse gas.

If the Earth had no atmosphere, thus no natural greenhouse effect, its average surface temperature would be about minus 18 degrees Celsius

1.2 GLOBAL WARMING

(zero degrees Celsius is the freezing temperature of water). At minus 18 degrees Celsius, little life would evolve on Earth's surface.

During Earth's 4.6-billion-year history, several processes released to the air all of the Earth's natural greenhouse gases, except for ozone. These processes included emissions (through volcanoes, fumaroles, and geysers) of greenhouse gases (water vapor, carbon dioxide, and methane) from the Earth's interior, bacterial metabolism (producing methane, carbon dioxide, nitrous oxide, and water vapor), bacterial photosynthesis (producing oxygen), and green-plant photosynthesis (producing oxygen). Ultraviolet sunlight cooked some oxygen to produce ozone, most of which formed high above the ground, in what is now called the stratospheric ozone layer. The formation of the ozone layer was critical for protecting the surface of the Earth from harmful ultraviolet sunlight, permitting life to move from underwater and underground to above the ground.

Natural greenhouse gases raised the temperature of the Earth substantially compared with the Earth without an atmosphere, permitting life to flourish on the Earth. Just before the start of the Industrial Revolution, around 1760, Earth's average temperature was about 15 degrees Celsius. That is 33 degrees Celsius higher than Earth's temperature without greenhouse gases (minus 18 degrees Celsius). This warming was due to the natural greenhouse effect. Of this temperature rise, about 66 percent was due to water vapor, about 25 percent was due to background carbon dioxide, and about 6.2 percent was due to background ozone, most of which is in the upper atmosphere.[6]

1.2.2 GLOBAL WARMING

Global warming is the rise in the Earth's globally averaged ground and near-surface air temperature above and beyond that due to the natural greenhouse effect, owing to human activity. The Earth's average global warming in the period 2011–2020 compared with the period 1850–1900 was about 1.09 degrees Celsius.[7] In 2023, globally averaged temperature rose to 1.36 or 1.48 degrees Celsius above the 1850–1900 mean, depending on the data set used.[8,9] Temperatures in 2024 were even higher, closer to 1.5 degrees Celsius above the 1850–1900 mean. Since these values are averages, some places have warmed more, whereas others have warmed less or cooled. For example, the Arctic has warmed by over 5 degrees Celsius. Many other high-latitude locations (parts of Canada, Northern Europe, and Russia) have warmed by 2–5 degrees Celsius. The North Atlantic Ocean has cooled slightly.

1 WHAT PROBLEMS ARE WE TRYING TO SOLVE?

1.2.3 CAUSES OF GLOBAL WARMING

Figure 1.1 reflects an approximate global warming of 1.4 degrees Celsius since 1750. Global warming is the net effect of four major warming processes partially offset by one major cooling process (Figure 1.1). The four major warming processes are anthropogenic greenhouse gas emissions, anthropogenic warming particle emissions, anthropogenic heat emissions, and the urban heat-island effect. The cooling process is anthropogenic cooling particle emissions.

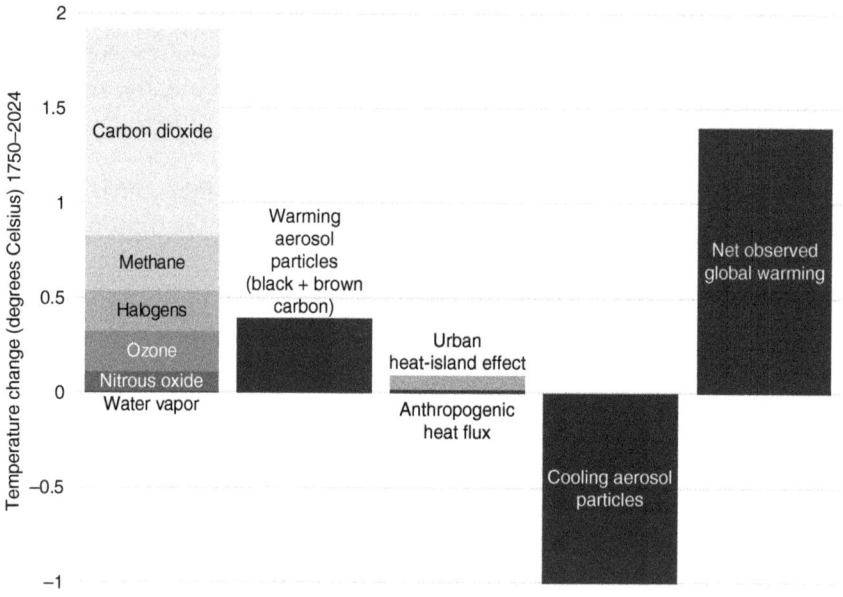

Figure 1.1 Estimated primary contributors to net observed global warming from 1750 to 2024.

Net observed warming is the sum of gross warming minus cooling. Gross warming is the warming due to greenhouse gases (left column), warming aerosol particles (second column), and the urban heat-island effect plus anthropogenic heat (third column). Cooling is due to cooling aerosol particles. Warming aerosol particles include black and brown carbon from fossil-fuel burning, biofuel burning, and open biomass burning. Cooling aerosol particle components include sulfate, nitrate, chloride, ammonium, sodium, potassium, calcium, magnesium, nonbrown organic carbon, and water. Of the gross warming (warming before cooling is subtracted out), about 45.7 percent is due to carbon dioxide, 16.3 percent is due to black plus brown carbon, 12 percent is due to methane, 9 percent is due to halogens, 8.8 percent is due to ozone, 4.3 percent is due to nitrous oxide, 3 percent is due to the urban heat-island effect, 0.7 percent is due to anthropogenic heat flux, and 0.23 percent is due to anthropogenic water vapor (updated from original source).[10]

1.2 GLOBAL WARMING

1.2.3.1 ANTHROPOGENIC GREENHOUSE GAS EMISSIONS

The primary anthropogenic greenhouse gases contributing to global warming are carbon dioxide, methane, halogens, ozone, nitrous oxide, and anthropogenic water vapor.

The primary anthropogenic sources of **carbon dioxide** today are fossil-fuel combustion, bioenergy combustion, open biomass burning to clear land, and chemical reaction during, for example, cement manufacturing, steel production, and silicon extraction. Owing to all of these sources, carbon dioxide in the air has increased from about 275 parts per million (ppm) to 429 ppm, or by 56 percent, between 1750 and 2025. One part per million of carbon dioxide means that for every one million molecules of total air, one molecule is carbon dioxide. Carbon dioxide has been increasing in the air, not only due to its emissions from human activity, but also because it stays in the air for a long time. The major removal mechanisms of carbon dioxide from the air are its dissolution into the oceans and other water bodies, and green-plant photosynthesis (the conversion of carbon dioxide and water vapor into oxygen and cell material by plants, trees, algae, and photosynthetic bacteria). However, these sinks remove the vast amounts of carbon dioxide in the air very slowly, over many decades.

The primary anthropogenic sources of **methane** are combustion of fossil gas, coal, oil, and bioenergy; combustion of open biomass; leakage to the air during fossil-gas, coal, and oil mining; and leakage of methane produced by bacteria to the air from landfills, rice paddies, livestock, and manure. Methane is removed from the air primarily by chemical reaction in the air itself and by bacterial metabolism at the surface of the Earth.

Halogens are a set of synthetic chemicals whose main uses are as refrigerants, solvents, degreasing agents, blowing agents, fire extinguishants, and fumigants. The first halogen was invented in 1928. Halogens enter the air when appliances or tubes sealing them in liquid form leak or are drained, and the liquid evaporates. Most halogens are **halocarbons**, which are chemicals that contain carbon and possibly hydrogen, but also either chlorine, bromine, fluorine, or iodine. The main types of halocarbons are the following: **Chlorofluorocarbons** (CFCs) are halocarbons containing carbon, chlorine, and fluorine. **Halons** are halocarbons containing carbon and bromine. **Perfluorocarbons** are halocarbons containing carbon and fluorine. **Hydrofluorocarbons** are halocarbons containing carbon, fluorine, and hydrogen. Some halogens, such as sulfur hexafluoride, have no carbon, so are not halocarbons. Because chlorofluorocarbons and halons contain stratospheric ozone-destroying chlorine and bromine, most countries outlawed them through international agreement starting with the 1987 Montreal Protocol. Hydrofluorocarbons and perfluorocarbons were developed as ozone-layer-friendly replacements. However, because many of them are greenhouse gases with long lifetimes in the air, such chemicals, while not directly damaging to the ozone layer, have the unintended consequence of enhancing global warming.

Ozone is the only greenhouse gas with no emission source. It forms chemically in the air. About 90 percent of ozone resides in the upper atmosphere (stratosphere), and the rest, in the lower atmosphere (troposphere). The troposphere is the layer of air between the ground and 8 kilometers above sea level at the north and south poles, and between the ground and 18 kilometers above sea level at the equator. The stratosphere is the layer of air just above the troposphere and extends to about 48 kilometers above sea level. Because of the substantial abundance of ozone in the stratosphere, the stratosphere is also called the ozone layer.

In the stratosphere, ozone (which has three oxygen atoms) forms from oxygen gas and ultraviolet light. Oxygen gas is the gas we breathe (when it is near the Earth's surface) and that our life depends on. It is made of two oxygen atoms bonded together. Ozone in the stratosphere forms chemically following the breakdown of oxygen gas by ultraviolet sunlight into two separate oxygen atoms. One atomic oxygen atom then combines with oxygen gas to form ozone. In the troposphere, ozone is produced following the breakdown, by ultraviolet sunlight, of nitrogen dioxide into atomic oxygen and atomic nitrogen. The atomic oxygen then combines with the oxygen gas to form ozone. The nitrogen dioxide is either emitted or comes from a chemical reaction between nitric oxide and certain reactive organic gases. Most emissions of nitric oxide, nitrogen dioxide, and reactive organic gases result from the burning of fossil fuels and bioenergy fuels by humans. Some comes from natural forest burning and bacterial metabolism. Some nitric oxide comes from lightning.

Ozone has a relatively short lifetime in the air. Most of its loss is due to chemical reaction. Since the Industrial Revolution, the mass of tropospheric ozone has increased by about 43 percent due to the worldwide increase in air pollution (the anthropogenic emissions of nitric oxide, nitrogen dioxide, and reactive organic gases). Since the late 1970s, stratospheric ozone has declined by about 5 percent due to the increased abundance of chlorofluorocarbons and halons in the stratosphere. Just as its abundance has grown rapidly in the troposphere due to increased air pollution, ozone's abundance and warming impact in the troposphere can decrease rapidly if air-pollution levels decrease. This is one reason that a strategy to eliminate air pollution can help to decrease global warming as well.

Nitrous oxide (laughing gas) is a colorless gas emitted naturally by bacteria in soils and in the oceans. Because it is long-lived, nitrous oxide can stay in the air for hundreds of years once emitted. It is a powerful greenhouse gas, so it causes substantial warming per molecule during this period. Humans have increased the abundance of nitrous oxide in the air through fertilizer use, burning agricultural waste, creating sewage, planting legumes (plants in the pea family), burning bioenergy fuels, burning forests, jet-fuel burning, nylon manufacturing, and manufacturing aerosol spray cans. Agriculture (fertilizer use, agricultural waste burning, and legume cultivation) is the largest source of human-emitted nitrous oxide today.

Anthropogenic water vapor comes from two main sources. The first is evaporation of water that is used to cool power plants and industrial facilities that run on coal, fossil gas, oil, biomass, or uranium. The second is emission of water vapor during the burning of fuels for energy. Water vapor emitted annually from these sources is only about one eight-thousand-eight-hundredth of the 500 million metric tonnes of water vapor emitted per year from natural sources. Nevertheless, this relatively small anthropogenic emission rate of water vapor contributes a modest 0.23 percent of global warming.[6]

1.2.3.2 ANTHROPOGENIC WARMING PARTICLE EMISSIONS

Dark aerosol particles may contribute more to today's global warming than any chemical aside from carbon dioxide (see Figure 1.1).[11–14] Dark particles, also called **warming particles**, contain primarily black and brown carbon. **Black carbon** is an agglomerate of pure carbon solid spherules attached to each other in an amorphous shape. The source of black carbon is incomplete combustion of diesel, gasoline, jet fuel, bunker fuel, kerosene, fossil gas, coal, crude oil, biogas, solid biomass, and liquid biofuels. Black carbon is often visible to the eye and appears black because it absorbs all wavelengths of sunlight, transmitting none to the eye. Black carbon particles convert the absorbed light to heat, raising the temperature of the particles. The heat is then reradiated to the surrounding air.

Black carbon and greenhouse gases warm the air in different ways from each other. Greenhouse gases are mostly transparent to sunlight. They warm the air by absorbing heat emitted by the surface of the Earth. They then reemit half of that heat upward and half downward, raising the ground and near-surface air temperatures. Black carbon particles, on the other hand, heat the air primarily by absorbing sunlight, converting the sunlight to heat, then reemitting the heat upward and downward. Like greenhouse gases, black carbon particles also absorb and reemit heat itself, but that process is important only at night and when their concentrations are high. For example, citrus farmers in the past burned crude oil in large iron "smudge" pots at night to create a black carbon haze that would trap heat emitted by the surface of the Earth to keep the citrus trees from freezing.

When other aerosol material, such as sulfuric acid, nitric acid, ammonium, brown carbon, and/or water coats the outside of a black carbon particle, the black carbon heats the air two to three times faster than without a coating because more light hits the larger particle, thus more light bends (refracts) into the particle. Inside the particle, this light bounces around until it hits the black carbon core and is absorbed by it.

Black carbon not only warms the air, but it also evaporates clouds and melts snow. When black carbon enters a cloud, it absorbs sunlight that is bouncing around in the cloud, converts the sunlight to heat, then emits the heat to the cloud, warming and evaporating the cloud. When black carbon falls on snow or sea ice, the black carbon similarly absorbs sunlight, converts the sunlight to heat, then emits the heat to the ice or snow, melting it.

Thus, for four reasons (its strong absorption when pure, its stronger absorption when coated, its ability to evaporate clouds, and its ability to melt snow and sea ice), black carbon is the second-leading cause of global warming after carbon dioxide. In fact, per unit mass in the air, black carbon causes over one million times more warming than does carbon dioxide.[11] However, because black carbon particles last only days to weeks in the air, their concentrations are much lower than are those of carbon dioxide, which stays in the air for decades. Nevertheless, because black carbon is continuously emitted, it always causes a strong warming.

Brown carbon is also a particle component that increases global warming and causes health problems. Whereas brown carbon is generally more abundant than black carbon, brown carbon is much less effective per unit mass at warming than is black carbon. As such, black carbon causes more overall warming than brown carbon does.

Whereas black carbon contains pure carbon, brown carbon contains carbon, hydrogen, and possibly oxygen, nitrogen, and/or other atoms. In other words, brown carbon is a type of **organic carbon** (which is a chemical containing carbon, hydrogen, and other atoms). Not all organic carbon is brown carbon. Brown carbon is the subset of organic carbon that absorbs short (blue) and some medium (green) wavelengths of visible light. The remaining long wavelengths (red) and some of the green wavelengths are mostly transmitted to the viewer's eye, making the particle haze appear brown. The more green light that is transmitted (the less that is absorbed), the more yellow the particles appear. Other organic carbon particles are often white or gray because they don't absorb much or any visible light.

The sources of brown carbon – the combustion of fossil fuels, bioenergy fuels, and open biomass burning – are also the sources of black carbon. However, the relative amount of brown or black carbon from a combustion source depends largely on the temperature of the flame. Hotter flames favor black carbon, whereas cooler flames favor brown carbon. For example, in smoldering biomass (a low-temperature flame), the ratio of brown to black carbon is about eight to one. In diesel combustion (a high-temperature flame), the ratio is about one to one.

Because black and brown carbon particles together cause such a large warming per molecule and have such short lifetimes in the air, reducing their emissions is the fastest way of slowing global warming.[13] Because such particles both cause substantial human death and illness, reducing their emissions not only slows global warming but also immediately improves human health. Thus, two major reasons exist to eliminate black and brown carbon particle emissions: to slow global warming rapidly and to improve human health rapidly.

1.2.3.3 ANTHROPOGENIC HEAT EMISSIONS

Anthropogenic heat emissions are emissions of heat from the use of electricity, friction created by vehicle tires on the road, the combustion of fossil fuels and bioenergy fuels, nuclear reaction, and open biomass burning to clear land. Such heat emissions warm the air directly. Much of the hot air eventually rises, converting the heat energy into **gravitational potential energy**, which is energy embodied in air lifted to a certain height against gravity. Differences in gravitational potential energy between one location and another create winds, which carry with them kinetic energy. Thus, some anthropogenic heat emissions are converted

1 WHAT PROBLEMS ARE WE TRYING TO SOLVE?

to energy in the wind. The increases in both temperature and wind speed due to heat emissions cause liquid water in the oceans and at the ground surface to evaporate. Since water vapor is a greenhouse gas, the production of water vapor accelerates the impact of the original heat emissions.

In sum, most anthropogenic heat is converted to other forms of energy that persist in the air, oceans, and land. Overall, the impact of anthropogenic heat emissions is much less than that of greenhouse gases, which persist for tens to thousands of years and cause greater overall warming than anthropogenic heat. Anthropogenic heat may contribute to about 0.7 percent of gross global warming to date (see Figure 1.1).[6] As such, eliminating fossil-fuel and bioenergy combustion and nuclear reaction, which WWS does, reduces anthropogenic heat emissions.

1.2.3.4 THE URBAN HEAT-ISLAND EFFECT

The urban heat-island effect is the temperature increase in urban areas due to the covering of soil and replacing of vegetation with impervious surfaces, such as concrete and asphalt. Covering surfaces reduces evaporation of water from soil and plants. Because evaporation is a cooling process, eliminating evaporation warms the surface. Built-up areas also have sufficiently different properties of construction materials that they enhance urban warming relative to surrounding vegetated areas. Worldwide, the urban heat-island effect may be responsible for about 3 percent of gross global warming (warming before cooling is subtracted out) (Figure 1.1).

1.2.3.5 COOLING PARTICLE EMISSIONS

Cooling particles are light-colored aerosol particles that cool the Earth's surface by reflecting sunlight to space and by thickening low clouds, which are largely reflective. Cooling particles contain primarily sulfate, nitrate, chloride, ammonium, sodium, potassium, calcium, magnesium, nonbrown organic carbon, and water. Because cooling particles tend to be more soluble in water than are warming particles, cooling particles allow water vapor to condense readily on them, enhancing cloudiness, thereby cooling the climate. Warming particles, on the other hand, tend to heat clouds, helping to burn them off. As with warming particles, cooling particles last only days to weeks in the air and cause major air-pollution health damage. As with warming particles, eliminating cooling particle emissions will improve human health dramatically. However, eliminating cooling particles will also raise global temperatures. This is why a strategy of eliminating all greenhouse gases, warming particles, and cooling particles simultaneously through a transition to WWS is necessary to solve both air-pollution and global-warming problems together.

1.2.4 IMPACTS OF GLOBAL WARMING

Global warming has already caused the world significant financial loss, and the cost is expected to grow to over $30 trillion per year by 2050.[4] Losses arise due to coastline erosion and flooding (from sea-level rise), fishery and coral reef damage, species extinction, illness and death due to heat stress and heat stroke, more malaria and dengue fever, agricultural loss, more famine and drought, more wildfires and air pollution, increased climate migration, and more severe weather and storminess (e.g., more intense hurricanes, tornadoes, and hot spells).

Higher temperatures increase air pollution in cities where the pollution is already severe.[15,16] Higher temperatures and the resulting drier wood and longer dry season also increase the risk, length, and intensity of wildfires, which themselves cause air pollution, loss of life, and structural damage. For example, during November 2018, three major wildfires in California, enhanced by drought and unusually high November temperatures, killed dozens of people, displaced hundreds of thousands more, rendered several thousand people homeless, and produced dangerous levels of air pollution throughout the state for over two weeks.

Similarly, global warming has already caused a lot of damage by increasing hurricane duration, size, wind speed, and storm surge. Global warming has further caused crops to fail in many parts of the world, triggering mass migrations. Such migrations are already occurring from the Middle East and North Africa to Europe, and from Central America to the United States, for example.

1.2.5 STRATEGIES FOR REDUCING AIR POLLUTION AND GLOBAL WARMING TOGETHER

Because all aerosol particles together are the leading cause of air-pollution mortality, reducing both cooling and warming particles is necessary from a public health perspective. However, Figure 1.1 indicates that cooling

1 WHAT PROBLEMS ARE WE TRYING TO SOLVE?

particles cause more cooling than warming particles cause warming globally. As such, if emissions of all warming and cooling particles are eliminated together without eliminating the remaining sources of heat, global warming will worsen. Similarly, since cooling particles mask about 40 percent of gross global warming, eliminating only cooling particles will increase net warming by about 70 percent (Figure 1.1).

One strategy to address global warming and human health simultaneously is to eliminate only warming particles. The downside of this strategy is that it permits most global warming and air pollution to continue. Instead, Figure 1.1 suggests that the best strategy for addressing human health and climate simultaneously is to eliminate greenhouse gases, cooling particles, and warming particles simultaneously. This will also reduce most anthropogenic heat and water vapor emissions.

This book is about how best to implement that strategy – eliminating all anthropogenic emissions of greenhouse gases, warming particles, and cooling particles at the same time. The solution is to transition the world's energy to 100 percent wind, water, and solar plus storage for all energy and to eliminate all nonenergy emissions.

1.3 ENERGY INSECURITY

Energy insecurity is a third major problem that needs to be addressed on a global scale. Several types of energy insecurity are of concern.

1.3.1 ENERGY INSECURITY DUE TO DIMINISHING AVAILABILITY OF FOSSIL FUELS AND URANIUM

One type of energy insecurity is the economic, social, and political instability that results from the long-term depletion of nonrenewable energy supplies. Fossil fuels and uranium are limited resources and will run out at some point. As fossil-fuel supplies dwindle, their prices will rise. Such price increases will first hit people who can least afford them – those with little or no income. Such people will suffer, since they cannot pay to warm their homes sufficiently during the winter, to cool their homes sufficiently during the summer, or for enough vehicle fuel.

Higher energy prices will also increase the cost of food and ultimately lead to economic, social, and political instability. The end result may be chaos and civil war.

A solution to this problem is to transition to an energy system that is sustainable – one in which energy is at less risk of being in long-term short supply. Such a system is one that consists of **clean, renewable energy**, which is energy that is replenished by the wind, the water, and the sun. Solutions that do not solve this problem are fossil-fuel power plants, with or without carbon capture, and almost all nuclear power plants, because they rely on fuels that will disappear over time.

1.3.2 ENERGY INSECURITY DUE TO RELIANCE ON CENTRALIZED POWER PLANTS AND OIL REFINERIES

A second type of energy insecurity is the risk of electricity or fuel loss due to the reliance on large, centralized electric-power plants, oil refineries, and fossil-gas storage facilities. If a city or an island relies on centralized power plants, and if one or more plants or the transmission grid goes down, electricity to a large portion of the city or island may be unavailable for an indefinite period. Such an event can arise from severe weather, a power-plant failure, or terrorism. An accidental fire or act of terrorism at an oil refinery or fossil-gas storage facility can similarly disrupt local and regional oil and gas supplies.

For example, on September 14, 2019, a terrorist attack on two Saudi Arabian oil-processing facilities knocked out the production of five million barrels of oil per day, or 5 percent of the world's – and half of Saudi Arabia's – daily oil production. Oil and gas refineries and storage facilities worldwide are continuously at risk of being attacked, and many become targets during conflict. Whereas decentralized power generation and storage facilities provided by WWS do not decrease the risk of attack to zero, they decrease the risk significantly due to the difficulty in damaging hundreds to thousands of smaller individual units rather than one or two larger ones.

> **Transition highlight**
> On September 18, 2017, Hurricane Maria hit Puerto Rico and knocked out power to its 1.5 million people for almost 11 months. The hurricane toppled 80 percent of the island's utility poles and transmission lines. With ten oil-fired power plants, two fossil-gas power plants, and one coal plant, the island's energy supply was all but wiped out by the loss of transmission. The long delay restoring power to homes and businesses occurred because of the need to rebuild most of the transmission system. A more distributed energy system with rooftop solar photovoltaics, distributed onshore and offshore wind turbines, and local battery storage would have allowed hospitals, fire stations, and homes to maintain at least partial power during the entire blackout period and would have reduced the time required to restore power to most customers. In fact, in

early 2019, the main utility in Puerto Rico proposed to divide the island into eight interconnected microgrids dominated by solar and batteries. If one microgrid goes down, the other seven will still function. On April 11, 2019, Puerto Rico went further by passing a law to go to 100 percent renewable electricity by 2050.

Another problem with large, centralized power plants is that they don't serve the 760 million people worldwide without access to electricity,[17] and they poorly serve another 2.1 billion people who have access to only dirty solid fuels (dung, wood, crop residues, charcoal, and coal) for home cooking and heating.[3] Burning solid fuels fills homes with smoke that causes short- and long-term illness to hundreds of millions of people and death to 3.2 million people worldwide each year.[3] Similarly, centralized power plants cannot provide electricity to remote military bases. Those bases obtain their electricity from diesel transported long distance and used in diesel generators. For example, in 2009, seven liters of diesel fuel were burned during the transport of each liter of diesel used to produce electricity in US military bases in Afghanistan.[18] Many soldiers died during the transport of the fuel.

Because WWS technologies are mostly **distributed** (decentralized) technologies, they can be used in microgrids to provide electricity when it is otherwise unavailable. A **microgrid** is an isolated grid that provides electricity to an individual building, hospital complex, community, military base, or data center, for example. A microgrid may be either far from a larger grid or wired to a larger grid but disconnected from it. A WWS microgrid may consist of any combination of solar PV panels, wind turbines, geothermal electricity generators, batteries, electrolyzers for producing hydrogen, hydrogen storage, hydrogen fuel cells for producing electricity and heat, battery-electric-vehicle chargers, electric heat pumps, and energy-efficient appliances. Electricity in a microgrid may also be used to purify wastewater, desalinate salty water, and/or grow food in a container farm or a greenhouse.[19] When used in a microgrid, WWS can bring electricity to people without it or to a disaster zone. Microgrids can also provide electricity to computer clusters or data centers not connected to the grid.

In sum, a transition to WWS facilitates the creation of microgrids and more distributed energy sources. Both factors reduce the chance that severe weather, power-plant failure, or terrorism will deny people energy. Fossil-fuel power plants, with or without carbon capture, and nuclear power plants do not solve this insecurity problem because these plants are mostly large and centralized. In addition, fossil fuels almost always require fuel to be imported to a region. With a clean, renewable energy microgrid, this problem is eliminated since all energy is produced locally from natural sources, namely, wind, water, and sunlight.

1.3.3 ENERGY INSECURITY DUE TO RELIANCE ON FUEL SUPPLIES SUBJECT TO HUMAN INTERVENTION

A third type of energy insecurity is the risk associated with fuel supplies that can be manipulated or that fluctuate substantially in price. Such risks often arise when one country relies on another country to supply its energy. For example, many countries, particularly island countries, must import coal, oil, and/or fossil gas to run their energy system. Similarly, prior to the 2022 war in Ukraine, over 35 percent of the European Union's fossil gas was imported from Russia. During the war, bans placed on Russian fuel decreased the flow substantially such that, by January 1, 2025, Russia provided only about 2.5 percent of European Union fossil gas. Japan imports over 75 percent of its oil, primarily from the Middle East. Israel imports over 90 percent of its oil, primarily from Azerbaijan, Brazil, and Kazakhstan. Importing fuel not only results in higher fuel prices, but it also creates reliance of one country on another. This reliance may be tested in times of international conflict. In some cases, a country that controls the energy may withhold it through a ban, an embargo, or price manipulation, or just may not be able to supply it anymore. Similarly, fossil-fuel and uranium supplies, even within a country, can be held up due to a labor dispute or civil war.[20]

Fossil-fuel and nuclear power plants are particularly prone to this problem because they rely on fuels that must be supplied continuously, either from across international borders or domestically. In many cases, especially for island countries, the fuels must be transported long distance.

A clean, renewable WWS energy system built within a country avoids this type of energy insecurity. This is mainly because WWS requires no mined fuels (oil, fossil gas, coal, or uranium) to run. Instead, WWS relies only on natural energy sources. Eliminating mined fuels eliminates the energy insecurity associated with them.

Transition highlight
On Friday, October 18, 2024, the island country of Cuba (population 11 million) was plunged into darkness as the main power source on the island, a 330-megawatt oil-powered centralized power plant, shut down due to a combination of low fuel supply, increased electricity demand during a heat wave, and aging infrastructure. The low fuel supply was due

largely to decreased shipments of oil from Venezuela, Cuba's main oil supplier, as well as from Russia and Mexico. Venezuela had been struggling to produce enough fuel to export, and Russia was embroiled in a war with Ukraine. In addition, Hurricane Milton the week before had prevented the transfer of oil from ships offshore to the island. The blackout caused an almost complete shutdown of commerce, services, and internet. This blackout illustrates the confluence of two energy-security problems associated with fossil fuels: the use of centralized power plants and the reliance on imported fuels. WWS addresses both limitations by relying on mostly decentralized energy produced within a country.

Although a country that supplies 100 percent of its own energy with WWS minimizes the risk of energy insecurity due to international conflict and price manipulation, a benefit arises when adjacent countries trade WWS electricity between each other. Such trading, in the absence of conflict, reduces the overall cost of energy and improves the reliability of the overall energy system.

1.3.4 ENERGY INSECURITY DUE TO FUELS THAT HAVE MINING, POLLUTION, OR CATASTROPHIC RISK

A fourth type of energy insecurity is the risk associated with byproducts of energy use. For example, facilities and vehicles that burn fossil fuels and bioenergy fuels produce air pollution that kills millions of people worldwide each year. The perpetual mining of fossil fuels and uranium also causes health damage to miners and damage to land, vegetation, and animals. Underground coal mining, for instance, leads to black-lung disease in many miners. Underground uranium mining increases lung-cancer rates in miners, who breathe in polonium, a carcinogenic radioactive decay product of radon. Radon is a gaseous radioactive decay product of uranium itself. Nuclear reactors also produce radioactive waste that must be stored for hundreds of thousands of years. In addition, nuclear reactors run the risk of a reactor core meltdown. Finally, the historic spread of nuclear energy to dozens of countries has contributed to the proliferation of nuclear weapons in several of these countries.

A transition to clean, renewable energy avoids these risks to health, the environment, and public safety. The continued use of fossil fuels and bioenergy fuels, with or without carbon capture, and of nuclear power, prolongs these energy-security problems.

2

WWS SOLUTIONS FOR ELECTRICITY GENERATION

The solution to air pollution, global warming, and energy insecurity is, in theory, simple and straightforward: Electrify or provide direct heat for all energy; obtain the electricity and heat from only wind, water, and solar sources; store energy, transmit electricity over long distances; and reduce energy use. This chapter first explores the main components of a wind–water–solar system and then focuses on the WWS electricity-generating technologies that will replace traditional energy sources, thereby eliminating all global emissions from energy.

2.1 COMPONENTS OF A WWS SYSTEM

Figure 2.1 summarizes the main components of a 100 percent wind–water–solar energy, storage, transmission, and equipment system that maintains grid stability. It includes WWS electricity and heat generation; hydrogen generation; electricity, heat, cold, and hydrogen storage; transmission and distribution; energy efficiency; and appliances and machines that use WWS electricity.

What is meant by electrifying or providing direct heat for everything? Almost all energy worldwide is used for electricity, transportation, heating and cooling of buildings, and industry. In a 100 percent WWS world, all transportation will be converted to battery-electric vehicles or hydrogen-fuel-cell-electric vehicles, where the hydrogen is produced from WWS electricity (green hydrogen). Electric heat pumps will power most of the air and water heating and air-conditioning for buildings, including for showers, clothes washing, clothes drying, and dishwashing. A portion of heating and cooling will come from centralized (district heating) facilities and be distributed through water pipes to buildings. The remaining heat and cold will be produced within buildings themselves. Heat from geothermal reservoirs and sunlight will provide some air and water heating for buildings, primarily through district heating.

2 WWS SOLUTIONS FOR ELECTRICITY GENERATION

WWS Generation
WWS electricity generation
  Wind turbines
  Solar photovoltaics
  CSP[1] plants
  Geothermal plants
  Hydro plants
  Tidal, ocean-current turbines
  Wave devices

WWS heat generation
  Solar collectors
  Geothermal systems

WWS Grid
Transmission/distribution
  AC/HVAC/HVDC[2] lines
  Distribution lines
  Grid management
    Software
    Demand response

Grid interconnection
  Among WWS generators
  Among regions/countries

WWS Storage
Electricity storage
  Batteries
  CSP[1] storage
  Pumped hydro storage
  Hydropower reservoirs
  Flywheels
  Compressed air
  Gravitational storage
  Grid-hydrogen storage

District heat storage
  Water tanks
  Soil (borehole storage)
  Water pits
  Aquifers

District cold storage
  Water tanks
  Ice
  Aquifers

Building heat storage
  Water tanks
  Thermal-mass materials

Industrial heat storage
  Firebricks

Nongrid-hydrogen storage
  Hydrogen storage tanks

WWS Equipment
Building & district air/water heating
  Electric heat pumps

Building and district cooling
  Electric heat pumps

Industrial heat
  Arc/induction furnaces; electric crackers
  Resistance furnaces, kilns, and boilers
  Electron-beam/dielectric heaters
  Electric heat pumps; solar heat

Hydrogen generation/compression
  Electrolyzers; compressors

Transportation vehicles
  Battery-electric
  Hydrogen-fuel-cell-electric

Some appliances/machines
  Electric-induction cooktops
  Electric leaf blowers/lawn mowers
  Heat-pump washers/dryers/dishwashers

Some green hydrogen industrial processes
  Ammonia manufacturing
  Steel manufacturing

Efficiency/reduced energy use
  Insulate/weatherize buildings
  LED[3] lights; efficient appliances
  Telecommute; use public transport

[1] CSP is concentrated solar power
[2] AC is alternating current electricity; HVAC is high-voltage alternating current electricity; HVDC is high-voltage direct current electricity
[3] LED is light-emitting diode

Figure 2.1 Main generation, transmission, storage, and use components of a 100 percent WWS system to power the world for all purposes.

For industrial processes, firebricks, heated by electric-resistance heating, will store and provide as much medium- to high-temperature heat (150 to 2,000 degrees Celsius) as possible, replacing fossil fuels and bioenergy fuels for such heating and avoiding the need for most electric heat generators. Remaining medium-to-high-temperature heat will be produced with electric-arc furnaces; electric-induction furnaces; electric-resistance furnaces, kilns, and boilers; electric crackers; electron-beam heaters; dielectric heaters; and solar heaters. Low-temperature heat (below 150 degrees Celsius) will be produced by electric heat pumps and solar heaters.

2.1 COMPONENTS OF A WWS SYSTEM

Energy use in general will be reduced by capturing and recycling waste heat and cold, improving insulation, using more energy-efficient appliances, and creating more pedestrian and bike-friendly cities, for example.

All of the electricity and heat in this new paradigm will be provided by WWS sources. Energy not used right away will be stored, either as electricity, heat, cold, or hydrogen. Electricity will also be transmitted from its source to its end use through short- and long-distance electricity transmission and distribution lines. In cities, some heat and cold will be transported by hot- and cold-water pipes. The WWS electricity-generation technologies for each city, state, province, and country will include some combination of onshore and offshore wind turbines, solar photovoltaics on rooftops and in power plants, concentrated solar power plants, geothermal power plants, conventional and run-of-the-river hydroelectric power plants, tidal turbines, ocean-current turbines, and wave devices. WWS heat not produced from electricity will be generated from solar heat collectors and geothermal heat systems.

Types of storage will include electricity, heat, cold, and hydrogen storage. Major electricity storage options include pumped hydropower storage, existing hydroelectric dams, CSP coupled with thermal-energy storage, batteries, grid-hydrogen storage, flywheels, compressed-air storage, and gravitational storage with solid masses. Major low-temperature heat storage media include soil, water, and heat-absorbing materials. High-temperature heat will be stored in firebricks. Major cold storage media include water and ice. Stored hydrogen will be used primarily for long-distance, heavy transport; steel and ammonia manufacturing; grid-electricity backup; and electricity and heat production in microgrids. Hydrogen will be produced by splitting water with WWS electricity (electrolysis). Thus, it will be green hydrogen. In some systems, storage will be colocated with electricity or heat generation to reduce cost. For example, batteries may be colocated with residential rooftop PV systems to reduce the need for grid electricity when electricity prices are high. Reducing the use of grid electricity also reduces the use of transmission lines, reducing the chance of a wildfire caused by a transmission-line spark.

The size of a WWS electricity-generating technology is generally defined in terms of its nameplate capacity. A **nameplate capacity** (also called rated capacity, generating capacity, or plant capacity) is the maximum instantaneous discharge rate of electricity from an electricity-producing machine's generator, as determined by the manufacturer of the machine. Whereas a **motor** converts electricity to mechanical motion, a **generator** is just a motor running in reverse, converting mechanical motion to electricity. Nameplate capacities are given in units of power. The base unit of power is the **watt**. When 1 watt of power is produced, 1 **joule** of energy is created (discharged) by an electricity generator per second. Thus, the nameplate capacity of a wind turbine is the maximum rate of energy discharge from the turbine's generator. In other words, a wind turbine that has a nameplate capacity of 1 kilowatt (1,000 watts) can discharge no more than 1,000 joules of energy per second in the form of electricity from its generator.

Energy storage is defined in terms of both power and energy. The **maximum charge or discharge rate of storage** is the maximum power (rate of change of energy) going into or out of storage, respectively. The **maximum storage capacity** is the maximum energy that can be stored, and equals the maximum discharge rate multiplied by the number of hours of storage at that rate. Alternatively, the number of hours of storage is the time required to discharge the maximum energy stored at the maximum discharge rate. Next, WWS electricity-generating options are discussed.

2.2 ONSHORE AND OFFSHORE WIND ELECTRICITY

Wind turbines convert the kinetic energy in the wind into electricity. The energy that arises due to the movement of air or water is **kinetic energy**. In most wind turbines, a slow-turning turbine blade spins a shaft connected to a gearbox. Progressively smaller gears in the gearbox convert the slow-spinning motion (3–20 rotations per minute for modern turbines) to faster-spinning motion (750–3,600 rotations per minute), just as shifting from a higher gear to a lower gear on a bicycle allows one to pedal faster to travel the same distance. A fast-spinning motion is needed to convert mechanical energy to electrical energy in a wind turbine's generator.

Some modern wind turbines, called **direct-drive** turbines, are gearless, with the shaft connected directly to the generator. To compensate for the slow spin rate of a gearless turbine's shaft within the generator, the generator must be larger and heavier than it would otherwise be with a geared turbine. However, because direct-drive turbines avoid the use of a gearbox, they are simpler, require less maintenance, and produce less noise than geared turbines. Because each has advantages, both direct-drive and geared turbines are manufactured today.

The **hub height** of a horizontal-axis wind turbine (the most common type) is the height above the ground or ocean surface of the horizontal axis that the turbine spins around. The power output of a wind turbine increases with increasing turbine hub height because wind speeds generally

2.2 ONSHORE AND OFFSHORE WIND ELECTRICITY

increase with increasing height above the ground or ocean surface in the lower atmosphere, and instantaneous wind power output is proportional to the cube of the instantaneous wind speed. As such, taller turbines capture faster winds.

Wind farms are often located on flat open land, in mountain passes, on ridges, and offshore. Individual turbines as of 2025 have ranged from less than 1 kilowatt (1,000 watts) to 26 megawatts (million watts) nameplate capacity. Computer simulations of a 35-megawatt turbine are in the works.

Small individual wind turbines, with nameplate capacities of 1–10 kilowatts, are often used to produce electricity in the backyard of an individual home or within a city street canyon. These local turbines do not produce much total electricity over a year but, depending on wind speed, the amount is often sufficient to offset most of a homeowner's electricity usage.

Onshore wind farms usually serve the electric grid and contain a few to dozens of mid-sized wind turbines (1–10 megawatts in nameplate capacity) to provide electricity for part of a town or a city.

Offshore wind farms are also connected to the electric grid and usually contain a few to dozens of mid- to large-sized turbines (3–26 megawatts in nameplate capacity). Wind speeds offshore are generally faster than onshore because a relatively smooth ocean surface results in little friction to slow down the winds. Over land, vegetation, buildings, mountains, rocks, and other surface protrusions act to slow the winds.

> **Transition highlight**
> One particular 20-megawatt turbine designed for offshore use has a 160-meter hub height above the ocean surface. Its blade diameter is 292 meters.[21] Thus, the height of its furthest vertical extent is 306 meters, or 102 percent of the height of the Eiffel Tower. At a mean annual wind speed of 8.5 meters per second, typical for offshore wind, it can provide electricity for up to about 12,500 households that each draw 6,000 kilowatt-hours per year. Another turbine, rolled out by Dongfang Electric Corporation during October 2024 in Fuzhou, China, has a nameplate capacity of 26 megawatts, a hub height of 185 meters, and a blade diameter of 350 meters. At a mean annual wind speed of 8.5 meters per second, such a turbine can provide electricity for about 15,000 households that each draw 6,000 kilowatt-hours per year.

Offshore wind turbines have either bottom-fixed foundations or floating foundations. Bottom-fixed foundations are used primarily in water depths down to 50 meters. However, a new design allows bottom-fixed foundations that extend 90 meters down from the ocean surface to the ocean floor.[22] Floating wind turbines avoid the need for a foundation. They have a floating platform secured to the seabed by cables and can be placed in water of any depth.

High-altitude wind-energy capture with turbine-containing kites flying from 500 meters to 10 kilometers above the ground has also been pursued, although it has not been commercialized to date.

Because the wind does not always blow and, when it does blow, its speed changes uncontrollably over time, winds are variable in nature. As such, wind-turbine electricity output also varies with time; thus, wind is called a **variable WWS resource**. Another term commonly used to describe variability is **intermittency**. However, all energy resources are intermittent due to scheduled and unscheduled maintenance. Variable resources are those with energy outputs that both vary with the weather and are affected by maintenance. Because wind output is variable, combining wind with batteries, hydroelectricity, and other types of electricity storage helps to match demand for electricity (which also varies with time) with supply. This is particularly necessary when wind produces a high percentage of the total electricity on a grid, but less so when it produces a low percentage.

2.3 WAVE ELECTRICITY

Winds passing over water create surface waves. The faster the average wind speed, the longer the wave is sustained, the greater the distance the wave travels, and the taller the wave becomes. **Wave devices** capture energy from ocean waves and convert that energy into electricity.

Because wave electricity output varies with time, wave electricity is also a variable electricity source. However, wave output is less variable than is offshore wind output, even at the same location. The reason is that waves form from winds dragging the water over a long distance. The variability in wind speed at a given location along a wave's path has little impact on the wave, which already has momentum due to upstream winds. As such, waves at a given point represent the impact of winds accumulated over a long distance, whereas winds at one point are instantaneous and more variable.

One type of wave device is a free-floating device that bobs up and down with the waves, creating mechanical energy that is converted to electricity in a generator housed inside the device. The electricity is then sent through an underwater cable to shore. Most of the body of the device is submerged under water. With some designs, the entire device is just below the water line so that it can't be seen from the coast.

Another type of wave device has an arm connected to a pier, jetty, breakwater, floating platform, or fixed platform. The other end of the arm is connected to a floater that rises and falls with a single wave's motion. Because the device is connected to a structure, it is not free-floating in the open ocean. The pumping motion due to the rising and falling floater is transmitted, through fluid pressure, to a power station on the platform that it is connected to. The fluid pressure is used to spin a hydraulic turbine, whose rotating motion is converted to electricity in a generator. Because the turbine and generator are on land, this type of wave device is easier to maintain than one completely offshore. However, a shore-based wave device can be placed in fewer locations than a free-floating device, which can be placed anywhere over the ocean where waves occur.

Despite the development of a variety of wave-energy technologies, the wave-energy industry is still less mature, in 2025, than the wind and solar photovoltaics industries. As a result, wave-energy costs are still higher than the costs of wind and solar.

2.4 GEOTHERMAL ELECTRICITY AND HEAT

Geothermal energy is energy extracted from hot water or steam below the Earth's surface. In both cases, the heat originates from hot rocks or soil. Rocks and soil are both heated in four ways. These include by **conductive** (atom-by-atom or molecule-by-molecule) transfer of energy from the center of the Earth (where temperatures are 4,300 degrees Celsius) to near the Earth's surface; the conduction of heat from volcanoes; the decay of radioactive elements in the Earth's crust; and the conduction, through soil, of sunlight that hits the Earth's surface. Most high-temperature rocks are near volcanoes, which are mostly near tectonic plates. Low-temperature warm rocks and soil exist underground everywhere. Even when the ground is frozen, soil 6 meters deep is warm enough to heat buildings.

Whether geothermal heat can be used to provide heat for a building or electricity depends on temperature. Low-temperature heat (below 120 degrees Celsius) is useful for heating buildings. It is extracted by ground-source heat pumps, pipes that capture native (naturally occurring) hot water or steam from hot springs, and pipes circulating water or air that capture heat from soil.

High-temperature heat (120 to 400 degrees Celsius) is needed to generate electricity. Temperatures in the Earth's crust, away from volcanoes, geysers, and hot springs, naturally increase by 15–30 degrees Celsius per kilometer of depth. Thus, to generate electricity away from naturally occurring conventional hot underground areas, drilling 4–8 kilometers and under impermeable rock is usually needed. This is the idea behind **enhanced geothermal systems** (EGS), which are geothermal electricity-generating systems that use heat from deep in the Earth, away from naturally occurring conventional heat sources. The main advantage of enhanced geothermal systems is that they can be sited in many more locations than conventional geothermal systems. A second advantage is that they can tap into an effectively unlimited heat reservoir deep in the Earth. The United States, for example, has the potential for about 40 gigawatts of conventional geothermal electricity generation in 13 states, but also for about 5,500 gigawatts of enhanced geothermal electricity generation across all 50 states.[23] During October 2024, the United States approved its first major (2 gigawatts) enhanced geothermal plant, in Utah, with an expected operation year of 2028.

Geothermal electricity-generating plants produce a steady supply of electricity. As such, they are **baseload** plants, which are electricity-generating plants that produce a constant supply of electricity for an extended period to meet loads on the grid. **Loads** are demands for electricity, such as for powering light bulbs or refrigerators. Enhanced geothermal plants are useful for providing not only baseload grid electricity but also constant electricity for a computer cluster or data center through a microgrid disconnected from the grid.

Prior to the 1900s, steam and hot water from the Earth were used to provide heat for buildings, industrial processes, and domestic water. The first use of geothermal heat to produce electricity was by **Prince Piero Conti** of **Larderello**, Tuscany, Italy. In 1904, he lit four light bulbs by using high-temperature steam from a geothermal field near his palace to drive a steam engine attached to a generator. A **steam engine** uses steam to push and pull a piston up and down. This piston is attached to a rotating cylinder. A generator converts the rotating motion to electricity.

In 1911, Conti installed the first geothermal power plant, which had a nameplate capacity of 250 kilowatts. This plant grew to 405 megawatts by 1975. The second electricity-producing plant in the world was built at the Geysers Resort Hotel, California, in 1922. This plant was originally used only to generate electricity for the resort, but it has since been developed to produce a portion of electricity for the state of California.

Today, the three major types of geothermal plants for producing electricity are dry steam, flash steam, and binary. Dry and flash steam plants operate when the geothermal reservoir temperature is 180–370 degrees Celsius or higher. In both cases, two boreholes are drilled – one for cool,

2.4 GEOTHERMAL ELECTRICITY AND HEAT

liquid water to flow down to hot rocks below, and a second for steam alone (in the case of dry steam) or liquid water plus steam (in the case of flash steam) to flow up. Due to the high temperature of the water after it is heated by the rocks, the water flows convectively up, so no pump is needed to push the water up from the bottom of the reservoir. As the steam or steam–hot-water mixture passes through the plant, the heat is converted to electricity, and the resulting cold liquid water is sent back down into the first hole.

In a **dry steam plant**, the pressure of the steam rising up the first borehole powers a turbine, which drives a generator to produce electricity. About 70 percent of the steam recondenses after it passes through a condenser, and the rest is released to the air as water vapor. Because several chemicals, including carbon dioxide, nitric oxide, sulfur dioxide, and hydrogen sulfide in the geothermal reservoir steam do not recondense along with water vapor, these gases are emitted to the air as well.

In a **flash steam plant**, the liquid water plus steam from the geothermal reservoir enters a water tank held at low pressure, causing some of the water to vaporize ("flash"). The vapor then drives a turbine. About 70 percent of this vapor is recondensed. Again, the remainder escapes with carbon dioxide and other gases. The liquid water is injected back into the ground.

Binary geothermal plants are developed when the geothermal reservoir temperature is 120–180 degrees Celsius. Water rising up the second borehole is enclosed in a pipe and heats, through a heat exchanger, a low-boiling-point organic fluid, such as isobutane or isopentane. The evaporated organic gas turns a turbine that powers a generator to produce electricity. Because the water from the reservoir remains in an enclosed pipe when it passes through the power plant, and is reinjected into the reservoir, binary systems emit virtually no carbon dioxide or other pollutants. Due to the lower water temperature in a binary plant versus a steam plant, a pump is usually needed in a binary plant to push water from the reservoir up the second borehole.

Prior to 2000, almost all new geothermal electricity plants were dry steam or flash steam. Since then, about 90 percent of plants built have been binary. Because they can operate in lower-temperature reservoirs, binary plants can be situated in many more locations than can steam-based plants. Most future enhanced geothermal system plants will also likely be binary plants because, due to the high cost of drilling, their boreholes will be drilled only to depths needed to obtain temperatures usable for a binary plant. Other plants would require deeper boreholes.

2.5 HYDROELECTRICITY

Hydroelectricity (**hydropower**) is produced by water flowing downhill through a water turbine connected to a generator. Most hydropower is produced from water held in a reservoir behind a large dam. This type of hydropower is referred to as large, or conventional, hydropower. A hydropower dam requires a reservoir behind it, which results in the flooding of a large area of land. The largest conventional hydropower plant in the world in 2025 was the Three Gorges Dam on the Yangtze River in China. It opened in 2004 and has a nameplate capacity of 22.5 gigawatts (billion watts). The second-largest plant was the Baihetan Dam on the Jinsha River in China, which is an upper stretch of the Yangtze River. It opened in 2021 and has a nameplate capacity of 16 gigawatts. Some other large hydropower plants include the 14-gigawatt Itaipu Dam on the Parana River bordering Brazil and Paraguay, the 13.86-gigawatt Xiluodu Dam in China, and the 10.2-gigawatt Guri Dam on the Caroni River in Venezuela.

A growing portion of hydroelectricity is produced by water flowing down a river directly through a turbine, or by water that is diverted through pipes and through a turbine near the edge of a river before returning to the river. Both types of hydroelectricity are **run-of-the-river hydropower**. The advantage of run-of-the-river hydro over conventional hydro is that large amounts of land are not flooded behind a dam. As a result, run-of-the-river hydro is less useful for storage. However, run-of-the-river hydro, with a modest storage pond behind it, as with conventional hydro, can provide electricity within 15–30 seconds of a need. The largest run-of-the-river hydro plant worldwide is the 11.2-gigawatt Belo Monte plant on the Xingu River in Brazil.

A conventional hydro plant consists of a dam, a water storage reservoir behind the dam, penstocks, sluice gates, a powerhouse, and a downstream water outlet. A **penstock** is a pipe, channel, or tunnel through which water flows from the storage reservoir to a water turbine. A **sluice gate** is a gate to stop or control water flow between the storage reservoir and the penstock. A **powerhouse** is a building containing water turbines, generators, and power transmission cables. When water passes through a **water turbine** in the powerhouse, the turbine's blades spin, rotating a metal shaft connected to the generator, which converts the mechanical rotating motion to electricity.

A run-of-the-river hydro facility consists of a water turbine and generator, but no reservoir, except for a small holding pond in some cases, and no pipes, except when the water is diverted a short distance away from the river, through the turbine, and back to the river.

2.5 HYDROELECTRICITY

Only a fraction of dams with reservoirs worldwide contain hydroelectric equipment. For example, the United States had about 92,000 dams in 2025, but only 2,200 (about 2.4 percent) of these also had turbines to generate electricity.

A hydropower plant can be run as a baseload plant, a load-following plant, or a peaking power plant. Whereas a baseload power plant produces a constant supply of electricity for an extended time, a **load-following plant** runs continuously but ramps its power production up and down to meet 5- to 15-minute-average changes in demand on the grid. Hydropower plants, when run as load-following plants, run continuously and adjust their output every 5 to 15 minutes to meet such changes. A **peaking power plant** generates electricity within seconds to a few minutes to meet specific peaks in electricity demand that other electricity sources, such as wind turbines and solar panels, cannot meet immediately. Some fossil-gas plants serve no other purpose than to meet peaks in demand. A hydropower plant, on the other hand, is flexible enough that it can run as a baseload, load-following, or peaking plant as needed. However, a hydropower plant's output is often limited by competing uses of the water that it holds and by how much water it can release downstream at a given time.

In a hydropower plant, neither the powerhouse nor the penstock needs to be located inside the dam. The penstock can be built to channel water around the dam to a powerhouse in front of or to the side of the dam. This may be desirable in cases where turbines are added to a hydropower plant, years after the dam is built, in order to increase the plant's peak discharge rate (nameplate capacity) while keeping the annually averaged water stored in the reservoir constant. Adding turbines is useful for plants that are run in load-following or peaking mode. Adding turbines to a hydropower plant is called **uprating**. Uprating a hydropower plant may be "one of the most immediate, cost-effective, and environmentally acceptable means of developing additional electric power."[24]

The average power output from a hydropower facility is limited not only by the nameplate capacity of its generators but also by the annually averaged amount of water available in the reservoir to run through the turbines. A measure of the practical average power output of a hydropower facility that accounts for both the water availability and the turbine nameplate capacity is called the installed capacity of the hydropower plant.

The **installed capacity** of a hydropower plant is the smaller of the average power that can be produced by available water in a hydropower reservoir and the nameplate capacity of the turbine generators in the plant itself.[25,26] For other types of power-producing technologies, such as wind turbines and solar panels, the installed capacity is simply the nameplate capacity of the technology.

2.6 TIDAL AND OCEAN-CURRENT ELECTRICITY

Tidal currents (**tides**) are back-and-forth currents in the ocean caused by the rise and fall of the ocean surface due to the gravitational attraction among the Earth, the moon, and the sun. The rising and sinking motion of the ocean surface forces water below the surface to move horizontally as a current. Because tides run about six hours in one direction before switching directions for another six hours, they are predictable.

An **ocean current** is a continuous flow (in one direction) of seawater driven by winds or by temperature and salinity gradients in the ocean. The Gulf Stream current, for example, is a warm, fast-flowing current driven by winds. It runs from the Gulf of Mexico, past the tip of Florida, up the US coast, to the Newfoundland coast, then to the North Atlantic ocean, where it splits into two other currents. Ocean currents, in fact, run along all major coastlines in the North and South Pacific Oceans, the North and South Atlantic Oceans, and the Indian Ocean. For example, the current running along the west coast of North America from Washington State down to Southern California is the California Current. The current running northward along the west coast of South America is the Humboldt (Peru) Current.

A **tidal turbine** captures the kinetic energy of an ebbing and flowing tidal current in the ocean, the continuous flow of an ocean current, or the flow of a river current. Because tidal, ocean, and river currents are all relatively steady, tidal turbines usually supply baseload (constant) electricity to the grid when they are not undergoing maintenance.

Tidal turbines can be mounted on the seafloor or hang from under a floating platform. They can also be placed in rivers to capture the energy of continuously flowing river water or of tidal water that ebbs and flows in a river. One such tide flows from the Atlantic Ocean, through Chesapeake Bay, near Virginia, then up the Potomac River, past Washington, DC.

Like a wind turbine, a tidal turbine consists of a blade that spins a turbine. The turbine provides rotational energy to a generator that converts the rotational energy into electricity to be transmitted to shore. A tidal turbine's rotor, which lies under water, may be exposed to the direct flow of water or placed within a narrowing duct that increases the speed of the water flowing toward it.

2.7 SOLAR PHOTOVOLTAIC ELECTRICITY

Transition highlight
One type of tidal turbine, called a tidal kite, has 12-meter-wide wings and is shaped like an airplane. It is tethered to the seabed and steered autonomously in a figure-eight pattern. Water passes through its blades 40 meters underwater, causing them and a rotor to spin. The spinning rotor is connected, inside the device, to a generator, which produces up to 1.5 megawatts of electricity. The electricity is sent through a cable to shore, where its first use will be to power part of the Faroe Islands.[27]

2.7 SOLAR PHOTOVOLTAIC ELECTRICITY

A **solar photovoltaic** panel consists of an array of cells containing a material that converts sunlight into direct current (DC) electricity. The DC electricity can be used directly to charge a battery. Or an inverter can convert the DC electricity into alternating current (AC) electricity, which is either used immediately in a building or sent to the electricity grid for people to use. Photovoltaics are often mounted just above the ground in large (utility-scale) power plants. Smaller PV units are also mounted on or built into roofs or walls of buildings. They are also ground-mounted on hillsides, in yards, and in vacant lots. Some are mounted on carports, parking lots, parking structures, and balconies. These small-scale systems, which service residential, commercial, industrial, and government buildings directly, are referred to as **distributed-PV** systems.

There are two main types of distributed-PV systems: **behind-the-meter** (BTM) and **in-front-of-the-meter** (FOM) systems. FOM systems are part of the main transmission and distribution system, just like utility-PV systems are. However, FOM systems are smaller than 20 megawatts in nameplate capacity, whereas utility-PV systems are larger than that. Second, FOM systems connect to distribution lines, whereas utility-PV systems connect to transmission lines. Because FOM systems connect to distribution lines, they serve buildings directly, minimizing the need for more transmission lines. However, distribution lines are connected to transmission lines, so FOM PV can also feed electricity to the transmission system. FOM-PV systems are, thus, subject to the same market and grid connection rules as utility-PV systems.

BTM systems are also smaller than 20 megawatts, but usually a few to tens of kilowatts in size. They serve buildings directly, but if the BTM system is connected to the grid, any excess electricity produced by the system may be sent back to the grid. If not enough BTM PV electricity is available to serve a building, the grid can then supply electricity to the building. Because the meter that determines electricity use for a building only reports the incoming electricity from the grid and the outgoing electricity back to the grid, but not the electricity consumed by the building from the BTM system, BTM systems are behind-the-meter systems. If a BTM system is not connected to the grid, the system is run in isolation as a microgrid.

BTM PV systems are often colocated with battery storage. A BTM system first provides electricity to a building it services. Any excess electricity is then stored in batteries. Any remaining electricity after that is sent to the grid if the BTM PV system is grid-connected. If the BTM system is not connected to the grid, the remaining electricity is involuntarily lost, or **curtailed**. Thus, if everything else is the same, BTM PV reduces immediate grid electricity demand by supplying electricity directly to buildings, avoiding the need for grid electricity for those buildings.

Finally, PV may be built into the tops or sides of cars, trucks, buses, trains, ships, and airplanes to power some of the electrical needs in those vehicles. This type of PV is **mobile PV**. Often, the PV panels installed on vehicles can bend.[28]

PV technology has evolved sufficiently that transparent PV can now be built into glass windows.[29,30] Large solar PV arrays that are thin and flexible enough to be rolled up and transported from place to place have also been commercialized[31] as have certain foldable PV panels.[32,33] Such **transportable PV** panels are useful for disaster relief, remote military bases that must be moved regularly, and communities that need a rapid but nonpermanent source of electricity. In utility-PV plants, PV panels are either mounted at a fixed tilt or on trackers that rotate to follow the sun. For most distributed PV systems, they are mounted with a fixed tilt.

> **Transition highlight**
>
> Solar PV systems have been built to float on lake and ocean surfaces and to rest over canals. During 2021, Indonesia built a 145-megawatt-nameplate-capacity floating solar PV farm. The country has potential for 28 gigawatts of floating PV across 375 lakes and reservoirs. In India a 3.6-kilometer stretch of the Narmada irrigation canal in Vadodara is covered with 33,816 PV panels generating 10 megawatts of peak power. The system not only saves cropland from being covered with PV panels, but it also reduces water evaporation from the canal.[34]

The presence of clouds, snow, desert dust, air pollution, shadow-casting buildings and trees, or very high temperatures can reduce the efficiency of a solar PV panel. Rainfall naturally cleans panels. However, during long periods with no rain, PV output in moderately polluted locations can drop by 10 to 20 percent. In areas with substantial wind-blown desert dust, output can drop by up to 40 percent.[35]

When a tree blocks direct sunlight, a thin cloud can enhance sunlight incident upon a panel by bouncing light from the cloud, around the tree, onto the panel. Similarly, light can bounce off snow onto a panel, increasing panel output. Further, low temperatures can increase panel output compared with high temperatures.

Because the sun disappears during the night and is often blocked by clouds, air pollution, or obstacles during the day, sunlight is a variable energy source, and solar PV electricity is an intermittent electricity source. However, solar PV can be combined with batteries or other types of electricity storage to provide steadier (baseload) output and, in many situations, load-following or peaking output.

2.8 CONCENTRATED SOLAR POWER

With **concentrated solar power**, mirrors or reflective lenses focus (concentrate) sunlight onto a collector containing a fluid in order to heat the fluid to a high temperature. The heated fluid (usually synthetic oil or molten salt) is then used to boil water. The resulting steam passes through a steam turbine connected to a generator, which produces electricity. Up to 30 percent of the energy in the collected heat is converted to electricity.

Like solar PV, CSP is a variable and intermittent electricity source. However, some types of CSP generators allow the heat to be stored for many hours so that CSP electricity can be produced during the night or when heavy clouds are present. Thus, CSP combined with storage can be used to provide baseload, load-following, or peaking power.

A disadvantage of CSP compared with solar PV, though, is that CSP requires direct sunlight to operate effectively, whereas solar PV can use either direct or diffuse sunlight. **Direct sunlight** is the beam of light that comes directly from the sun that hits a PV panel or CSP mirror. **Diffuse sunlight** is light that bounces (scatters) out of the direct solar beam due to gas molecules, aerosol particles, and cloud drops along the path of the direct beam. The light scatters in many directions and multiple times because it is intercepted and redirected by many gas molecules, aerosol particles, and cloud drops. Some of the light eventually bounces onto the solar panel or mirror. Light that appears under a thick blanket of cloud is almost all diffuse light because the cloud blocks the direct beam.

Deserts, which have few clouds, are usually the best places for direct sunlight, thus, CSP. Solar photovoltaics can operate anywhere on Earth, including at the north and south poles, even if skies are almost always cloudy, because they can convert both diffuse and direct light to electricity.

One type of CSP collector is a set of long **parabolic trough** (U-shaped) mirror reflectors, which focus light onto a pipe containing oil. The hot oil flows to a chamber, where the heat is transferred to water, which boils to produce steam that is sent to a steam turbine connected to a generator to produce electricity.

A second type of collector is a **central tower receiver** with a field of mirrors surrounding it. In the central tower, the focused light heats a circulating thermal storage fluid, such as a **molten salt** mix, to 500–600 degrees Celsius. The molten mix often consists of sodium nitrate and potassium nitrate. The hot salt flows to a heat exchanger, where the heat is transferred to water, which boils to produce steam that is sent to a steam turbine connected to a generator to produce electricity. After passing through the steam turbine, the steam is piped to a condenser, where it recondenses to a liquid. From there, the cool liquid water is piped back to the heat exchanger to be boiled again. Thus, the water circulates in a closed loop. Meanwhile, after exchanging its heat with water in the heat exchanger, the molten salt temperature is now down to about 260 degrees Celsius. The "cold" salt is then moved to a cold holding tank until it is recirculated to the central tower. Thus, the molten salt also circulates, but in a separate closed loop.

In many parabolic trough and central tower CSP plants, some of the hot fluid (oil or molten salt mix) is stored in an insulated thermal storage tank before it is used to boil water in the heat exchanger. The purpose is to delay the boiling of water, and thus electricity production, until little or no solar electricity is generated by direct sunlight, such as at night or when heavy clouds are present.

A third type of CSP technology is a **parabolic dish-shaped** (like a satellite dish) reflector that reflects light onto a receiver while rotating to track the sun. The receiver transfers the heat to hydrogen in a closed loop. The expansion of hydrogen against a piston or turbine produces mechanical power used to run a generator to produce electricity. The power conversion unit is air-cooled; thus, water cooling is not needed. Parabolic dish CSP is not coupled with thermal-energy storage.

CSP plants need to be cooled by either air or water. The use of air cooling, which is desirable in water-constrained locations, reduces overall CSP plant water requirements by 90 percent at a cost of only about 1–5 percent less electric-power production.[36]

2.8 CONCENTRATED SOLAR POWER

Because the components of CSP plants are made of abundant raw materials, material shortages are not expected to limit the mass production of CSP plants. For example, CSP plants consist primarily of mirrors, receivers, and thermal storage fluid. Mirrors are made mostly of glass with a reflective silver layer on the back of the glass. Receivers are stainless steel tubes with an outer surface that absorbs sunlight and that is surrounded by an outer antireflective glass tube. None of these materials has limited availability.

The largest CSP farm in the world in 2025 was the 700-megawatt Mohammed bin Rashid Al Maktoum Solar Park in the United Arab Emirates. It includes 600 megawatts of parabolic troughs, a 100-megawatt central tower receiver, and 15 hours of molten salt storage. The second largest was the Noor/Ouarzazate station in Morocco, which also consists of parabolic troughs, a tower, and storage. The Ivanpah facility in California was third, at 392 megawatts, and consists of three central towers surrounded by mirrors.

3

WWS SOLUTIONS FOR ELECTRICITY STORAGE

Hydro, geothermal, tidal, and ocean-current electricity production can be steady for long periods; thus, these generators can provide baseload (constant-output) electricity. However, wind, solar PV, and wave electricity outputs vary during the day and by season. As such, these electricity sources provide variable output. Given that solar and wind may end up supplying 90 percent or more of all WWS energy generation worldwide, on average, it is important to have electricity storage technologies available to provide backup when solar and wind are unavailable. Storage also allows excess WWS electricity and heat generated during the day, for example, to be used at night. Major electricity storage options include existing hydroelectric dams, pumped hydroelectric storage, batteries, concentrated solar power coupled with thermal-energy storage, flywheels, compressed-air energy storage, gravitational storage with solid masses, and green-hydrogen storage. This chapter discusses these technologies.

3.1 HYDROELECTRICITY RESERVOIR STORAGE

Conventional hydroelectricity (hydropower or just "hydro") plants have built-in storage since the water that generates the electricity for them is stored in a reservoir behind a dam. Whereas reservoir water can be drained through a turbine to create electricity as needed, the reservoir is charged only naturally through rainfall and runoff. Thus, electricity storage associated with a conventional hydro plant differs from that of battery storage. A battery can be charged when excess electricity is available and discharged when electricity is needed. A conventional hydro plant can often produce electricity precisely when it is needed, but it cannot control when rain and runoff will refill the reservoir. The only option for controlling recharge is to turn the hydropower plant into a pumped hydro storage facility.

Hydro plants can be run as baseload plants, load-following plants, or peaking plants. When they are used for load following and peaking, hydro plants can meet minute-by-minute changes in electricity demand on the grid faster than fossil-gas, coal, or nuclear power plants are able to. The **ramp rate** of a power plant is the speed at which it ramps up from zero to maximum power. The ramp rate of a hydropower plant is fast. For example, a hydropower turbine can generate full power from no power within 15–30 seconds. That is the time required for the water to flow gravitationally from the sluice gate, after it opens, through the penstock, to the turbine. Once the water reaches the turbine, the turbine spins and immediately produces electricity in an attached generator. As such, the ramp rate of a hydropower plant is 100 percent in 15–30 seconds. In comparison, the ramp rate of a fossil-gas open-cycle turbine is 100 percent in 300 seconds (5 minutes); that of a fossil-gas combined-cycle turbine is 100 percent in 600– 1,200 seconds (10–20 minutes); and those of coal and nuclear plants are both 100 percent in 20–100 minutes.[37]

Because hydropower plants can produce electricity within 15–30 seconds of a need, they are often used to provide backup power on an electricity grid. At a high continuous electricity discharge rate, hydropower plants can meet a high continuous demand for hours to days, depending on the reservoir water content, the nameplate capacity of installed turbines in the plant, and downstream flow restrictions. When used only to meet gaps in supply, such plants may sit idle for days to weeks before being used several days in a row for two to three hours per day.

3.2 PUMPED HYDROPOWER STORAGE

A related type of storage that can discharge electricity quickly is **pumped hydropower storage** ("pumped hydro"). Pumped hydro storage is used primarily to fill in short-term (minutes to a day) gaps in electricity supply.

Pumped hydro consists of two reservoirs – an upper one and a lower one. The lower one can be the ocean, a natural lake, a human-made lake or reservoir, or a continuously running river. The upper reservoir can be a natural lake or a human-made lake or reservoir. One proposed pumped hydro system consists of Lake Mead, the reservoir behind Hoover Dam, and a pumping station 32 kilometers downstream along the Colorado River.[38]

The idea behind pumped hydro is that when excess electricity is available or when electricity prices are low, a motor pumps water uphill through pipes, from a pumping station in the lower reservoir, to the upper reservoir. When electricity is needed, water drains downhill through a water turbine connected to a generator that generates electricity.

The efficiency of pumped hydro (ratio of electricity delivered to the sum of electricity delivered and electricity used to pump the water uphill) is about 80 percent. Pumped hydro's ramp rate is the same as that of conventional hydro, from zero to 100 percent power in 15 to 30 seconds. As such, pumped hydro storage ramps up faster than do fossil-gas plants (100 percent power in 5–20 minutes), or nuclear plants (100 percent power in 20–100 minutes).[37]

> **Transition highlight**
> Excluding conventional hydro reservoirs, about 95 percent of all electricity storage built on Earth in 2022 was pumped hydro storage. Worldwide, about 616,000 potential pumped hydro sites exist. These sites can store an estimated 23 million gigawatt-hours of electricity for the electric grid. This represents over 100 times the electricity needed to back up a 100 percent renewable electricity system worldwide.[39]

The largest pumped hydro facility worldwide is the 3.6-gigawatt Fengning facility in China's Hebei province, which became operational on August 11, 2024. It can store up to 40 gigawatt-hours of electricity. At a peak discharge rate of 3.6 gigawatts, the 40 gigawatt-hours of stored electricity can be extracted continuously for about 11 hours. The second-largest facility is a 3.003-gigawatt facility in Bath County, Virginia, in the United States. That facility stores 24 gigawatt-hours.

3.3 STATIONARY BATTERIES

Batteries in a 100 percent WWS world will be used primarily in battery-electric vehicles, electronic devices (e.g., phones, watches, computers, flashlights), moveable equipment (e.g., leaf blowers, lawn mowers, chainsaws), and stationary electricity storage systems. Batteries are useful because they generate electricity almost immediately on demand. As such, a battery-electric vehicle accelerates to a high speed much faster than does an internal-combustion-engine vehicle. Similarly, batteries can provide needed electricity to the grid much faster than can coal, fossil-gas, nuclear, or even hydro facilities. For example, the ramp rate of batteries from 0 to 100 percent output is about 20 milliseconds, whereas that of an open-cycle fossil-gas turbine is 5 minutes, or 15,000 times slower. Stationary batteries can provide load-following or peaking power for a large electricity grid or for an isolated microgrid. Wall-mounted or floor-mounted battery packs are commercially available for homes, commercial buildings, microgrids, and city electric grids.

A **battery** is an electrochemical cell that converts chemical energy into electricity. A **battery pack** is a collection of such cells. Each cell consists of

3.3 STATIONARY BATTERIES

two half-cells divided by a separator. Each half-cell consists of a current collector, an electrode, and an electrolyte solution.

The **current collector** is a metal cap in each half-cell through which electrons move either forward or backward. In both cells, the current collector is attached to an **electrode**. The current collector in one cell is called the positive current collector (**cathode**). In a **lithium-ion battery**, the electrode attached to the positive current collector is often made of lithium cobalt oxide, which is the source of lithium ions (Li^+) that move back and forth between the half-cells. The current collector in the other cell is called the negative current collector (**anode**), and the electrode attached to it is often made of pure carbon (graphite). The graphite's purpose is to receive and bond to lithium ions and electrons.

The **electrolyte solution** contains a lithium salt dissolved in an organic solvent. The liquid solution facilitates the passage of lithium ions between the two half-cells. The **separator** allows lithium ions, but not electrons, to pass between the two half-cells.

During battery charging, a charger plugged into the wall is wired to both current collectors. The charger provides energy to existing electrons flowing through the circuit. Some electrons initially flow through the circuit from the cathode to the anode side because some of the lithium cobalt oxide molecules on the cathode side dissociate into a positively charged lithium ion, a negatively charged electron, and neutrally charged cobalt oxide, and the electron takes the path of least resistance through the circuit. Once on the anode side, the energized electrons pull like a magnet at the the lithium cobalt oxide on the cathode side to break it apart. Free positively charged lithium ions are then drawn magnetically from the cathode side, through the electrolyte solution and separator, to the negatively charged electrons accumulating on the anode side. There, the lithium ions combine with the electrons and graphite to form lithium-graphite, neutralizing charges.

Meanwhile, electrons released at the positive current collector (cathode side) can't pass through the separator. Instead, they take the path of least resistance by flowing through a wire connected to the charger, completing the circuit. When no more lithium ions are available to flow from the positive to negative current collectors, the battery is fully charged.

During battery discharging, a wire is connected from the negative current collector (anode) to an electronic device (e.g., a light). Another wire is connected from the device to the positive current collector (cathode). The device, when turned on, draws electrons from the negative current collector. The electrons pass through the wire to the device. The device consumes some of the energy held by the electrons, but the electrons themselves are conserved and continue to flow to the positive current collector and into the adjacent electrolyte solution. The negative charge buildup due to electrons accumulating at the positive current collector (cathode) induces positively charged lithium ions to dissociate from the lithium-graphite present near the negative current collector, and to flow, through the separator, to the positive current collector, where they recombine with electrons and cobalt oxide to reform lithium cobalt oxide. When all lithium ions have migrated from the negative to the positive current collector, the battery is depleted and needs recharging.

Two parameters important for determining battery capabilities are their **round-trip efficiency** (ratio of electricity discharged from the battery to the electricity used to charge the battery) and the number of times they **cycle** (fully discharge plus charge) before the battery is no longer useful. Battery round-trip efficiencies are generally 85–90 percent but can be lower or higher depending on the quality of charger used. A typical car battery goes through 500–3,000 cycles before degrading. However, some newer lithium batteries degrade by only 35 percent after 28,000 full charge/discharge cycles over eight years.[40] If these cycles occurred once daily, the battery would theoretically last 77 years. However, it is unlikely a battery can be cycled daily for 77 years because batteries degrade even when they are not used, limiting their life to 30–40 years.

Traditional lithium-ion batteries contain cobalt. The largest cobalt mines today are in the Democratic Republic (DR) of the Congo, China, Canada, and Russia. China produces the most refined cobalt. A major concern with mining cobalt has been the use of child labor in the DR Congo's cobalt industry. A concern with all mining is environmental damage. Alternatives to lithium-ion batteries include **lithium-iron-phosphate** (LFP) batteries. These batteries use LFP instead of lithium cobalt oxide in their cathode. Lithium-ion batteries store more energy per unit mass of battery than LFP batteries. However, LFP batteries are more stable and thus less fire-prone. Plus, LFP batteries can go through more cycles before degrading than lithium-ion batteries. Many battery-electric vehicles already use LFP batteries.

> **Transition highlight**
> Lithium-based battery costs have decreased dramatically since their 1991 introduction. In 1991, lithium-ion cell prices were ~$7,500 per kilowatt-hour of energy storage.[41] One **kilowatt-hour** of energy is equivalent to 3.6 million joules of energy. By 2010, the lithium-ion cell price had dropped to about $450 per kilowatt-hour and, by 2018,

to $181 per kilowatt-hour.[41] LFP battery prices have historically been a bit higher than lithium-ion battery prices. Nevertheless, from March of 2022 to 2023 to 2024, LFP cell prices dropped from $118 to $108 to $53 per kilowatt-hour in China.[42] Thus, lithium-based battery cell prices dropped by at least 99.3 percent from 1991 to 2024.

Battery pack prices are generally 20–40 percent higher than cell prices. LFP battery pack prices in China were $76 per kilowatt-hour in March of 2024.[42] Battery packs are used in battery-electric vehicles directly. If they are used for distributed or grid-scale electricity storage, an additional installation cost arises. The price of a battery pack plus installation is about twice the cell price. Thus, the stationary battery storage price in March 2024 in China was about $106 per kilowatt-hour of storage. Installed stationary battery pack prices need to be about $60 per kilowatt-hour worldwide by 2035 to enable a low-cost large-scale transition to 100 percent WWS.[43I] The rapid decline in price over just the past two years suggests this price will be met even before 2030.

> **Transition highlight**
> Battery adoption has grown rapidly. For example, by October 31, 2024, California boasted 11.46 gigawatts of battery peak output among its grids. Residential and commercial buildings had another 1.35 and 0.58 gigawatts, respectively, for a total of 13.39 gigawatts statewide. Most batteries could discharge at their peak output for 4 hours, so they stored up to 53.6 gigawatt-hours of electricity. More than half of California's batteries were added in just one year. By October 31, 2024, Texas also had 9.3 gigawatts of batteries; the whole US had over 21 gigawatts.

Batteries aside from standard lithium-ion ones just described have been developed. These include the following.

Iron-air batteries (also called reversible rust batteries) have one cell containing iron oxide (rust) and another containing pure metallic iron. During charging, an electric current breaks down iron oxide into metallic iron and oxygen. The oxygen is released to the air. During discharging, metallic iron combines with oxygen from the air to reform iron oxide and produce electricity. The main advantage of an iron-air battery is the low cost of the main material (iron) and the system as a whole. The estimated battery pack cost in 2024 was $20 per kilowatt-hour of energy storage,[44] which compares with ~$76 per kilowatt-hour for LFP batteries. Good things about iron-air batteries are that iron is a common, safe, and durable element, and the batteries can be scaled and recycled easily. However, because iron-air batteries are much larger than lithium-based batteries with the same storage capacity, iron-air batteries are not useful for transportation, only for stationary electricity storage.

Transition highlight
In 2024, the first plant for manufacturing iron-air batteries opened in Weirton, West Virginia, through a startup called Form Energy.[44] Their batteries will first be used to provide 85 megawatts (million watts) of peak power and 8,500 megawatt-hours of grid electricity storage in Lincoln, Maine. Thus, this battery storage facility is anticipated to allow 100 hours of storage at the peak discharge rate of 85 megawatts. This compares with 4 hours of storage for lithium-based batteries. However, lithium-based batteries can also achieve 100 hours of storage when they are concatenated together. For example, 25 four-hour batteries can provide 100 hours of storage at the peak discharge rate of one battery. Batteries with fewer hours of storage have higher peak discharge rates than batteries with more hours of storage, for the same energy storage capacity.

Basalt-stone batteries consist of two insulated steel tanks, each filled with pea-sized basalt stones. One tank is cold, and the other, hot. Excess renewable electricity from the grid is used to compress, and thereby heat, air from the cold tank. The lower the temperature of the cold tank, the less the energy needed to raise the temperature by a fixed amount. This is because compressing hot air takes more energy than compressing cold air. The compressed hot air is then transferred to the hot tank with an electric heat pump, which then expels cold air into the cold tank. The hot tank heats to 600 degrees Celsius, whereas the cold tank cools to –30 degrees Celsius. Due to the insulation, the hot and cold tanks can maintain their temperatures for several days. To store more energy, more stones are added to each tank. When electricity is needed, more cold air from the cold tank is compressed and sent to the hot tank, forcing hot air from the hot tank to move to an expander. As air expands and cools in the expander, the air flows through an expansion turbine to generate electricity. The cold air is then returned to the cold tank. Because basalt is inexpensive, the cost of this battery is also anticipated to be low. The roundtrip efficiency of the battery is 55–60 percent.[45] However, if some of the lost heat is sent to a district heating system, then the efficiency increases to 90 percent. The estimated future cost of basalt-stone batteries is about $11 per kilowatt-hour.[45]

Sodium-sulfur batteries contain sodium and sulfur in the two half-cells. During discharging, sodium in the sodium half-cell gives up an electron. The electron travels through an external circuit that has an electricity-using device (such as a light bulb) connected to it and then to the sulfur side of

3.3 STATIONARY BATTERIES

the battery. The positive sodium ion simultaneously passes through a membrane to the sulfur side, where it reacts with the sulfur and the electron to form a sodium-sulfur compound. During charging, the reverse process occurs. Sodium-sulfur batteries have high energy densities and high charge and discharge efficiencies, and they last many cycles. Because they are made of inexpensive material, they are also low cost, which makes them ideal for stationary electricity storage. Due to their high energy density (similar to lithium-based batteries), they can also be used for transportation. In fact, one of the first applications of the sodium-sulfur battery was in the 1991 Ford Ecostar demonstration battery-electric vehicle, which never went into commercial production. Sodium-sulfur batteries operate at high temperature (290–360 degrees Celsius). One such battery operating at these temperatures is installed in western Bulgaria. Its peak discharge rate is 250 kilowatts, and it stores 1,450 kilowatt-hours of energy, giving it 5.8 hours of storage at the peak discharge rate.[46] It is expected to fully cycle once per day for 20 years (thus, 7,300 cycles in total). Research is ongoing to allow the batteries to be used at medium temperatures (120–300 degrees Celsius) and room temperature.[47,48,49]

Aluminum-ion batteries use aluminum instead of lithium to flow between cells. An aluminum ion, Al^{3+}, carries three times the charge of a lithium ion, Li^+. This allows aluminum-ion batteries to be smaller than lithium-ion batteries. However, the higher charge of an aluminum ion also results in more interaction of aluminum ions than lithium ions with other chemicals in the battery, reducing the efficiency and storage capacity of aluminum-ion versus lithium-ion batteries. Aluminum-ion batteries are less flammable than lithium-ion batteries. Aluminum is also a more common and less-expensive raw material than lithium. An aluminum-ion battery paired with a graphene electrode may charge up to 60 times faster than a lithium-ion battery and hold up to three times the charge of other aluminum-based battery cells.[50] Another electrode material, organic redox polymer, may also increase the storage capacity of aluminum-ion batteries.[51]

Salt-water batteries are similar to lithium-ion batteries, except that with salt-water batteries, sodium ions, instead of lithium ions, migrate. A concentrated saline solution (salt water or sodium sulfate mixed with water) is used as the electrolyte to conduct the sodium ions. Salt-water batteries are nonflammable, nonexplosive, and have no toxic chemicals, so they are easily recycled. In addition, salt-water batteries can run for 10,000 cycles or more. However, salt-water batteries have low power densities, and the voltages in them are one-third of those in lithium-ion batteries, so they require a large volume, meaning they can't be used in vehicles and are suitable primarily for stationary power storage. In addition, their cost is currently higher than that of lithium-based batteries.

Vanadium-flow batteries consist of two separated tanks, each with vanadium ions of a different oxidation state dissolved in a liquid electrolyte. The fluid in each tank is pumped to a membrane, across which ions are exchanged, creating a current that flows between the two tanks through an external wire. During electricity charging, ions flow in one direction through the pipe. During discharging, they flow in the opposite direction. Vanadium-flow batteries can be run for 20,000 cycles, are recyclable, and are safe.[52] Disadvantages are that they have low energy densities (and thus require a large volume compared with lithium-based batteries and can be used only for stationary electricity storage) and are more expensive than lithium-based batteries.[52]

3.4 CONCENTRATED SOLAR POWER WITH STORAGE

Parabolic-trough and central-tower concentrated solar power (CSP) plants work by heating a fluid (either oil for a parabolic trough or a molten salt mixture for a central tower), which passes next to but is separated from water in a heat exchanger in order to boil the water. The resulting steam runs a steam turbine to generate electricity. For both the parabolic-trough and central-tower plants, the hot fluid can first be stored in an insulated tank before it is sent to the heat exchanger. This allows electricity production to be delayed until either nighttime, days of heavy cloud cover, or times when electricity demand is high or when no other WWS electricity source is available. After the steam passes through the turbine, it is routed to a condenser, which liquefies the steam and sends the liquid water back to the heat exchanger in a closed loop to be reheated. After the molten salt exchanges its heat with water in the heat exchanger, the molten salt is also cold and is sent to a holding tank before it is passed back through either the parabolic-trough or the central-tower receiver to reheat.

Whereas CSP heat can be stored in a tank containing molten salt, the heat can alternatively be stored in a phase-change material. A phase-change material is an initially solid material that absorbs a large amount of heat before it melts and that releases a large amount of heat when it resolidifies. Phase-change materials require less volume and thus store heat at less cost than molten salt.[53]

The storage associated with CSP is usually sized to last up to 15 hours at the peak discharge rate (nameplate capacity) of the CSP generator before

3.4 CONCENTRATED SOLAR POWER WITH STORAGE

the storage is depleted. Storage of up to 15 hours allows for 24 hours per day of electricity production from CSP plus storage.

Transition highlight
The Gemasolar CSP plant in Seville, Spain, was the first central tower CSP plant to include storage. It consists of 2,650 mirrors, a 140-meter-tall tower, and 8,500 tonnes of molten salt that flows through the tower.[54] When cold (290 degrees Celsius) liquid salt is pumped up to the top of the tower from a cold storage tank, it is heated to 565 degrees Celsius by the concentrated light from the mirrors before the hot salt is moved to the hot storage tank. When electricity is needed, the hot salt is passed next to but separated from water to produce steam, which generates electricity through a 20-megawatt generator. Gemasolar first produced electricity 24 hours per day in July of 2011. The plant has provided electricity continuously (24 hours per day) for stretches of up to 36 days. During cloudy and winter days, a CSP-with-storage plant also produces electricity at night, but for fewer hours.

The **capacity factor** of a CSP plant is the annually averaged power produced by the generator in the plant divided by the maximum possible power produced by the generator (its nameplate capacity). With no storage, the capacity factor of a CSP plant is limited by the instantaneous electricity generated, averaged over a year. Excess heat not used immediately is wasted because the generator can produce only a limited amount of electricity. A typical capacity factor of a CSP plant without storage in a sunny location is around 25 percent. With storage, however, the capacity factor increases to around 65 percent or higher. This is accomplished first by adding more mirrors and molten salt (for a central-tower CSP plant) and storing the salt after it is heated. During the night, or when sunlight is weak, the stored hot salt is used to boil water. The additional steam is used to produce electricity in the absence of sunlight, increasing the plant's annual electricity output and thus its annually averaged capacity factor.

The capacity factor of a CSP plant can be increased even further if wasted heat that is not turned into electricity is captured and used to produce heat for industrial processes or used as a source of heat for heat pumps.

Finally, CSP plants with storage can help to keep the electric grid stable. With storage, CSP plants can ramp from zero to 100 percent power production in about 10 minutes,[55] which is faster than the ramp rate of a coal or nuclear plant (20–100 minutes) and similar to that of a fossil-gas plant (5–20 minutes).[37] As such, CSP with storage can help meet peaks in electricity demand, with 100 percent WWS.

3.5 FLYWHEELS

A **flywheel** is a spinning wheel or disk, usually made of steel or carbon fiber, that rotates around an axis. The axis is perpendicular to the ground, as it is with a spinning top, so that gravity acts equally on all sides of the spinning wheel. A flywheel stores energy as rotational energy, then converts that energy to electricity as needed. A flywheel is an electric motor, an energy storage device, and a generator, all in one. When excess electricity is available, it powers an electric motor to rotate the flywheel up to a high speed. Of the energy added to a flywheel, a small portion is used to keep the flywheel rotating. The rest is maintained as stored energy. If the flywheel is run in a vacuum, air resistance is zero, so frictional loss is minimized. Another way to minimize frictional loss is to use an electromagnetic bearing or a permanent magnet. This allows the spinning rotor to float.

When electricity is needed, the motor turns into a generator, which converts rotational energy into electricity. When electricity is produced, the flywheel slows but does not stop. A flywheel can store more and more rotational energy until its rotor shatters. Steel flywheels are limited to around 3,000 rotations per minute. High energy density carbon fiber flywheels can spin up to 60,000 rotations per minute. Aside from possible breakage, flywheels require little maintenance and have a long life (about 20 years) with no impact on the environment except for the impact of obtaining the materials used to build them.

A flywheel accumulates and stores kinetic energy produced by either steady or intermittent electricity over any period of time and releases that energy as electricity at a fast rate over a short period. As such, flywheels are ideal for storing excess electricity from intermittent solar and wind. However, flywheels developed to date have stored only small amounts of energy (e.g., 3–25 kilowatt-hours) and have relatively high loss rates (1.5–3 percent per hour). As such, the stored electricity must be used quickly; otherwise, it will all be lost in 33–66 hours.

On the other hand, flywheels can begin to discharge quickly (within 4 milliseconds) and can discharge at a high rate (10–100 kilowatts). Thus, two useful applications of a flywheel are to provide short-term peaking power for the electric grid and to charge electric vehicles quickly. For example, a flywheel that has a peak discharge rate of 100 kilowatts can add 20 kilowatt-hours to an electric vehicle in 12 minutes.

Flywheels have historically been expensive due to their high electricity loss rates. However, with increasing use of recyclable and stronger materials, flywheel costs have declined, lifetimes have increased (to 25 years/50,000 cycles), and temperature ranges of usability have expanded (now from −29 to 60 degrees Celsius).[56] As a result, flywheels are now deployed commercially in hundreds of residential homes.[56]

3.6 COMPRESSED-AIR ENERGY STORAGE

Compressed-air energy storage (CAES) is another technology that can be used to accumulate intermittent renewable electricity in storage over a long period and then to resupply that electricity to the grid when needed. With compressed-air storage, excess intermittent electricity is used to compress air. When electricity is needed, the compressed air is expanded in an expansion turbine. The turbine's rotating shaft is connected to a generator, which converts the rotational energy to electricity.

Compressed air can be stored in an underground cavern, a salt dome, an aquifer, or a closed vessel. Whereas compressed-air storage has been studied extensively, only a few utility-scale facilities have been built, including one in Germany (1978); two in the United States – Alabama (1991) and Texas (2012); one in Ontario, Canada (2019); one in Sardinia, Italy (2022); and three in China – Jiangsu (2022), Zhangjiakou (2022), and Shangdong (2024). The round-trip efficiencies (electricity output divided by electricity input) of these plants have ranged from 42 to 70 percent.

> **Transition highlight**
> In one example, the Hydrostor storage facility in Ontario, Canada, a compressor compresses air within a former salt cavern deep under the town of Goderich. Excess electricity from the grid is used to power the compressor. When electricity is needed, the air is expanded and runs through a 10-megawatt turbine connected to a generator. Sufficient electricity is stored to provide five hours of electricity to 2,000 homes.

The locations of large-scale underground compressed-air storage facilities are limited geographically. However, a small compressed-air storage system can be connected to an electricity-producing device, such as a wind turbine. The wind turbine operates normally to produce electricity in a generator. Electricity from the generator at hub height is then used to power a motor that compresses air that is stored in an air-storage tank. When electricity is needed for the grid, the compressed air from the storage tank is expanded and converted back to electricity in an expander-driven generator. The motor, storage vessel, and second generator are all housed at ground level, below the turbine.[57]

3 WWS SOLUTIONS FOR ELECTRICITY STORAGE

Transition highlight
Another variation of compressed-air storage is the use of excess renewable electricity to cool and then compress air until it condenses as a liquid. The high-pressure, cold liquid is stored in a container until electricity is needed. At that point, the air is warmed until it reevaporates. The expanding air then drives a turbine to generate electricity. This type of storage is referred to as **liquid-air energy storage**. The process is made efficient by storing the heat released during condensation and using that heat to help reevaporate the liquid air. The process does not require special materials, such as lithium used in batteries. A 50-megawatt/300 megawatt-hour storage facility based on this concept, which was developed by inventor Peter Dearman, is being built in Carrington, Manchester, UK, and is expected to be complete by 2026.[58]

3.7 GRAVITATIONAL STORAGE WITH SOLID MASSES

A form of electricity storage similar to pumped-hydro storage, except that it uses solid materials instead of water, is **gravitational storage with solid masses**. In one version of this storage, excess electricity from the grid is used to power an electric motor in a crane to lift cement blocks against the force of gravity and stack them, one at a time, on a tower. When electricity is needed, the crane grabs each block and slowly lowers the block toward the ground. The downward motion uncoils the hoist chain holding the block. The rotating motion during the uncoiling is translated to a rotating motion inside the same electric motor that lifted the block, turning the motor into an electric generator.[59] The electricity produced by the generator is sent to the grid.

Transition highlight
One company that is building a gravitational storage system is using a large crane to lift and lower concrete blocks.[60] An advantage of this system is its long lifetime, estimated to be at least 35 years. The efficiencies for charging and discharging storage are stated both to be 90.5 percent, resulting in a round-trip efficiency (electricity consumed per unit electricity input) of 82 percent. The one-way efficiencies break down into a crane/pulley efficiency of 96 percent, a gear efficiency of 98 percent, a motor/generator efficiency of 98.2 percent, a variable frequency drive efficiency of 98.7 percent, and a transformer efficiency of 99.3 percent.[61] So long as the crane is holding a block, electricity can be produced almost instantaneously when needed.

In another version of gravitational storage, excess grid electricity is used in an electric motor to move a train carrying rocks or concrete blocks up a hill. When electricity is needed, the train rolls down the hill, almost instantly producing electricity from a generator (motor running in reverse). The round-trip efficiency is stated to be about 86 percent, similar to that of moving mass vertically.[62]

In a third version of this technology, excess grid electricity is run through a motor to lift sand or gravel in a container up a mountain slope via a cable. Containers of sand or gravel are stored at the top of the mountain. When electricity is needed, the containers are lowered down the slope via the cable. The rotating motion of the coil at the top of the mountain produces electricity in the generator. This type of storage is referred to as **mountain gravity energy storage**.[63] It can store electricity for use on the timescale of seconds, minutes, days, weeks, or months. It is most efficient in locations with tall, steep mountain slopes.

In a fourth version of this technology, concrete blocks or large stones are lowered and raised within abandoned fossil-oil and gas wells.[64] The United States alone has over 3.2 million abandoned wells, many of which leak gas. The idea is first to plug the base of each well and then to use its 1,000-meter depth for storage. When excess electricity is available, it is used in a motor to lift a concrete block up the well. When electricity is needed, the concrete block is lowered, and the pulley lowering it is connected to the motor, which is now run in reverse as a generator to generate electricity.

3.8 GREEN-HYDROGEN STORAGE

Green-hydrogen storage involves the production of hydrogen from clean, renewable WWS electricity with an electrolyzer, followed by the compression and storage of the hydrogen in a tank. The electricity is reproduced on demand by running the hydrogen through a fuel cell. Green-hydrogen storage on its own is more expensive than battery storage per unit of power (kilowatt) discharged but less expensive per unit of energy (kilowatt-hour) stored.[65]

The reason for the high cost of green-hydrogen storage per unit power output is the need for several pieces of equipment in addition to water as a fuel. The equipment needed includes a rectifier, an electrolyzer, a compressor, a storage tank, and a fuel cell. A **rectifier** converts AC electricity from the grid to DC electricity that is used to run the electrolyzer. The **electrolyzer** uses DC electricity to split two water molecules ($2H_2O$) into two molecular hydrogen ($2H_2$) molecules and one oxygen molecule (O_2). The **compressor** compresses the hydrogen molecules for storage in the storage tank. The **fuel cell** converts the stored hydrogen plus oxygen from the air back into water while producing DC electricity and heat in the process. A battery, on the other hand, requires only a rectifier and the battery.

The reason green-hydrogen storage costs less per unit energy stored than a battery is that, with green-hydrogen storage, only a larger storage tank is needed to increase energy storage, and storage tanks represent only a small portion of green-hydrogen storage's overall cost. On the other hand, to increase battery storage, a bigger battery is needed, and the battery itself is the most expensive component of battery storage cost.

Transition highlight
A study across 145 countries transitioning all energy sectors to 100 percent WWS found that, in some countries, using only existing conventional hydro and batteries resulted in the lowest-cost energy system.[65] In the remaining countries, the combination of conventional hydro, batteries, and green-hydrogen storage was lowest cost. In no case was hydropower plus green-hydrogen storage lowest cost. In addition, combining green-hydrogen storage for grid electricity with storage of hydrogen for nongrid purposes (steel and ammonia manufacturing and long-distance transport) reduced cost compared with separating grid from nongrid-hydrogen storage.

4

WWS SOLUTIONS FOR TRANSPORTATION

Transportation is the conveyance of people, animals, or goods from one place to another. Types of transportation include skateboards, bicycles, motorcycles, passenger vehicles, sport-utility vehicles, small trucks, large trucks, semi-trucks, buses, forklifts, cranes, tractors, bulldozers, asphalt pavers, backhoe loaders, wheel loaders, cold planers, compactors, excavators, harvesters, graders, off-road vehicles, trains, ferries, motorboats, yachts, ships, helicopters, airplanes, battle tanks, infantry fighting vehicles, armored personnel carriers, armored combat support vehicles, mine-protected vehicles, light armored vehicles, light utility vehicles, amphibious vehicles, and more. All of these, except skateboards and bicycles, currently move primarily by burning gasoline (petrol), diesel, biodiesel, methanol, ethanol, liquefied natural gas, bunker fuel, or jet fuel in an internal combustion engine. These fuels are all derivatives of fossil fuels or biofuels. The cleanest and most efficient method of replacing these fossil-fuel and biofuel vehicles is to convert them to battery-electric vehicles or hydrogen-fuel-cell-electric vehicles. Such vehicles emit no chemicals from the tailpipe aside from, in the case of hydrogen-fuel-cell-electric vehicles, water vapor. If the electricity going into the battery or used to produce hydrogen is from WWS, then the vehicles are emission free in their energy production as well. This chapter discusses these two WWS solutions. Hydrogen combustion is not supported because it results in chemical air pollutants containing hydrogen, nitrogen, and oxygen. Ammonia combustion is not supported because it results in similar chemicals plus unburned ammonia, which is one of the most dangerous chemicals in photochemical smog due to its propensity to dissolve in water and form particulate matter.

4.1 BATTERY-ELECTRIC VEHICLES

Battery-electric vehicles store electricity in batteries, then draw power from the batteries to run an electric motor that moves the vehicle. Battery-electric vehicles produce zero tailpipe emissions. However, in many countries today, a good portion of electricity used for battery-electric vehicles comes from fossil-fuel power plants. In a 100 percent WWS world, though, all electricity for battery-electric vehicles will come from zero-emission WWS sources.

Some argue that if electricity for a battery-electric vehicle comes from a polluting source, then the battery-electric vehicle is just as bad for health as a fossil-fuel vehicle, which emits pollution continuously from its tailpipe. However, that is not the case. The reason is that the intake fraction of pollution from street traffic is 15–30 times that of pollution from power plants.[66] The **intake fraction** is the ratio of the mass of pollutant inhaled by a population to the mass of pollutant emitted by a source. In other words, people, on average, breathe in a much greater fraction of vehicle exhaust than power-plant exhaust. As a result, taking pollution off the street and moving it to a power plant reduces health impacts significantly, although it does not eliminate health impacts. With 100 percent WWS, though, all powerplant emissions are also eliminated.

In addition to producing tailpipe emissions, fossil-fuel vehicles produce emissions from the mining, transporting, and processing of oil to produce gasoline, diesel, jet fuel, and bunker fuel. In a WWS world, such emissions are eliminated.

Fossil-fuel- and biofuel-powered vehicles also emit aerosol particles during the use of brake pads. In such vehicles, the car slows when a brake pedal is pushed. This pushes a brake against a turning tire disk, reducing the rotation rate of the tire, slowing the vehicle. The resulting friction creates heat and causes some particles to break off the brake pads and float into the air. Electric vehicles eliminate most brake-pad emissions because such vehicles slow down primarily through regenerative braking, although drivers still occasionally need to use their brakes and thus their brake pads.

Regenerative braking works as follows. To slow a car, the driver's foot is taken off the accelerator pedal. Doing so causes rotational energy from the rotating wheels to feed into a generator (the car's motor running in reverse) to "regenerate" electricity, which is then stored in the vehicle's supercapacitors, which are similar to batteries (see Chapter 9). Thus, regenerative braking, which converts rotational energy into electricity, slows the car, avoiding the need for the use of brake pads, avoiding brake-pad particle emissions.

Regenerative braking was invented in 1886 by the Sprague Electric Railway & Motor Company, founded in 1884 by **Frank Sprague** (1857–1934). Sprague had worked for Thomas Edison since 1883, but because Sprague wanted to develop motors, which Edison was less interested in, he

4.1 BATTERY-ELECTRIC VEHICLES

left Edison's company in 1884 to start his own. Sprague's first application of regenerative braking was to electric streetcars that his company put in place in Richmond, Virginia, in 1888. Sprague's company also developed regenerative braking for elevators.

All vehicles today emit particles due to tire wear. However, such particles are mostly larger and fewer in number than combustion particles. Because tire particles are generally large, most drop out of the air faster and do not penetrate deep into people's lungs like combustion particles do. Finally, for all vehicles, energy is used to mine materials for the vehicle, build the vehicle, and recycle the vehicle at the end of its life. In a WWS world, such energy will come from clean, renewable WWS sources, with zero emissions.

In sum, transitioning to battery-electric vehicles substantially reduces air pollution and its health problems. Providing the electricity needed to manufacture and run a battery-electric vehicle with WWS reduces pollution even more.

4.1.1 EFFICIENCY OF BATTERY-ELECTRIC VEHICLES

Another advantage of today's battery-electric vehicles compared with gasoline vehicles is that battery-electric vehicles can travel three to five times as far per unit of energy input from the vehicle charger as a gasoline vehicle travels per unit of energy in gasoline.

For example, only 17–20 percent of the energy in gasoline goes toward moving a gasoline passenger vehicle. This is the **tank-to-wheel efficiency** of the vehicle. The rest of the energy (80–83 percent) is lost as waste heat.

On the other hand, 64–89 percent of the electricity from a plug outlet moves a battery-electric passenger vehicle, and the rest is waste heat. This is the **plug-to-wheel efficiency** of a battery-electric vehicle. This efficiency accounts for the efficiency loss of charging the vehicle and other factors.

As such, a battery-electric vehicle requires much less energy input than does a fossil-fuel vehicle. Put differently, a conversion from fossil-fuel vehicles to battery-electric vehicles reduces energy demand for transportation fuel by a factor of between three and five.

The plug-to-wheel efficiency of a battery-electric vehicle is determined as follows. First, a permanent-magnet electric motor has an efficiency of 89–96 percent. An alternative, an induction electric motor, has an efficiency of 84–94 percent.[67] Thus, the range of efficiencies of electric-car motors is 84–96 percent. The Tesla *Model S*, *Model X*, *Model 3*, and *Model Y* all use permanent-magnet motors as of 2024.

In addition, efficiency losses occur due to converting electricity from the grid to chemical energy in a battery. Such vehicle-charging losses can range from 4 to 20 percent of the electricity going into the battery, depending on the current and voltage used to charge the vehicle. Another 1–2 percent of energy is lost during conversion of the battery's chemical energy back to DC electricity. In addition, 2–3 percent is lost with the combination of converting that DC electricity back to AC electricity in an inverter, adjusting the voltage for use in the electric motor, and using power electronic controls in the vehicle. Inverter losses are only about 1 percent of energy.

Accounting for all these losses gives the overall plug-to-wheel efficiency of a battery-electric passenger vehicle as 64–89 percent, with an average of 77 percent.

Transition highlight
A 2023 Ford *F-150* four-wheel-drive extended-range electric pickup truck requires 480 watt-hours of electricity to move the vehicle one mile (1.61 kilometers). This translates to 931.3 kilometers traveled per gigajoule of electrical energy. A 2023 four-wheel-drive Ford *F-150* flex-fuel pickup truck, which runs on either gasoline or an ethanol-gasoline blend, travels 14 miles (22.5 kilometers) per gallon of the blended fuel. This translates to 252.3 kilometers traveled per gigajoule of energy in the blended fuel. Thus, the electric *F-150* travels 3.7 times the distance of the flex-fuel vehicle for the same energy input. Accounting for the price of electricity and of an ethanol-gasoline blend in Iowa gives a fuel-cost savings alone of driving the electric *F-150* of $26,500 over 15 years if both trucks are driven 15,000 miles per year.[68]

A battery-electric vehicle can increase its efficiency, and thus its range, by using regenerative braking. Regenerative braking converts rotational energy into electricity, slowing the vehicle.

Transition highlight
Regenerative braking can not only extend battery-electric vehicle range substantially, but it can also eliminate, in at least one special case, the need to recharge an electric vehicle entirely. The largest stand-alone electric vehicle in the world in 2019 was an enormous electric dump truck operating in a quarry in Biel, Switzerland. The empty truck used electricity to climb from the bottom to the top of the quarry. At the top, it was filled with ore, more than doubling the truck's weight compared with its uphill trip. The additional weight created so much kinetic energy that the electricity produced from regenerative braking during the downhill trip exceeded the electricity consumed during the uphill trip. As such, the truck did not need recharging during its daily operation.[69]

4.1 BATTERY-ELECTRIC VEHICLES

In a WWS world, battery electricity will be used not only to power light-duty vehicles, but also short- and medium-range semi-trucks, short-range aircraft, some military vehicles, and short-distance boats and ships. The history of battery-electric transportation is discussed next.

4.1.2 HISTORY OF BATTERY-ELECTRIC TRANSPORTATION

Today, battery-electric vehicles are being developed for ground, water, and air transport. Here, the evolution of battery-electric vehicles is explored for each of these transport types.

4.1.2.1 GROUND VEHICLES

Passenger Cars

Scotland's **Robert Anderson** built the world's first battery-electric vehicle, which was also the world's first horseless carriage, sometime between 1832 and 1839. To accomplish this feat, he affixed a battery and a motor to a carriage. At the time, batteries were not chargeable, so the carriage's battery needed to be replaced after each use. Thus, Anderson also developed the first swappable battery system. Rechargeable battery-electric vehicles were not possible until 1859 when the French physicist, Gaston Planté (1834–1889) invented the first chargeable battery, a lead-acid battery.

Around 1884 in Birmingham, UK, Thomas Parker used lead-acid batteries in his invention of the world's first electric car that used chargeable batteries. This car had room for two people, a top speed of 19.3 kilometers per hour, and a range of 64.4 kilometers. Parker later invented an electric tram that ran on the streets of Birmingham and an electric locomotive for use both above and below ground.

Meanwhile on January 29, 1886, the German engineer Karl Benz filed for a patent in Germany for the world's first gasoline-powered car. Gasoline is a refinery product of crude oil discovered by accident in 1861 during the refining of crude oil to kerosene. Until 1879, when Benz recognized that gasoline was useful for running combustion engines, gasoline was used only as a solvent or discarded. Benz drove his two-seat, uncovered gasoline car in public for the first time on July 3, 1886. Its top speed was 16 kilometers per hour, thus slightly slower than Parker's electric car.

In 1887, William Morrison of Des Moines, Iowa, developed the first battery-electric car in the United States. It was built by combining a carriage produced by the Des Moines Buggy Company, 24 batteries that he made

4 WWS SOLUTIONS FOR TRANSPORTATION

himself, and an electric motor. In 1888, he showcased the car, which had a top speed of 32.2 kilometers per hour and an overall range of 81 kilometers, in the Des Moines city parade.[70]

In 1894, Pedro Salom and Henry G. Morris of Philadelphia patented several battery-electric street cars and boats. One battery-electric carriage they developed, called the Electrobat, traveled 40 kilometers at a top speed of 32 kilometers per hour. Salom and Morris turned this idea into a startup, which they sold to Isaac L. Rice, who in turn founded the Electric Vehicle Company in New Jersey. By the early 1900s, the company had produced more than 600 electric taxi cabs used in New York City, Boston, Baltimore, and other cities. In New York City, the cab batteries were replaced at a battery-swapping station. Conflicts among investors caused the company to fold in 1907.

In 1898, Ransom Olds, who had founded the Olds Motor Vehicle Company in 1897, decided to manufacture electric cars, which were less complicated and more reliable than gasoline cars of the day, which he also produced. Olds retrofitted a part of his factory in Detroit, Michigan, to mass-produce battery-electric vehicles. However, on March 9, 1901, a worker's mistake sparked a fire that destroyed the entire factory and Olds' electric-car dream. Only one electric car survived the blaze.

In 1899, the Belgian Camille Jenatzky developed an early electric race car, shaped like a torpedo. It was the first electric vehicle to break the 100-kilometer-per-hour speed barrier.

By 1900, electric cars had become so popular that they comprised 38 percent (33,842 cars) of the US automobile fleet, a higher share than gasoline cars (22 percent – 19,600 cars) but a share slightly lower than that of steam-powered cars (40 percent – 35,600 cars).[71] Steam-powered cars are propelled by a steam engine, which is the same type of engine used to move a locomotive. However, in a steam-powered car, the steam is produced by burning kerosene to boil water in a small boiler (about 0.6 meters in diameter) within the car. Kerosene is a refinery product of crude oil. In a locomotive, coal is used to boil water. In a steam engine, steam pushes a piston up, and condensation of the steam (by spraying liquid water into it) creates a vacuum to pull the piston down. This up-and-down motion is translated into a circular motion with a rod and crank connecting the pistons to a cylinder. The rotating cylinder moves the car's wheels.

The public wanted electric cars because they were quiet, easy to drive, required few repairs, and did not produce pollution. Unfortunately, once Henry Ford began mass-producing, in 1908, the *Model T*, which cost one-third of the price of a battery-electric car, the electric-car market collapsed.

4.1 BATTERY-ELECTRIC VEHICLES

The expansion of gas stations and the greater range and top speed of gasoline cars through the 1920s sealed the fate of all battery-electric vehicles for the next several decades.

> **Transition highlight**
> It is interesting to wonder what might have happened if oil had never been discovered. No gasoline car or steam-powered car running on kerosene would have been invented. Battery-electric vehicles may have dominated car, truck, and boat transportation. More effort might have gone into improving batteries and electric vehicles, so battery-electric vehicle technologies might have improved rapidly in the early 1900s. Over the next 100 years, enormous amounts of health- and climate-affecting air pollutants from tailpipes would have then been avoided. Of course, the electricity would have come mostly from burning coal and wood. In addition, many of the industrial byproducts of crude oil, such as plastics and some medicines, would not have been invented. On the other hand, more environmentally friendly alternatives to such products could have been invented instead. Overall, if oil had never been discovered, it is likely that the transition away from fossil fuels today would have been much faster than it is.

Following the death of the electric-car industry in the early 1900s, it wasn't until 1973 that General Motors developed a new-generation tiny two-seat urban battery-electric-car prototype. The car was never sold commercially. Only after the Arab Oil Embargo of October 1973 to March 1974 did interest in funding the research and development of battery-electric vehicles revive. This resulted in the American Motor Company producing electric delivery jeeps for a US Postal Service test program in 1975. These vehicles, though, were still limited in their range (64 kilometers) and top speed (72 kilometers per hour). More work was needed. In 1976, the US Department of Energy began to fund the research and development of battery-electric and hybrid gasoline/battery-electric vehicles.

In the early 1990s, interest in battery-electric vehicles rose again in the US due to tougher internal-combustion-engine vehicle emission standards enacted under the 1990 US Clean Air Act Amendment and by the California Air Resources Board. As a consequence, General Motors developed the *EV1* electric car, which had a range of 129 kilometers and good acceleration. Because the cost to produce it was high, the *EV1* was never commercialized, and General Motors scrapped its development in 2001.

In 1997, Toyota mass-produced the first commercial hybrid gasoline/battery-electric vehicle, the *Prius*. It was released worldwide in 2000. In 1999, Honda also released a gasoline/battery-electric hybrid, the *Insight*.

The commercial availability of both vehicles motivated research and development of a new generation of pure battery-electric vehicles.

Arguably the most important event in the history of battery-electric transportation was the announcement in 2006 by a startup in California's Silicon Valley, **Tesla Motors**, that it would produce an all-electric luxury sports car. Tesla Motors was founded on July 1, 2003, in San Carlos, California, by **Martin Eberhard** and **Marc Tarpenning**. **Elon Musk** was the major initial funder of Tesla Motors. **Ian Wright** and **J. B. Straubel** joined Tesla during the next few months. All five are considered cofounders of the company. Straubel was the chief technical officer. Elon Musk took over as chief executive officer of the company during October 2008. Tesla Motors was named after inventor **Nikola Tesla** (1856–1943), who designed and improved electric motors, electric generators, and the alternating current (AC) electricity grid.

Tesla Motors (later changed to Tesla, Inc.) focused on an all-electric car rather than on a hybrid car because of the simplicity and reliability of using only electricity rather than electricity plus gasoline. They built a sports car first to change the public perception of electric vehicles. At the time, the public had a poor image of electric vehicles, associating them with short range, low speed, and clunkiness.

The sports car built was the Tesla *Roadster*. The first one was delivered during February 2008. Between 2008 and 2012, 2,450 *Roadsters* were sold in over 30 countries. The car had a peak range of 244 miles (393 kilometers) on a single charge and an acceleration of 0 to 60 miles per hour (96.6 kilometers per hour) in 3.7 or 3.9 seconds, depending on the model. It had a Lotus body that looked sleek. The amazing thing about the Tesla *Roadster* was not the number of cars sold but how it changed people's view of electric vehicles. Instead of electric cars being stereotyped as transport for tree huggers over short distances, they were now considered sleek, beautiful, powerful, and capable of being used for at least 90 percent of travel. However, the price of the *Roadster* was out of reach for most people on Earth.

Tesla, though, had already made plans to capitalize on the popularity of the *Roadster* by developing a series of lower-priced electric passenger cars. These included the *Model S* (five-door hatchback sedan), first delivered in 2012; the *Model X* (three-row crossover sport utility vehicle), first delivered in 2015; the *Model 3* (midsize four-door sedan), first delivered in 2017; and the *Model Y* (compact sport utility vehicle), first delivered in 2020.

After Tesla created a new market for battery-electric vehicles, many other car manufacturers began developing prototypes and then vehicles for sale. By 2024, over 40 electric-car models were available for sale in the

4.1 BATTERY-ELECTRIC VEHICLES

United States alone. The longest-range vehicle was the Lucid *Air* (range of 830 kilometers, or 516 miles), followed by the Tesla *Model S* (647 kilometers, or 402 miles), and the Hyundai *Ioniq* 6 (581 kilometers, or 361 miles). The least expensive cars per dollar of range were the Hyundai *Ioniq 6 SE Long Range* ($65 per kilometer of range), the Hyundai *Kona Electric* ($77.6 per kilometer), the Nissan *Ariya* ($81 per kilometer), the Nissan *Leaf* ($83 per kilometer) and the Tesla *Model Y* Long Range ($84.5 per kilometer). Only one car, the Nissan *Leaf*, had a base price of less than $30,000.[72]

By the end of 2023, about 28.2 million pure battery-electric cars had been sold worldwide, including 9.9 million in 2023 alone.[73] Of the 28.2 million battery-electric cars on the road, 16.1 million were in China, 6.7 million were in Europe, 3.5 million were in the United States, and 1.9 million were in the rest of the world. In the United States as of May 2024, five times the number of battery-electric cars were registered in California as in Florida, the state with the second-most registrations. Texas and Washington State were ranked third and fourth, respectively.[74] While this is progress, 28.2 million cars represent only 1.9 percent of the 1.475 billion vehicles on the road worldwide (1.1 billion passenger cars and 375 million trucks and buses).

Transition highlight
In September 2024, the number of battery-electric cars in Norway (754,300) surpassed the number of gasoline cars (753,900), although the number of diesel cars (around 1 million) still exceeds both these figures.

Bikes, Scooters, and Motorcycles

A logical use of batteries and motors for transportation is in bicycles, scooters, and motorcycles. A battery and electric motor added to a bicycle (**e-bike**) enables riders to either pedal or rely on the bike's motor. An **e-scooter** and **e-motorcycle** replace fossil-fuel-powered versions. The range of an e-bike and e-scooter is 24 to 145 kilometers (15–90 miles), depending on battery size and terrain. The range of an e-motorcycle is 80–724 kilometers (50–450 miles). By the end of 2024, about 44 million e-bikes, 104 million e-scooters, and 58 million e-motorcycles had been sold cumulatively worldwide. Several countries rely on e-scooters and e-motorcycles as their main form of transport.

Pickup Trucks

Several companies have moved beyond cars, developing electric pickup trucks, semi-trucks, and buses. Electric pickup trucks available commercially in 2024 included the Rivian *R1T*, Tesla *Cybertruck*, Ford *F-150 Lightning*, RAM *1500 REV*, Alpha *Wolf*, Fisker *Alaska*, Chevrolet *Silverado EV*, GMC *Hummer EV Pickup*, and the Scout *Terra*.

> **Transition highlight**
> The 2025 Rivian *R1T* pickup truck has a maximum range of 660 kilometers (410 miles) on a single charge when fitted with a large battery pack of 149 kilowatt-hours. So, if the truck is charging at a rate of 20 kilowatts (typical for a relatively fast home garage charger), the time to fully charge it is 7.45 hours.

Semi- and Other Commercial Trucks

Semi-trucks are used mostly to move commercial goods. Battery-electric semi-trucks available commercially in 2024 included the Tesla *Semi*, the Volvo *Semi*, the Daimler/Freightliner *eCascadia Semi*, the Nikola Tre *Electric Semi*, the Bollinger *B4 Semi*, the Farizon *Homtruck Semi*, and the Volkswagen MAN *eTGX Semi*. The Tesla *Semi* maximum range was 805 kilometers (500 miles).[75] The ranges of the other semis were 443, 370, 531, 298, 500, and 800 kilometers (275, 230, 330, 185, 311, and 497 miles), respectively.

In 2023, about 54,000 new battery-electric commercial trucks of all kinds were sold worldwide: 38,200 in China, 10,800 in Europe, 1,200 in the US, and 3,700 in the rest of the world.[76] This brought the cumulative number of such commercial trucks to over 330,000 worldwide, with about 92 percent in China, 5.6 percent in Europe, 0.57 percent in the US, and 1.83 percent in the rest of the world.[76]

In September 2024, REE Automotive announced, together with Roush Industries, the production start of 5,000 commercial *P7* trucks per year in Detroit, Michigan, that are "autonomous-ready" and will run on battery electricity or hydrogen-fuel-cell electricity. In November 2024, Madison, Wisconsin, added two battery-electric garbage trucks to its fleet of 62 battery-electric buses and 100 battery-electric city-owned passenger cars. Cambridge, Massachusetts, similarly added a battery-electric garbage truck and had three more on order.

Buses

Another important vehicle, especially in cities, is the bus. In 2023, about 50,000 new battery-electric buses were sold worldwide: 30,000 in China, 8,000 in Europe, 1,000 in the US, and 10,000 in the rest of the world.[76] Cumulatively, 635,000 electric buses in operation were worldwide by the end of 2023, with about 90 percent in China.[76]

4.1 BATTERY-ELECTRIC VEHICLES

Transition highlight
One major bus manufacturer is BYD (Build Your Dreams). They are headquartered in China and have bus manufacturing plants in several countries. BYD manufactured its first bus in 2010. One of their most popular buses is the BYD K9, also known as the BYD ebus. It uses a lithium-iron-phosphate battery and has a range of 253 kilometers (157 miles) on a single charge, under typical urban street conditions.

Trains, Funiculars, Cable Cars, Trams, Streetcars, Trolleys, and Light Rail

Another area of progress is with electric trains. Electric trains eliminate the belching smoke from trains once powered by steam engines fueled by coal. Electric trains use an electric motor, which is efficient, converting 84–96 percent of input electricity into motion. They also use regenerative braking, thereby generating electricity when they slow down and reducing emissions of pollution particles that arise when brake pads or brake blocks are applied to a train's wheels. An electric train obtains its electricity from either an overhead line, a third rail, or onboard batteries.

Built in 1802, in Shropshire, UK, the first railway locomotive worldwide ran on coal. Only in 1837 did Scottish inventor **Robert Davidson** (1804–1894) build a prototype electric locomotive, which ran on swappable batteries. He then built an improved version in 1841, a train that could travel at a speed of 6.4 kilometers per hour for 2.4 kilometers while towing 5.4 tonnes. Unfortunately, coal-locomotive workers destroyed the train out of fear that they would lose their jobs to this new technology.

Aside from Davidson's prototype, only a few electric locomotives have run solely on batteries since then. In 1917, the Kennecott Copper Mine in Latouche, Alaska, used two battery-powered locomotives to haul dirt. From 1968 to 2009, the Toronto Transit Commission operated a battery-electric locomotive on the Toronto subway. The London Underground also regularly uses battery-electric locomotives for maintenance work. Almost all other electric trains have run on an electrified third rail or an electrified overhead wire.

Possibly the first nonpolluting (both at the vehicle and at the energy source) nonanimal-powered, nonsail-boat commercial public transportation vehicle in history was the funicular. A **funicular** is a railway on a steep slope with two parallel tracks and one car on each track. The two cars are connected by a cable that loops over a pulley at the top of the track. The cars move in concert. As one slides down, it pulls the other up. The word funicular derives from the Latin word *funis*, which means rope.

Transition highlight

The first known funicular in the world was the Prospect Park Incline Railway on the United States side of Niagara Falls. It was built in 1845 but closed in 1907 following a deadly accident that killed one person. Each car carried 15–20 passengers from the top to the bottom of the falls. The whole system was covered. Water was originally used to make one car heavier than the other. Enough water was added to a water tank under the floor of the car at the top of the falls to make the car sufficiently heavy to pull the bottom car up as it rolled down the incline. When the first car reached the bottom, the water was emptied from its tank, and water added to the tank of the second car, which was now at the top. In this way, this railway system operated with zero emissions from either the vehicle or its source of power. The use of water to move the cars was replaced by electricity once the Niagara Falls hydroelectric facility opened in 1895. With the change, the whole system was still pollution free. Several funiculars using water were built in the 1800s, including in Giessbach, Switzerland (1879) and Braga, Portugal (1882). Most other funiculars were powered by steam engines running on coal.

Another nonpolluting (but only at the vehicle), nonanimal-powered commercial public transportation vehicle developed was the **cable car**, invented in San Francisco, California. In 1869, **Andrew Smith Hallidie** (1836–1900) witnessed an accident in San Francisco in which horses pulling a horsecar up a steep, wet cobblestone road (Jackson Street, between Kearny and Stockton Streets) were killed when the horsecar slid backwards, pulling them down. He thought that there must be a safer way to transport people and goods up the steep hills of San Francisco.

Hallidie had a history, starting in 1856, of producing and improving wire ropes for mines, aerial trams, and bridges. He had also built bridges himself. Guided by his experience, Hallidie invented the cable car, which was first tested on August 2, 1873. The first cable-car line in San Francisco was on Clay Street. It began regular service on September 1, 1873, and was immediately successful. Between 1873 and 1890, 23 cable-car lines were built in the city. However, the 1906 earthquake and fire in San Francisco destroyed most of them. Despite a rapid rebuild, by 1912, only eight lines remained. Due to competition by electric streetcars and buses, only three cable-car lines remain in San Francisco today, and they are used primarily for tourism. Two operate between Hallidie Plaza and Fisherman's Wharf, each via a different path, and one operates between Market Street and Van Ness Avenue.

Cable cars move on steel rails. An open slot between the tracks contains a cable that is attached to the car and extends, inside the track, all the way to a central powerhouse full of large wheels. Turning each wheel winds a

4.1 BATTERY-ELECTRIC VEHICLES

cable for an individual car, pulling the car. Cable cars often have grips on both ends so they do not need to be turned around. The cars have no way of moving themselves. They are pulled solely by the cables. As such, cable cars emit zero pollution at the car. Whereas the powerhouse wheels used to pull the cable cars were originally turned using steam from coal burning, this method was eventually replaced with the use of electric motors.

In 1879, 42 years after Davidson invented the electric locomotive and 6 years after Hallidie invented the cable car, the German inventor **Werner Siemens** (1816–1892) operated the first nonbattery-powered electric passenger train, in Berlin, Germany. It consisted of three cars pulled by a locomotive, which traveled up to 13 kilometers per hour. It obtained its electricity from an electrified third rail between the tracks. Over a four-month period, it carried 90,000 passengers around a 300-meter-long circular track.

In 1881, Siemens opened the world's first electric tram line, in Lichterfelde, Germany, near Berlin. It was powered by an electrified third rail. **Tram** generally means the same, in different parts of the world, as **streetcar**, **trolley**, or **light rail**. Trams are smaller and lighter than regular trains and run on a track through public streets. They are powered by electricity from an overhead wire or a third rail. Trams differ from cable cars, which are pulled by underground cables. The first tram with an electrified overhead line appeared near Vienna, Austria, in 1883.

Tram (streetcar) technology improved following inventions by Frank Sprague and the Sprague Electric Railway & Motor Company. In 1886, Sprague's company invented both a constant-speed, nonsparking motor and regenerative braking. The company used these technologies in the world's first large electric streetcar system, completed in Richmond, Virginia, in 1888. The streetcars used electrified overhead lines. In 1889, Boston, Massachusetts, added streetcars, as did other cities. In San Francisco, the first electric streetcar began operating in 1892. Following the 1906 earthquake and fire that destroyed most cable-car lines, electric streetcars took over all but eight cable-car routes.

The first subway in the world to run on electricity was installed in London in 1890. Its source of electricity was an electrified third rail. Electric subways in Budapest, Hungary (1896) and Boston (1897) followed.

Electric (using an electrified third rail or overhead wire) trains, subways, and streetcars are now ubiquitous worldwide. In China, for example, more than 120,000 kilometers (75 percent) of all railway track had been electrified by 2024.[77] In India, it was about 66,000 kilometers (96 percent), while in Japan, almost all railway tracks had been electrified. Electrification rates in the European Union were 56 percent, and in South Korea, 78 percent. In the US, though, railway-track electrification rates were only 1 percent. A challenge going forward is to electrify all railways for all distances everywhere worldwide and power that electricity with WWS.

Agricultural and Construction Machines

Agricultural and construction machines have only recently joined the transition to electricity. In 2017, Solectrac manufactured the first two commercial electric tractors, the *Compact Electric Tractor* (22-kilowatt-hour battery pack) and the *eUtility* (28-kilowatt-hour battery pack). As mentioned earlier, in 2019, the largest stand-alone electric vehicle in the world was completed, the electric dump truck for moving rocks and soil, operating in a quarry in Biel, Switzerland. In 2021, Monarch Tractor developed the first autonomous (no driver needed) electric tractor. Unfortunately, during the first few months of 2022, only 0.02 percent (57) of all new tractors sold in the US were battery-electric. Although electric tractors have lower fuel and maintenance costs and produce less noise than combustion tractors, the main limitation to the growth of electric tractors has been an insufficient charging infrastructure. This is being addressed with bigger and swappable batteries. In 2024, for example, Volkswagen introduced an electric tractor with a 32-kilowatt-hour swappable battery pack for farmers in Rwanda to rent. Also in 2024, Volvo introduced a battery-electric excavator and wheel loader to replace diesel versions at construction sites.

4.1.2.2 WATER VESSELS

Batteries are not limited to land-based vehicles. Several battery-electric marine vessels have been developed for short-distance water transport. In 1893, the Electric Launch Company began manufacturing battery-electric boats in the United States. The company first built 55 boats, each 11-meters long, to ferry over one million passengers during the 1893 World's Columbian Exposition in Chicago. By 1900, the number of battery-electric pleasure boats worldwide exceeded the combined number of coal- and gasoline-powered ones. Around 1910, though, battery-electric boat adoption plummeted as coal- and gasoline-powered boats improved their range and lowered their costs.

> **Transition highlight**
> One battery-electric boat built by the Electric Launch Company was a pleasure-craft for Anheuser-Busch cofounder, Adolphus Busch, that was delivered to him in 1912. This 16.8-meter all-mahogany boat, *Chief Uncas*, was used

by the Busch family during summers on Otsego Lake, near Cooperstown, New York. The boat was converted to gasoline in 1950 but then restored to battery electricity in 2012, with 16 batteries powering two propellers.

In a renaissance of electric marine transport, Norway's Norled began operating the all-electric ferry *Ampere* in 2015. The 80-meter-long ferry moves cars and people 34 times per day between the villages of Lavik and Oppedal, Norway, which are 5.7 kilometers apart. The ferry's battery storage capacity is 1,000 kilowatt-hours.

In 2017, India commissioned a 50-kilowatt-hour, 75-passenger solar-powered ferry, *Aditya*, that operates between Vaikkom and Thanvanakkadavu (2.5 kilometers apart), in the state of Kerala. It has 20 kilowatts of solar PV on its roof. Also in 2017, Denmark retrofitted the 1,250-passenger, 240-car *Tycho Brahe* ferry to make it all-electric (4,160 kilowatt-hours of battery storage). This ferry operates between Denmark and Sweden. In 2017, Finland retrofitted its oldest ferry (operating since 1904), *Elektra*, to be all electric with 1,060 kilowatt-hours of battery storage. The ferry operates between Nauvo and Parainen, in the Turku archipelago.

In 2019, Alabama began operating the battery-electric Gee's Bend ferry (270 kilowatt-hours of battery storage), which carries 15 vehicles and 132 passengers. In July 2020, Damen Shipyard delivered five 50-passenger electric ferries, each with 120 kilowatt-hours of battery storage, to Copenhagen, Denmark. In October 2020, two tourist ferries, each with 316 kilowatt-hours of battery storage, began operating at the base of the Niagara Falls in New York.

In early 2021, Norway launched a 4,300-kilowatt-hour-battery-capacity, 139.2-meter-long ferry that operates between Moss and Horten, 10 kilometers apart. The ferry holds 600 passengers and 200 cars. In July 2021, the Hollands Shipyards Group provided Norway's Brevik Fergeeselskap company a battery-electric ferry with a battery storage capacity of 1,300 kilowatt-hours and room for 98 passengers and 16 cars.

In August 2024, the all-electric ferry, *Ellen*, with 4.3 megawatt-hours of battery storage, completed its maiden voyage off southern Denmark's coast. The 57-meter-long ferry carries up to 200 passengers and 30 vehicles a distance of 40 kilometers. Incat Tasmania is constructing an enormous (40 megawatt-hour) 130-meter-long, 2,100-seat, 225-car battery-electric ferry for use in Argentina. The ferry will be delivered in 2025. In December 2024, San Francisco announced the purchase of three 150-seat battery-electric ferries that will be in operation by 2027.

Several recent developments have occurred with respect to nonferry water transport. In 2017, China developed an all-electric coal-carrying container ship with 2,400 kilowatt-hours of batteries and a range of 80 kilometers. In May 2020, Damen Shipyards Group unveiled the world's first all-electric dredger, a 530-tonne, 62-meter-long vessel. The dredger was delivered to a client in Egypt in August 2021.

> **Transition highlight**
> In 2017, a 30-meter hybrid (battery-electric/hydrogen-fuel-cell-electric) catamaran, the *Energy Observer*, was built. It was powered by solar PV, which covered all available surface area of the vessel. During sunny days, the PV provided electricity for the two electric motors, desalination equipment, and appliances on the boat. The solar electricity also charged the batteries and produced hydrogen through an electrolyzer. During the night and on cloudy days, either the batteries or fuel cells running on the stored hydrogen provided electricity for the motors. Sufficient hydrogen was available to run the vessel for six days if no other electricity source was available.

In August 2020, Türkiye launched the world's first electric tugboat, with 2,900 kilowatt-hours of battery storage, for use in the waters near Istanbul. In December 2020, Vietnam built an all-electric tugboat with a battery capacity of 2,800 kilowatt-hours, for use in New Zealand. In 2021, the German AIDA Cruises added a 10,000 kilowatt-hours battery pack to a 300-meter-long cruise ship. Also in 2021, Japan developed two electric propulsion tankers, each with a 4,000-kilowatt-hour battery pack, to operate in Tokyo Bay.[78] In November 2021, the *Yara Birkeland* became the world's first all-battery-electric cargo ship to travel autonomously (with no crew and guided remotely). It is 80 meters long and has 7 megawatt-hours of battery storage. It travels between ports off the coast of Norway. In December 2021, the world's first battery-electric bunker tanker (a small tanker ship used to load fuel oil onto ocean-going vessels) was launched in Tokyo Bay. It has a 3,480-kilowatt-hour battery system.

> **Transition highlight**
> In August 2021, the Swedish boat maker Candela launched a battery-electric hydrofoil powerboat, *C-8*, that seats eight and cruises at 40.7 kilometers per hour for up to 106 kilometers. Its top speed is 50 kilometers per hour. In September 2024, the *C-8* became the first battery-electric boat to travel between two countries in the Baltic Sea. It went 278 kilometers on a round trip between Stockholm, Sweden, and the Finnish autonomous region of Aland, with two charging stops along the way. The trip's electricity cost was 40 to 50 euros. An accompanying gasoline boat's fuel cost was 750 euros. Thus, the battery-electric boat fuel cost was about 95 percent lower. On October 29, 2024, a Candela

4.1 BATTERY-ELECTRIC VEHICLES

30-seat battery-electric hydrofoil ferry, the *P-12 Nova*, began service in Stockholm. The ferry reduces energy use by 80 percent and commuting time by half compared with a diesel ferry, by reducing water friction. The same ferry model will enter service to cross Lake Tahoe in 2025, reducing the use of cars to reach the same destination.

In January 2022, China completed a trial voyage of the world's largest pure battery-electric cruise ship, the *Yangtze River Three Gorges 1*. The ship is 100-meters long and designed to carry up to 1,300 passengers on the Yangtze River. It has a battery capacity of 7,500 kilowatt-hours. Singapore launched its first fully electric cargo ship, the *Hydromover*, in November 2023. The ship is 18.5-meters long, carries up to 25 tonnes of cargo and 10 passengers, has a 422 kilowatt-hours battery, and has a 74-kilometer range.

In January 2024, the first US battery-electric tugboat, the *eWolf*, was delivered to the Port of San Diego. It is 25-meters long, has a 6,200 kilowatt-hours battery pack, and eliminates the use of 30,000 gallons of diesel per year. Also in 2024, Damen introduced a battery-electric vessel to transfer maintenance crews up to 46 kilometers to an offshore wind platform, where the vessel is recharged for the return trip. In May 2024, China launched a 50-megawatt-hour, 120-meter-long battery-electric container ship that carries 10,000 tonnes of cargo. In September 2024, an 81-meter French ship, *Anemos*, became the first cargo ship in over 100 years to cross the Atlantic powered mostly by sail. It delivered 1,000 tonnes of cognac and champagne from France to New York City. The cloth sails were controlled by computer. The ship has backup diesel generators, but they were used only sparingly, so it emitted one-tenth of the pollution of a fossil cargo ship. In November 2024, the first battery-electric Thames-River tour boat was under construction in Cornwall, UK. It seats 250 and is expected to carry 500,000 passengers per year starting in summer 2025.

4.1.2.3 AIRCRAFT

Aircraft emit about 2–3 percent of global carbon dioxide emissions, mostly high in the air.[79] Aircraft also emit black and brown carbon particles and many gaseous pollutants. The particles and water vapor emitted by aircraft produce contrails, which also affect climate. Whereas aircraft may cause about 1.3 percent of near-surface global warming, they may also cause about 4 percent of upper-tropospheric global warming and 6 percent of Arctic warming.[80] As such, transitioning aircraft to battery-electric or hydrogen-fuel-cell-electric propulsion is critical for helping to address climate damage. Whereas short-haul aircraft can be battery-electric, medium- and long-haul aircraft will likely need to be hydrogen-fuel-cell-electric.[81]

In 2009, the Swiss Company, Solar Impulse, built the first electric-powered aircraft, *Solar Impulse 1*, which contained solar cells and batteries. In 2010, a single pilot flew the plane for 26 hours, including for 9 hours at night, opening the world's eyes to the potential for electric flight. *Solar Impulse 2*, a second plane with more solar cells and batteries than the first, was built in 2014 and flown, from March 9, 2015, to July 26, 2016, around the world in stages.

> **Transition highlight**
>
> In 2015, the Israeli electric-aircraft manufacturer Eviation began developing an electric plane, *Alice*. The plane was intended to fly either nine passengers or 1,200 kilograms of cargo. The most recent version of the plane has an estimated range of 463 kilometers on a single charge (based on current battery technology) and flies at a top speed of 480 kilometers per hour. The plane has 3.7 tonnes of lithium-ion batteries, which comprise 60 percent of the plane's takeoff weight, and two motors. *Alice* is expected to operate commercially by 2027.[82]

In 2017, China test-flew a two-seat battery-electric plane, *RX1E*. In October 2019, a four-seat version, *RX4E*, with 70-kilowatt-hours of battery storage, and a range of 300 kilometers (186 miles), was flown. China certified the plane for mass production in December 2024. In September 2024, an updated two-seat plane, *RX1E-A*, with a range of 240 kilometers (149 miles), was also flown over Beijing.

In 2019, Harbour Air, a North American seaplane airline, retrofitted one of its internal-combustion-engine planes with batteries installed by MagniX to form the *eBeaver*. The plane was tested on December 11, 2019. In April 2021, Harbour Air announced it would retrofit all 40 of its seaplanes with batteries and propulsion systems.

In June 2019, the Slovenian company Pipistrel announced it was building a battery-electric vertical-takeoff-and-landing (eVTOL) aircraft, the *801*, that seats a pilot and four passengers. It has a range of 97 kilometers (60 miles) and a top speed of 282 kilometers per hour (175 miles per hour). Such aircraft will replace helicopters. In April 2020, Jaunt Air Mobility introduced its own eVTOL electric aircraft, *Journey*, also with five seats and a top speed of 282 kilometers per hour (175 miles per hour). The aircraft is 63 percent quieter than a helicopter.

In May 2020, a nine-seat Cessna 208B Grand Caravan, retrofitted with a battery-electric propulsion system by MagniX and AeroTEC, flew for 30

4.1 BATTERY-ELECTRIC VEHICLES

minutes above central Washington State. This test flight was of the largest electric plane up to that date. In July 2021, United Airlines and its regional partner Mesa Airlines announced the purchase of 200 electric aircraft with 19 seats, the *ES-19*, from Sweden's Heart Aerospace. The aircraft has a range of 400 kilometers (248 miles) and will be available in 2026. In August 2021, New Zealand's Sounds Air announced it will order three of the same aircraft and that it is striving to be an all-electric airline by 2030.

In September 2021, the startup Wright Electric announced it was testing a 2-megawatt motor for use in battery-electric and hydrogen-fuel-cell-electric aircraft. Ten motors will provide enough power for a 140- to 170-seat single-aisle plane. In November 2021, Wright Electric announced it will commercially produce a 100-seat battery-electric plane with a range of 740 kilometers (460 miles) by 2027. In December 2021, a Boston company, Regent, announced it will test-fly an electric boat-plane hybrid, called a seaglider, over Tampa Bay. The seaglider rests at a dock and floats like a boat in areas with no waves. It then takes off from the water, hovering over waves using underwater wings, or hydrofoils. The first commercial version of the seaglider will carry 12 passengers.

In June 2023, the US Federal Aviation Administration certified the first eVTOL aircraft in the US, produced by Joby Aviation. The aircraft seats five and has six electric propellers, a range of 241 kilometers (150 miles), and a top speed of 341 kilometers per hour. In January 2024, the Dutch company Elysian released plans to build a 90-seat battery-electric aircraft that can fly 800 kilometers (497 miles) on a charge, by 2033. In June 2024, the battery company CATL used a high-density battery to test-fly a 4-tonne plane. Based on the result, CATL expects to build an 8-tonne plane that carries 9 to 12 passengers flying 2,000 to 3,000 kilometers, by 2028.

4.1.3 LITHIUM AND NEODYMIUM MINING

Today's batteries for battery-electric vehicles contain lithium. Lithium itself is not a toxic element. However, lithium-ion batteries also contain nickel, cobalt, aluminum, manganese, titanium, and/or phosphorus,[83] some of which are toxic if breathed in high enough concentrations. On the other hand, these chemicals are not generally inhaled, as they are confined to the batteries that contain them. Further, the quantities of those toxic chemicals are less than the quantities of lead, nickel, or cadmium in lead-acid or nickel-cadmium batteries. If lithium-ion batteries are thrown into a landfill, toxic chemicals will leak from them over time, contaminating groundwater. However, lithium-battery recycling, which is already occurring, avoids this problem.[84]

Nevertheless, most mining for lithium causes environmental degradation. Such damage, though, should be put in perspective. Batteries that now last 15–20 years require one-time mining at the beginning of their life. Battery recycling allows the lithium to be reused. Gasoline and diesel vehicles, on the other hand, require continuous mining of oil forever, so the mining damage is orders of magnitude greater than with lithium mining. One study found that 1 gigawatt of wind nameplate capacity replacing a coal power plant on the Texas grid reduces total mining by 25 million tonnes over 20 years.[20] That accounts for changes in the quantity of ores for steel, copper, and rare earth elements needed and for all soil and waste moved in the process.

Further, a traditional lithium mine becomes cleaner as it begins to run on WWS energy. For example, in July 2020, a Texas company announced it will use solar PV to provide 100 percent of the electricity required in the annual average for its Round Top Mountain rare-earth-element- and lithium-mining project in Hudspeth County, West Texas. Scientists have also developed a technique to extract lithium with no new mining.

> **Transition highlight**
> Environmental damage due to some lithium mining can be averted almost entirely. Existing geothermal electricity production in the Salton Sea, California,[85] and the Upper Rhine River, Germany,[86] entails lithium-rich brine water (water rich in salt) being pulled from geothermal wells and recycled back into the wells after high-temperature heat is extracted for geothermal electricity generation. Efforts are currently underway to extract lithium from such brines using the electricity and heat from the geothermal plant itself. If the plants are binary geothermal plants, such efforts will result in no greenhouse gas emissions, no air-pollution emissions, and no additional mining. Similarly, lithium can be extracted from the same brine that bromine is extracted from, as done in Arkansas.[87]

Powering mining operations with WWS is not limited to lithium mining. In April 2021, a gold mine in Mali began using solar PV plus batteries for its electricity. In August 2021, a company in Western Australia announced it would power a nickel mine with PV and another with PV plus batteries. A second company announced it would power an ilmenite mine in Madagascar with PV, wind, and batteries.[88] In September 2021, a potash mining company in Australia announced it would provide two-thirds of its power with a PV–wind-battery microgrid[89].

A question related to lithium-ion batteries is whether sufficient resource exists to provide enough lithium to power all electric vehicles, equipment, and stationary batteries needed for a 100 percent WWS world. Based on current world resource data,[90] enough lithium exists (105 million tonnes in 2023) to run over 13 billion battery-electric vehicles of substantial range. Currently, the world has 1.5 billion mostly fossil-fuel vehicles. Ideally, no more, and hopefully fewer, vehicles will be needed. Countries with the most lithium resources include, in decreasing order, Bolivia, Argentina, the United States, Chile, Australia, and China. In just one year, from 2022 to 2023, 7 million more tonnes of lithium were discovered, sufficient for 890 million new battery-electric vehicles (59 percent of the current total).

But lithium batteries will also be needed for stationary storage, cell phones, and other tools and appliances. Whereas more lithium will be discovered in the future, lithium is already being recycled, helping to reduce shortages.[84] Stresses on lithium supplies will also be alleviated by the fact that several batteries without lithium are available (see Section 3.3).

Another element used in many electric vehicles is neodymium. It is used in permanent-magnet alternating-current motors. However, the amount of neodymium needed per vehicle is an order of magnitude less than that required per wind turbine, and neodymium is not a limiting factor in wind-turbine production. Also, one type of electric motor, an induction motor, uses no neodymium. Finally, an alternative to neodymium, now being commercialized, is iron nitride. Iron-nitride magnets, when optimized, are stronger than neodymium magnets. They may also be more environmentally benign to mass-produce since they contain common elements. However, as of 2024, only one company was producing iron-nitride magnets, and in low numbers, because of the complexity of the manufacturing process.[91]

4.1.4 BATTERY FIRES

A concern about lithium batteries in battery-electric vehicles is their potential to catch fire. Batteries can catch fire if they are overcharged with equipment not designed for them, penetrated or crushed, exposed to water or extreme heat, or defective. To minimize the chance or impact of a fire, it is often suggested that vehicles (including e-bikes, e-scooters, and e-motorcycles) be charged in a ventilated area and with the charging equipment designed for them. Vehicle battery fires are difficult to put out, but they can be, with a lot of water applied over two hours. Despite battery-fire concern, US battery-electric vehicles are involved in only 25 fires per 100,000 vehicles sold, many fewer than the 1,530 fires per 100,000 gasoline vehicles sold and 3,475 fires per 100,000 gasoline–electric hybrid vehicles sold.[92]

4 WWS SOLUTIONS FOR TRANSPORTATION

4.2 HYDROGEN-FUEL-CELL-ELECTRIC VEHICLES

Hydrogen gas (H_2), or just hydrogen, has been present in the Earth's atmosphere since the formation of the Earth 4.6 billion years ago. Today, it is a well-mixed gas in the lower atmosphere, whose main natural source is bacteria living in the oceans and soils. Humans contribute to hydrogen's atmospheric burden too, as hydrogen is a byproduct of fossil-fuel combustion. In fact, automobiles are the largest human source of hydrogen into the atmosphere today.[93] Hydrogen is removed from the air primarily by bacteria in soil and the ocean.

Paracelsus "Paracelsus" (1493–1541), a Swiss physician and alchemist, born Theophrastus von Hohenheim, may have been the first to isolate hydrogen. He discovered that pouring sulfuric acid over the metals iron, zinc, or tin gave off a highly flammable vapor. In 1766, **Henry Cavendish** also made this determination and isolated hydrogen's properties.

Because hydrogen is much lighter than air, hydrogen has been used to lift passenger balloons and airships for flight as well as to lift party balloons. Combusted hydrogen has also been used to propel rockets and the space shuttles into space. Its use in the **Hindenburg** airship, which caught fire and crashed, killing 36 people on May 6, 1937, set back its use for transport for many decades. Even in that case, though, the fire damage was from the burning of the airship's shell rather than of hydrogen. Due to its low density, a hydrogen flame shoots upward, minimizing damage to a vehicle carrying it. On the other hand, jet fuel, diesel, and gasoline explode outward, destroying a vehicle and its contents.

4.2.1 GREEN-HYDROGEN PRODUCTION

Hydrogen can be produced synthetically by steam methane reforming, autothermal reforming of methane, methane pyrolysis, coal gasification, electrolysis, and photoelectrochemical water splitting. In a WWS world, only electrolysis and photoelectrochemical water splitting should be used, where the energy in both cases comes from WWS (**green hydrogen**). Section 8.8 discusses the nongreen-hydrogen techniques and why they should not be used.

The cleanest and simplest way to produce hydrogen is by **electrolysis**, where the electricity source is WWS. With electrolysis, electricity in an **electrolyzer** splits liquid water into pure hydrogen and oxygen gas. This process emits no other chemicals, if the electricity is from a WWS source. The hydrogen gas is compressed or liquefied, then stored in a tank for later use.

4.2 HYDROGEN-FUEL-CELL-ELECTRIC VEHICLES

Its best uses are in a fuel cell to regenerate electricity and heat, and in the manufacture of steel and ammonia. In July 2021, the Nordic and US- based steel company SSAB produced the world's first fossil-free steel, also called green steel, in Lulea, Sweden, using a green-hydrogen process. An electrolyzer powered by WWS electricity (mostly wind) produced the hydrogen. Electrolysis is not only simple, but it also eliminates the fuel mining, transport, processing, and emissions (of pollutants and carbon) associated with producing hydrogen from fossil fuels.

Three types of electrolyzers are the polymer-electrolyte-membrane electrolyzer, the alkaline-water electrolyzer, and the solid-oxide electrolyzer. The polymer-electrolyte-membrane electrolyzer is the most commonly used electrolyzer, but the solid-oxide electrolyzer may be the most efficient.[94]

The first two operational electrolyzers in the world were installed in 1999, one in Sweden and the other in Peru.[95] By the end of 2023, the number of electrolyzers installed worldwide had grown to at least 150, representing 1,100 megawatts of nameplate capacity and 189,000 tonnes of hydrogen produced per year. Of these, the United States had 5 electrolyzers; 3 were in Canada; 1 was in Central America; 13 were in South America; 3 were in Australia; 5 were in Japan; 2 were in Africa; 1 was in Taiwan; 16 were in Asia; 1 was in the Middle East; and over 100 were in Europe.

Transition highlight
An emerging method of producing hydrogen from WWS electricity is **photoelectrochemical water splitting**. With this technology, sunlight first produces electricity inside a PV cell. The electricity is then used within the same cell to split water into hydrogen and oxygen. As such, the PV cell and electrolyzer are merged into one device. Because this method involves a WWS electricity source and produces no pollutants, the hydrogen is green. This process reduces transmission-line electricity losses compared with PV electricity that is transmitted via the grid to an electrolyzer. However, the technology may not reach commercialization until about 2028.

4.2.2 HYDROGEN STORAGE

In a 100 percent WWS world, hydrogen will be used primarily in fuel cells to provide electricity to run an electric motor in mid- and long-distance planes and ships, some long-distance trucks and trains, and some military vehicles. Hydrogen will also be used for steel and ammonia manufacturing, some grid-electricity storage, and electricity and heat production in microgrids. Electrolysis, with electricity from WWS, will produce the hydrogen.

An advantage of producing and using hydrogen with 100 percent WWS is that excess WWS electricity can be used to produce hydrogen when more than enough WWS electricity is available to meet grid-electricity demand. Otherwise, excess WWS electricity might be curtailed (wasted), resulting in electricity prices higher than necessary. If excess WWS electricity is instead used to produce hydrogen, and if the hydrogen is stored, the cost of WWS generation drops to less than if the WWS electricity is curtailed.[65,96] Similarly, excess WWS electricity can be used to run electric heat pumps that produce heat or cold that is stored in water or soil for later use. Excess electricity can also be used to produce low-to-high-temperature heat that is stored in firebricks and then used for industrial processes.[43] As such, electricity, heat, cold, and hydrogen can work together to help power a 100 percent WWS economy at low cost.

At sea-level air pressure, hydrogen gas has less than one eight-thousandth the density of gasoline. Due to its low density, hydrogen must be compressed into a small volume, or its temperature must be dropped until it liquefies for it to be stored. Liquid (cryogenic) hydrogen's density is much greater than compressed hydrogen's density.

Compressed hydrogen can be stored in an onboard hydrogen-fuel-cell-electric vehicle storage container or a stationary hydrogen storage tank. Compressed-hydrogen storage containers for vehicles are made of either low-alloy steel or a composite material, or both. The steel can become brittle if it is exposed to hydrogen over a long time, which is why composites are often used. Stationary hydrogen storage tanks are made of low-alloy steel, a composite material, or stainless steel and carbon steel encapsulated within prestressed concrete.[97]

Liquid hydrogen is obtained by reducing hydrogen's temperature to below its boiling point (−253 degrees Celsius at sea-level air pressure). About 13 percent of the energy stored in liquid hydrogen is needed to cool the hydrogen enough to liquefy it. Liquid hydrogen is stored in cold tanks at below the evaporation temperature of hydrogen. In a closed tank, even a small increase in temperature will evaporate enough hydrogen to explode the pressure by a factor of ten thousand. As such, liquid hydrogen tanks require lots of insulation to prevent the hydrogen from evaporating (boiling).

Aircraft, rockets, and space shuttles require much more energy, thus much more hydrogen, than do cars. Given their limited sizes and large energy needs, long-distance aircraft require hydrogen to be stored in its most dense form, liquid hydrogen. Liquid hydrogen was used in demonstration flights for the Soviet Union's commercial aircraft Tupolev Tu-154B, beginning on April 15, 1988. The aircraft was fitted with a

4.2 HYDROGEN-FUEL-CELL-ELECTRIC VEHICLES

thermally insulated fuel tank behind the passenger cabin that contained liquid hydrogen. Combusted liquid hydrogen powered a third engine. Combusted liquid hydrogen also powered the space shuttles and most rockets into space. Despite the fact that liquid hydrogen has only a quarter of the energy density of jet fuel, a given mass of hydrogen used in a fuel cell propels an airplane further than the same mass of jet fuel. As such, the volume of hydrogen fuel storage in a hydrogen-fuel-cell-electric plane is only 1.8 to 4.8 times the volume of jet-fuel storage in a jet-fuel plane for the two aircraft to go the same distance.

4.2.3 HYDROGEN FUEL CELLS

A **fuel cell** is a device that converts chemical energy from a fuel into electricity and heat. A hydrogen fuel cell converts hydrogen and oxygen to electricity, heat, and water vapor. Fuel cells are connected in series to form a **fuel-cell stack**. Stacks contain a few to hundreds of fuel cells. Fuel-cell stacks of 50- to 125-kilowatt nameplate capacity are used in most vehicles. Stacks of 1–200 megawatts are used in power plants.

In a 100 percent WWS world, hydrogen fuel cells will be used for long-distance transport and stationary electricity generation. In a hydrogen-fuel-cell-electric vehicle, the fuel cell combines hydrogen from an onboard storage tank with oxygen from the air to produce electricity, water vapor, and waste heat. The electricity runs an electric motor, which produces rotational motion to turn the vehicle's wheels or propellers. The water vapor and waste heat are expelled.

Hydrogen may also be used in stationary fuel cells to provide grid or microgrid electricity. However, using electricity to produce hydrogen and then, within a few hours, using the hydrogen to reproduce electricity is inefficient compared with storing electricity in a battery and then discharging the battery to reproduce electricity. On the other hand, when electricity needs to be stored both over a few hours and over tens of hours or longer, then a combination of batteries and hydrogen fuel cells is often more efficient than batteries alone for electricity storage.[65] The overall efficiency of using hydrogen for grid electricity can also be improved if heat released during fuel-cell use is captured and used.

> **Transition highlight**
> One useful application of a hydrogen fuel cell is to produce electricity and heat in a microgrid. In 2016, an off-grid apartment complex built in Brutten, Switzerland, included 127 kilowatts of solar PV cells, an electrolyzer, two hydrogen storage tanks, a fuel cell, and a ground-source heat

pump. Excess electricity from the PV was used to electrolyze water into hydrogen, which was stored until needed. During winter and at night, hydrogen was run through the fuel cell to generate electricity and heat. The heat was used to heat water. The ground-source heat pump ran on electricity from either the fuel cell or PV, and heated and cooled the air.

When evaluating battery-electric versus hydrogen-fuel-cell-electric vehicles, analysts often compare the plug-to-wheel efficiency of one versus the other. The plug-to-wheel efficiency of a hydrogen-fuel-cell-electric passenger car ranges from 34.6 to 46.6 percent. That is the percent of electricity used to produce hydrogen converted to motion. The rest is waste heat. This efficiency is lower than the plug-to-wheel efficiency of a battery-electric passenger car (64–89 percent), but higher than the tank-to-wheel efficiency of an internal-combustion-engine car (17–20 percent).

The plug-to-wheel efficiency of a hydrogen-fuel-cell-electric car depends on several factors. First, electricity is needed to produce the hydrogen by electrolysis. Second, electricity is needed to compress or liquefy the hydrogen for storage in a holding tank before it is moved to a vehicle fuel tank. Third, some of the hydrogen leaks between its production and injection into the vehicle fuel tank. Fourth, energy stored in hydrogen's chemical bonds is lost as heat in the fuel cell. Additional energy is lost in the fuel cell when water vapor produced by the fuel cell escapes into the air, carrying with it latent heat. Finally, some electricity produced by the fuel cell is lost in an inverter and in wires between the fuel cell and the electric motor, and more electricity is lost as heat in the motor.

The **electrolyzer efficiency** is the ratio of the energy stored in the chemical bonds of hydrogen produced by the electrolyzer to the amount of electricity used by the electrolyzer to produce the hydrogen. Efficiencies range from 73.8 percent for standard electrolyzers to 95 percent for a new-technology alkaline electrolyzer.[94,98] The **compressor efficiency** is about 88 percent. In other words, of the total energy needed to produce and compress one kilogram of hydrogen, 4–26.2 percent is lost due to electrolysis and another 12 percent is lost due to compression. The remainder is stored in the bonds of hydrogen molecules.

Transition highlight
In October 2024, China announced the development of a plant to produce, store, and use 40,000 tonnes per year of green hydrogen with a large (400-megawatt) electrolyzer. The electricity for the electrolyzer will come from wind and solar. The hydrogen will be used to power 600 hydrogen-fuel-cell-electric trucks and a 200 megawatts/1,600 megawatt-hour hydrogen grid-electricity storage system.

4.2 HYDROGEN-FUEL-CELL-ELECTRIC VEHICLES

Of the energy in hydrogen entering a vehicle, 30–40 percent is lost when the fuel cell converts the hydrogen to DC electricity. Another 15.4 percent is lost as heat used to evaporate water produced from hydrogen and oxygen in the fuel cell. This lost energy is stored as latent heat in the water vapor, which is released to the air and ultimately condenses to liquid cloud water in the atmosphere. Another 2–3 percent of energy is lost in wires, power electronics, and the DC-to-AC inverter. Another 4–16 percent of energy is lost converting electricity to motion in the motor.

Once hydrogen is produced, some may leak. Hydrogen is a tiny molecule, much smaller than methane, so it can leak from pipes unless the pipes are well sealed. In a 100 percent WWS system, most hydrogen will not be produced far away from where it is needed and will not be transported by pipeline or truck. Instead, most will be produced locally after electricity is transmitted long distance. This will eliminate hydrogen pipeline losses. Leaks, however, may still occur. An estimated 0.3 percent of hydrogen produced may leak between production and consumption. Thus, the efficiency of the fuel-cell system inside a hydrogen-fuel-cell-electric passenger vehicle itself is 41.4–55.7 percent. Accounting also for the electrolyzer efficiency, compressor efficiency, and hydrogen leaks, the overall plug-to-wheel efficiency of a hydrogen-fuel-cell-electric vehicle is about 34.6–46.6 percent.

Because hydrogen-fuel-cell-electric passenger vehicles are less efficient than battery-electric passenger vehicles, a hydrogen-fuel-cell-electric passenger vehicle requires the energy from many more wind turbines or solar panels to move the same distance than does an equivalent battery-electric vehicle. However, aside from the leaked hydrogen and the water vapor emissions from hydrogen-fuel-cell-electric vehicles, the operational emissions from both are zero.

With large vehicles, such as long-distance heavy commercial trucks, trains, ships, airplanes, and military vehicles, the tables turn and hydrogen-fuel-cell-electric vehicles become more efficient than battery-electric vehicles. The reason is that long-range battery-electric vehicles use more energy per unit distance than short-range vehicles just to carry the batteries. Battery-electric vehicles don't consume fuel so don't decrease their mass during a trip. They consume only massless electricity. Hydrogen-fuel-cell-electric vehicles also carry more mass (more hydrogen fuel and fuel cells and larger storage tanks) as their range increase. However, such extra mass is less than the mass of extra batteries, and hydrogen fuel itself is light and dissipates as the vehicle travels. As such, the overall mass of a long-range hydrogen-fuel-cell-electric vehicle is less than that of a battery-electric vehicle and decreases during a trip.

In sum, the heavier a vehicle and the further it must travel, the more likely a hydrogen-fuel-cell-electric vehicle is to overtake a battery-electric vehicle in terms of efficiency. Hydrogen-fuel-cell-electric trucks, in fact, become advantageous over battery-electric trucks for a vehicle range longer than 800 kilometers. Most long-haul US commercial truckers drive 950–1,050 kilometers per day with only a 30-minute break during the drive. Thus, US trucks ideally need a range of 1,100 kilometers or a refueling or recharging time of less than 30 minutes. In Europe, truck drivers can drive a maximum of eight hours per day, so the driving range is limited to closer to 800 kilometers per day.

Transition highlight
In 2024, the Daimler Mercedes-Benz *GenH2* semi had a range of 1,000 kilometers (621 miles) and required 10–15 minutes to refuel. The Tesla battery-electric semi had a range of 805 kilometers (500 miles),[75] and at a supercharger, could recharge 70 percent of its range in 30 minutes. Thus, a Tesla battery-electric semi could meet the 800-kilometer range threshold of most European long-distance truckers but not the 950–1,050-kilometer range threshold of US truckers. The Daimler hydrogen-fuel-cell-electric semi, on the other hand, could meet the range threshold of both US truckers and European truckers. Given rapid battery improvements, it is likely that both battery-electric and hydrogen-fuel-cell-electric semis will ultimately be used for long-distance trucking.

Aside from more efficiency, hydrogen-fuel-cell-electric vehicles have other advantages over fossil-fuel vehicles. For example, hydrogen-fuel-cell-electric vehicles eliminate all tailpipe particle and gas emissions, aside from water vapor. But, during combustion, fossil-fuel vehicles emit more water vapor than do hydrogen-fuel-cell-electric vehicles. Hydrogen production, transport, and storage also result in hydrogen leaks. However, fossil-fuel vehicles emit more hydrogen[93] as a combustion product than do hydrogen-fuel-cell-electric vehicles from leaks.[99] In sum, replacing fossil-fuel vehicles with hydrogen-fuel-cell-electric vehicles will reduce global hydrogen and water vapor emissions.[99]

Transition highlight
Leaked hydrogen will play no role in warming the climate. A concern has arisen that – because hydrogen chemically reacts in the air to destroy chemicals that break down two greenhouse gases, methane and lower-atmospheric ozone – hydrogen's leakage may contribute to global warming.[100] This concern is misplaced. The largest source of human-emitted hydrogen is automobile exhaust.[93] Even if all automobiles worldwide were replaced with hydrogen-fuel-cell-electric ones, the leakage rate of

4.2 HYDROGEN-FUEL-CELL-ELECTRIC VEHICLES

the hydrogen would need to exceed 3 percent for the emissions of leaked hydrogen to exceed the reduced emissions of hydrogen due to eliminating automobile exhaust.[99] In fact, less than 10 percent, and possibly less than 5 percent, of fossil-fuel vehicles will be replaced by hydrogen-fuel-cell-electric vehicles. The rest will be replaced by battery-electric vehicles. As such, the leakage rate of hydrogen would need to be 30 to 60 percent for hydrogen emissions not to decline with hydrogen-fuel-cell-electric vehicles. Such leakage rates are unrealistic given that all technologies carrying and storing hydrogen will be new. As such, leaked hydrogen will play no role in warming the climate.

For a battery-electric or hydrogen-fuel-cell-electric aircraft, ship, or military vehicle to replace a fossil-fuel vehicle, the former will ideally have an overall mass and volume similar to or lower than those of the fossil-fuel vehicle, and a range and power-to-weight ratio or thrust-to-weight ratio similar to or higher than those of the fossil-fuel vehicle. A vehicle's mass and volume affect not only its capabilities, but also how it interacts with roads, bridges, tunnels, runways, locks, canals, and ports. The power-to-weight ratio defines how fast a vehicle can accelerate to a top speed. For an airplane, the thrust-to-weight ratio defines its ability to take off within a runway's length. Those considering adopting new battery-electric or hydrogen-fuel-cell-electric vehicles are more likely to do so if the vehicle performance and range are similar to or better than those of existing fossil-fuel vehicle options.

The major components in a fossil-fuel vehicle that can be removed when building a battery-electric or hydrogen-fuel-cell-electric vehicle are the vehicle's engine, transmission, oil, coolant, automatic transmission fluid, fuel tank, and fuel. A battery-electric vehicle replaces these components with a battery pack, inverter, electric motor, gearbox (optional), and wiring. A hydrogen-fuel-cell-electric vehicle replaces these components with a hydrogen storage tank, compressed or liquefied hydrogen fuel, a fuel-cell stack, an electric motor, a gearbox, and wiring.

Transitioning long-haul aircraft, while challenging, can be done with near-future technology. A future hydrogen-fuel-cell-electric aircraft similar to a Boeing 747-8 could have the same thrust-to-weight and range as today's 747-8, but with 22 percent lower mass and 21 percent larger volume.[81] The larger volume is not an issue for aircraft. Long-distance ships and trains can similarly be transitioned to hydrogen-fuel-cell-electric versions. Military tanks and wheeled vehicles can be transitioned to either battery-electric or hydrogen-fuel-cell-electric vehicles.[81]

The main advantage of a hydrogen-fuel-cell-electric aircraft over a jet-fuel aircraft is that the former eliminates all aircraft exhaust emissions except for water vapor (any hydrogen leaks would occur before or during fueling of the aircraft). Thus, hydrogen-fuel-cell-electric aircraft eliminate exhaust emissions of particles containing black carbon, brown carbon, and sulfate; and of gases, such as carbon dioxide, carbon monoxide, nitrogen oxides, and sulfur oxides, along with a soup of organic gases. Due to the water-vapor emissions, contrails will still form behind hydrogen-fuel-cell-electric aircraft under the right temperature and humidity conditions in the background atmosphere. However, such contrails will form only on background particles rather than on exhaust plus background particles. As such, contrails from hydrogen-fuel-cell-electric aircraft will be about 70 percent thinner, thus dissipate faster, than contrails from jet-fuel aircraft.[10] Pure battery-electric aircraft emit zero water vapor so will eliminate all contrails.

4.2.4 HISTORY OF HYDROGEN-FUEL-CELL-ELECTRIC VEHICLES

Since the late 1990s, several types of hydrogen-fuel-cell-electric vehicles have been developed for land, sea, and air.

4.2.4.1 GROUND VEHICLES

In 1998, six fossil-fuel buses were retrofitted with Ballard hydrogen-fuel-cell-electric propulsion systems and dispatched equally in Vancouver and Chicago. They carried 200,000 passengers over three years. From 2003 to 2006, several hydrogen-fuel-cell-electric buses, manufactured by Hino Motors and Toyota Motors, were tested in Japan. From 2004 to 2006, Oakland California's AC Transit trialed three such buses and, in 2009, 12 more. Daimler dispatched three hydrogen-fuel-cell-electric buses to China in 2006, and 36 more in 2007.

In 2008, Honda leased the first hydrogen-fuel-cell-electric passenger car, the *FCX Clarity*, to consumers. Between 2008 and 2015, Honda leased 48 of the vehicles in the US, whose range was 386 kilometers (240 miles). The distribution was limited by the paucity of fuel-cell-charging stations. In 2014, Toyota unveiled its own hydrogen-fuel-cell-electric passenger car, *Mirai*, at the Los Angeles auto show. Through 2024, about 25,000 had been sold worldwide. The range of its extended-range 2024 model is 647 kilometers (402 miles). Refueling takes less than five minutes.

Transition highlight

In September 2018, the company Alstom began operating the first hydrogen-fuel-cell-electric train, *Coradia iLint*, in Germany. The train has a maximum speed of 140 kilometers per hour (87 miles per hour), has

4.2 HYDROGEN-FUEL-CELL-ELECTRIC VEHICLES

a range of 1,000 kilometers (621 miles), requires 15 minutes for refueling, and creates little noise. In August 2022, 14 trains began operating in Lower Saxony, Germany. In June 2023, the same train entered service in Quebec, Canada. In 2022, California ordered 29 hydrogen-fuel-cell-electric trains from the company Stadler. In 2002, East Japan Railways began testing a hydrogen-fuel-cell-electric train. In 2023, China converted an internal-combustion-engine locomotive into a hydrogen-fuel-cell-electric one that can run for almost eight days straight and be refueled in two hours. In 2023, Italy funded six projects to replace diesel trains with hydrogen-fuel-cell-electric ones by 2026.

In July 2021, Hyzon Motors delivered a 55-tonne hydrogen-fuel-cell-electric milk truck with a range of 520 kilometers (323 miles) for operation in the Netherlands. As of 2024, Nikola, Toyota, Hyundai, Honda, Volvo, General Motors, and Daimler, among others, were also manufacturing hydrogen-fuel-cell-electric commercial trucks. The 2024 Daimler *Mercedes-Benz GenH2* semi, for example, has a range of 1,000 kilometers (621 miles).

4.2.4.2 MARINE VESSELS

In March 2018, Yanmar and Toyota tested a 16.5-meter hydrogen-fuel-cell-electric boat, *Shimpo*, with a 60-kilowatt-hour battery pack. In 2021, SWITCH Maritime tested the world's first hydrogen-fuel-cell-electric ferry, *Sea Change*. It ran on compressed hydrogen gas, seated 75 passengers (no vehicles), had a 556 kilometers (345 miles) range, and was deployed in the San Francisco Bay in July 2024. In 2023, another hydrogen-fuel-cell-electric ferry, *MF Hydra*, began operating off the coast of Norway. This was the first marine vessel to run on liquid hydrogen; it seated 295 passengers, 8 crew, and 80 vehicles. In March 2024, India launched its first hydrogen-fuel-cell-electric catamaran ferry, which seated 50, used hydrogen and batteries to power its electric motors, and had 3 kilowatts of PV onboard.

4.2.4.3 AIRCRAFT

The first flight of a hydrogen-fuel-cell-electric aircraft occurred on September 29, 2016, when the four-seat *Hy4* propeller aircraft, developed by the German Aerospace Center (DLR) together with the companies H2Fly, Pipistrel, and Hydrogenics, took to the skies over Stuttgart, Germany. The *Hy4* was a hybrid with both hydrogen fuel cells and batteries.

The batteries were used for takeoff and landing; the fuel cells, for cruising. It had a maximum range of 1,500 kilometers (938 miles) and a top speed of 200 kilometers per hour (124 miles per hour). The empty weight of the aircraft was about 42 percent of the maximum weight. In December 2020, a new version of the plane with Cummins fuel cells and motor system was test flown over Stuttgart. In September 2023, a further updated version of the aircraft was flown using liquid hydrogen instead of compressed hydrogen gas, increasing the range of the aircraft.

In September 2020, the British-American company, ZeroAvia, conducted a test of a six-seat hydrogen-fuel-cell-electric plane, *Piper Malibu*, north of London, UK. This plane is a precursor to a 10- to 20-seat hydrogen-fuel-cell-electric plane with a range of 463 kilometers (288 miles) that they expect to deploy by 2025.

In December 2020, Airbus announced its development of a liquid hydrogen-fuel-cell-electric aircraft that seats 100, with six detachable pods along the wings, each with a hydrogen propulsion system, storage tank, and eight-bladed propeller. In June 2023, Airbus successfully tested a 1.2-megawatt hydrogen-fuel-cell-electric propulsion system for the pods. They plan to test-fly not only the 100-seat aircraft, but also an 853-seat A380 aircraft with pods, by 2026.

In December 2021, United Airlines announced it would order up to 100 hydrogen-fuel-cell-electric motors from ZeroAvia to retrofit 50 regional twin-motor aircraft with, by 2028. On July 2, 2024, United entered into a conditional purchase agreement with ZeroAvia for the motors. Also in December 2021, the California startup H2 Clipper announced it was developing a hydrogen-fuel-cell-electric cargo airship, which looks like a blimp and will carry 8 to 10 times the payload of a cargo plane at one-quarter of the cost. It will be 302 meters long and will cruise at 240 kilometers per hour (150 miles per hour). H2 Clipper is expecting to test-fly the airship in 2026.

On July 11, 2024, Joby Aviation test-flew a hydrogen-fuel-cell-electric vertical takeoff and landing (eVTOL) aircraft 842 kilometers (523 miles). The aircraft stores 40 kilograms of liquid hydrogen for a fuel cell to produce electricity for six electric motors. Batteries help with takeoff and landing. The aircraft is for use as an air taxi, starting in 2026.

4.2.5 PLATINUM FOR HYDROGEN-FUEL-CELL-ELECTRIC VEHICLES

One concern that arises with the large-scale production of hydrogen fuel cells for transportation is the potential shortage of the scarce metal, platinum, which is used in many types of fuel cells. However, a transition to

4.2 HYDROGEN-FUEL-CELL-ELECTRIC VEHICLES

100 percent WWS will eliminate all gasoline and diesel vehicles and their catalytic converters, which contain platinum. Catalytic converters convert 90 percent of the pollutant gases carbon monoxide, hydrocarbons, and oxides of nitrogen, in the exhaust stream of combustion vehicles to different gases (carbon dioxide, water, and molecular nitrogen). Since 5–10 percent or less of the vehicles that replace combustion vehicles will be hydrogen-fuel-cell-electric vehicles, and the rest will be battery-electric, neither of which requires a catalytic converter, the decreased platinum use in catalytic converters will far outweigh the increased platinum use in fuel cells. As such, platinum is not a limiting element in a WWS economy.

Transition highlight
If 10 percent of the world's 1.5 billion vehicle fleet in 2050 were converted to hydrogen-fuel-cell-electric vehicles, this would translate to about 150 million 50-kilowatt vehicles. Since each 50-kilowatt fuel-cell stack requires about 12.5 grams of platinum,[101] 150 million such vehicles require about 1,875 tonnes of platinum. In comparison, each catalytic converter uses about 3–7 grams of platinum, so 1.5 billion vehicles in the world hold 4,500 to 10,500 tonnes of platinum. As such, merely recycling 18–40 percent of the world's catalytic converters in existing vehicles would provide enough platinum for all hydrogen-fuel-cell-electric vehicles in this scenario. In addition, another 17,000 tonnes of platinum exist in known resources that can be extracted from ore worldwide.

5

WWS SOLUTIONS FOR BUILDINGS

A significant portion of a building's energy is used to provide air heating, water heating, air-conditioning, and refrigeration. Heat is also used for cooking, dishwashing, clothes washing, and clothes drying. Away from the tropics, demand for air heating in buildings is usually greatest during winter, whereas demand for air-conditioning is usually greatest during summer. In all locations, hot water and refrigeration are needed year-round, although energy demand for hot water peaks during cold months and demand for refrigeration peaks during warm months. In a 100 percent WWS world, air and water heating and air-conditioning in buildings will be provided by either district heating and cooling systems or individual heating and cooling systems. In both cases, heat and cold will be produced primarily with electric heat pumps, where the electricity comes from WWS sources and extracted from either the air, the ground, water, or a waste stream of hot or cold air or water. The heat and cold may be stored or used immediately. Additional heat may come from geothermal and solar sources. The remaining energy used in buildings is for electric appliances and gadgets, such as lights, televisions, refrigerators, computers, and phone chargers. This chapter discusses how WWS will power district and individual-building heating and cooling systems. It includes discussions of hot and cold storage options for both, and electric appliances and machines that will replace fossil-gas ones. The chapter also examines energy efficiency in buildings and techniques to reduce building energy use. Finally, it discusses a modern district heating and cooling system and an all-electric home.

5.1 DISTRICT HEATING AND COOLING

District heating is the name given to a heating system in which hot water in a centralized boiler is distributed in a closed loop through insulated pipes to multiple buildings for either air heating (through radiators), potable water heating, or both. Once the heat is transferred from the water to the building, the resulting cooler water is returned to the centralized boiler for

5.1 DISTRICT HEATING AND COOLING

reheating. **District cooling** is the production of cold water at a centralized facility (usually the same facility as for district heating) and distribution of the cold water by pipes to buildings to provide cold air and/or refrigeration.

The first district heating system worldwide began operating in Lockport, New York, in 1877. Inventor **Birdsill Holly** (1820–1894) was trying to find a way to heat his whole home with one central system rather than with individual heaters in each room. At the time, each individual room in a building was heated with its own coal stove or wood fireplace. Holly built a large coal-fired water boiler in his backyard to produce steam, which he sent by wooden pipe 150 meters underground to his home. He then sent the steam by different pipes to each room to distribute the steam relatively evenly. After successfully testing his idea in his home, he connected his system to his neighbor's home as well. When that worked, he started a company, the Holly Steam Combination Company, which built the world's first community-wide heating system, in Lockport, in 1877. Holly subsequently expanded his business to other cities and improved district heating technology by inventing measuring devices and regulators needed to run the systems. Holly is also well known for inventing the district fire-protection system, which is a system that propels water around a city at constant pressure to fire hydrants. Holly's inventions of the district fire-protection system and district heating have since spread worldwide.

Within five years, **Wallace C. Andrews** (1833–1899) brought Holly's district heating technology to New York City. On March 3, 1882, the New York Steam Company, owned by Andrews, sent high-temperature (up to 200 degrees Celsius) steam through a total of 8 kilometers of pipes to buildings in lower Manhattan, New York City, to heat them. Burning coal to boil water created the steam. Heat losses from coal combustion and through poorly insulated and leaky pipes were significant, but coal was plentiful. The district heating system in New York City allowed skyscrapers to be built without needing their own coal-fired boiler in the basement.

Steam district heating still exists in Manhattan, with Consolidated Edison providing steam to over 1,700 buildings. From 1882 to 1930, steam-based district heating powered by coal (**first-generation district heating**) operated in many US and European cities. One danger of using steam pipes is that pressure can build up, causing them to explode. To reduce pressure, some steam is vented through an exhaust pipe to the air, where the steam can be seen.

From 1930 to 1970, **second-generation district heating** emerged. This system was based on burning coal and oil to heat water, and sending the hot water through pressurized pipes to buildings. Input water temperatures exceeded 100 degrees Celsius, and pipe losses were less than with steam heating, partly due to the use of concrete ducts around the pipes for insulation. These systems were installed worldwide but particularly in Eastern Europe after World War II.

Third-generation district heating evolved from 1970 to the present. It uses prefabricated, insulated water pipes buried underground to reduce heat loss. The insulation allows input water temperatures to be below 100 degrees Celsius, which in turn reduces the energy needed to heat the water. The heat for water in these systems was originally provided by a combination of coal, fossil-gas, and biomass combustion and by recycling the heat of combustion from electricity generated by these fuels. More recently, heat has also been recycled from data centers and created from excess electricity produced by wind, solar, and geothermal plants. In some district heating systems, the heat is stored underground in soil, water pits, or aquifers instead of in above-ground boilers.

Third-generation district heating was developed and used first in Scandinavian countries. Today, 50–65 percent of building air and potable water heating in Denmark, Iceland, Sweden, Finland, Estonia, Latvia, Lithuania, Poland, Russia, and Northern China is from district heating. About 12.5 percent of all heat delivered in the European Union and about 8 percent delivered worldwide is from district heating.[102]

Fourth-generation district heating emerged in 2015 and is currently evolving. It combines district cooling together with district heating and lowers the temperature of the district heat to 30–70 degrees Celsius to minimize heat losses and the energy needed to produce heat.[103] The lower temperatures permit low-temperature heat sources (e.g., recycled waste heat and electric heat pumps used to heat centralized boilers) to deliver more heat at lower cost.[104]

District cooling involves a centralized chiller that stores water that has been cooled. As with district heating, the cold water is sent through an insulated piping network to buildings to cool air in the buildings. With fourth-generation district heating, electric heat pumps are used to heat water in boilers and cool water in chillers. The heat pumps extract heat or cold from the air, water, or the ground and move it indoors. If waste hot water from buildings, data centers, or manufacturing processes is used as the source of heat for a heat pump, the heat pump requires less electricity; thus, it runs more efficiently than if room-temperature water is used. Ideally, in a fourth-generation system, all electricity for a heat pump is obtained from 100 percent WWS. An example of a fourth-generation district heating system is the Stanford Energy System Innovations project (see Section 5.1.4).[105,106]

5.1 DISTRICT HEATING AND COOLING

Fifth-generation district heating is a concept that has not yet been realized. It is similar to fourth-generation district heating but adds to that a bidirectional (as opposed to just in one direction) exchange of hot and cold water between adjacent buildings to reduce overall energy requirements further.[104] This requires a heat pump within each building to ensure an optimal temperature for the water sent between buildings.

In a 100 percent WWS world, fourth- or fifth-generation district heating and cooling will be used in densely populated areas, such as cities, where centralized systems are most advantageous. It can also be used in college campuses, hospital complexes, military bases, and small remote communities. Rural and many suburban homes and buildings are less ideal for district heating and cooling due to the length of trenches and pipes needed and the resulting cost and energy losses through the pipes. Homes and buildings that are not on district heating and cooling loops will have their own domestic hot-water tanks and use heat pumps for water and air heating, and for air-conditioning.

With WWS, large electric heat pumps and solar and geothermal heat collectors will provide district heat. WWS electricity will power the heat pumps. Heat for the heat pumps will come from air, soil, water, or recycled waste heat. Waste heat from industry or buildings will be captured and piped back to the district heating center. The heat produced for district heating will either be used immediately, stored in water tanks, or stored underground for later use.

The same heat pumps used for district heating will provide the cold for district cooling. The source of cold for cooling will also be air, soil, water, or recycled waste cold (air or water). The cold will be used immediately or stored in water tanks, ice, or aquifers for later use. In some cases, the cold will be drawn from cold lake water and sent through a heat exchanger, where the cold water will absorb heat from the air inside a building, cooling the building.[107]

An advantage of district heating and cooling is that it uses large heat pumps, thus avoiding the need for many individual-building heat pumps. It also allows heat and cold to be produced only when grid-electricity prices are low, then stored for later use, rather than on-demand every hour of every day, including when prices are high. A disadvantage of district heating and cooling is it requires dedicated water piping to all buildings served. This can be costly if the buildings are far apart. Thus, district heating and cooling is most cost-effective in densely populated cities.

District heating and cooling requires storage of the heat and cold, respectively. The main short-term storage technology for heat is a water tank; the main short-term storage technologies for cold are a water tank and ice. Heat and/or cold can also be stored seasonally underground in soil (borehole storage), water pits, and aquifers. These types of storage are discussed next.

5.1.1 HEAT AND COLD STORAGE IN WATER TANKS

Tank thermal-energy storage is the most common type of heat and cold storage worldwide. It involves storing hot or cold water in a storage tank. The water can be heated or cooled before it reaches the tank or while in the tank. Large water tanks are used as part of district heating and/or cooling systems. Domestic hot-water tanks for homes and other buildings are also tank storage, but such tanks are much smaller than those used for district heating and cooling.

District heating tanks are made of concrete, steel, or fiber-reinforced plastic. Concrete tanks usually contain an interior polymer or stainless steel liner to minimize water vapor and heat diffusion out of the tank. They are also insulated on the outside. Steel tanks are insulated as well.

Hot-water tanks are called **boilers**. The hot water is used to heat the air or water in a building. The hot water is used for showering, bathing, cooking, hand washing, and drinking. Cold-water tanks are called **chillers**. The cold water is used for air-conditioning and, sometimes, refrigeration. Although community-scale boilers and chillers can be large, they are generally used to store heat and cold for only days to weeks. Underground thermal-energy storage is less costly per unit of energy than tank storage. It also has the advantage of providing seasonal heat and cold storage. More specifically, heat captured from the sun in summer can be stored underground until it is used in winter. However, unlike above-ground boilers and chillers, which can be built almost anywhere, underground storage can be built in fewer locations.

Water stored in tanks can be heated directly with solar or geothermal heat or with heat from an electric heat pump. Tank water can be cooled with a heat pump. Ideally, electricity used to heat water is excess electricity that is not needed on the grid. Using excess electricity to produce heat stored in a tank is an ideal way of decreasing the cost of electricity because it reduces the **curtailment** (wasting) of excess WWS electricity.

Using a heat pump to heat water in a boiler also reduces winter energy demand by 68–81 percent compared with burning fossil gas to heat the same water because heat pumps are much more efficient than fossil-gas heaters or even electric-resistance heaters. Thus, boilers powered by heat pumps reduce energy requirements significantly.

5.1 DISTRICT HEATING AND COOLING

Hot water stored in a water tank warms air in a building when the water passes through a radiator. **Radiators** are heat exchangers. **Heat exchangers** are devices that allow a heated liquid or gas to transfer its heat to another liquid or gas without the two fluids mixing together or otherwise coming into direct contact. The transfer occurs by conduction, where heat energy is passed from one atom to the next in the metal or other material that separates the two fluids. In the case of a radiator, water flows through pipes, often with fins to increase the surface area exposed to the air. Heat from the water in a radiator conducts through the metal in the radiator to the air, then the energized air molecules disperse throughout the room. Radiators can be placed in or against a wall, under a floor (**radiant floor heating**), or in a ceiling. They may or may not be accompanied by a fan. After hot water passes through a radiator, it is piped to subsequent radiators until it returns, cold, to the district heating and cooling center. Ideally, the waste cold is then extracted by a heat pump to cool down the water in the cold-water tank. Since cold is extracted from the piped water, that water is now warm and is sent back to the hot-water tank.

Similarly, cold water piped to a building is used to cool air in the building after the cold water passes through a heat exchanger. The heat exchanger moves heat from the warm air in the building to the cool water, which warms up. The warm water is then circulated back to the district heating and cooling center. The waste heat is then extracted by a heat pump to warm the water in the hot-water tank. Since heat is extracted from the piped water, that water is now cold and is sent back to the cold-water tank.

5.1.2 COLD STORAGE IN ICE

A type of thermal-energy storage that is effectively electricity storage is cold storage in ice. **Ice storage** has been used for decades in universities, hospitals, stadiums, and other large facilities. It involves freezing liquid water to produce ice when excess electricity is available or when electricity prices are low, then running water through coils embedded in the ice when cooling is needed. Water passing through the coils is chilled, and the cold water is piped to a building. In the building, the cold water is run through a radiator to air-condition the building. Thus, ice storage avoids the need for daytime electricity use for air-conditioning. Since summer electricity demand peaks during the late afternoon in many places, ice storage is a method of reducing peak electricity demand in these places. The main advantage of ice storage over battery storage is the lower cost of ice storage. Ice storage costs $35–$40 per kilowatt-hour of thermal storage versus the $100–$200 per kilowatt-hour system cost of battery-electricity storage. The round-trip efficiency (cooling energy output to electricity input) of ice storage is around 82.5 percent.[108]

5.1.3 UNDERGROUND HEAT AND COLD STORAGE

Whereas boilers are useful for meeting district or home heat storage for days to weeks, longer-term heat storage can be obtained from underground heat storage reservoirs filled primarily during the summer and drained during the rest of the year. Such seasonal heat storage is called **underground thermal-energy storage**. Three main types of underground thermal-energy storage are borehole (soil), water-pit, and aquifer thermal-energy storage.

5.1.3.1 BOREHOLE THERMAL-ENERGY STORAGE

A **borehole thermal-energy storage** system is a large, underground heat exchanger that stores solar heat collected during summer for use during winter and other seasons. It consists of an array of boreholes drilled into soil. A plastic pipe with a U-shaped bend at the bottom is inserted down each borehole. The space around each pipe in each borehole is then filled with highly conductive grouting material to increase the molecule-to-molecule conduction of heat from hot water that passes through the pipe to the soil surrounding the grouting material. Once hot water in the pipe has transfered its heat to the soil, the water is cold and is sent back to a solar collector, heat pump, or heat exchanger to collect more heat. Most ground heating occurs during summer. When hot water for buildings is needed during winter, cold water is sent down the pipe; heat that is stored in the soil transfers, molecule-by-molecule, through the grouting material and pipe wall to the water in the pipe, and the hot water flows to a boiler for distribution to the buildings.

A good example of a borehole storage system is the **Drake Landing Solar Community** in Okotoks, Alberta, Canada. Starting in 2004, 52 homes were constructed. Solar collectors were installed on the garage roof of each home. The collectors contained a glycol solution (mixture of water and nontoxic glycol) that absorbed solar heat, particularly during long summer days. The heated solution was transferred from the roofs, through insulated pipes, underground to a building where the heat from the solution was transferred, through a heat exchanger, to water stored in a short-term hot-water storage tank (boiler). The water temperature in the boiler was maintained above 40 degrees Celsius.

5.1 DISTRICT HEATING AND COOLING

Hot water in excess of what could be stored in the boiler was piped to the borehole field, which was about 35 meters in diameter. In this field, 144 holes were drilled to 35 meters depth, and each was filled with a pipe with a U-shaped bend at the bottom. Insulation was placed on top of the pipes, and a grassy field, used as a community park, was grown on top of the insulation. The boiler delivered hot water through a separate pipe to each hole in the borehole field. Each pipe extended to the bottom of its respective borehole. The heat from the water in each pipe was transferred by conduction to the surrounding soil, raising the soil temperature up to 80 degrees Celsius by the end of summer. Cold water in each pipe returned upward and was sent back to the boiler to be heated again.

During summer and autumn, all home heating needs in the community were met by solar heat supplied to the boiler, and excess heat was sent to the borehole field. During winter and other times that the boiler alone could not satisfy all building heat demands, cold water was piped through the borehole field to collect heat and bring it back to the boiler. The hot water from the boiler was then distributed to the 52 homes for air and domestic water heating. With this system, up to 100 percent of winter heat demand was satisfied by summer heat collection. The overall efficiency of the system, which is defined as the fraction of heated fluid entering underground storage that is ultimately sent by pipe to homes and used for air or water heat, was about 56 percent.[109]

In sum, Drake Landing was a district heating system with a community hot-water tank, but where much of the hot water used in winter was stored underground starting in summer. The system was operational from 2007 to 2024, when it had to shut down because of a lack of replacement parts and expertise to continue maintenance of the facility. However, the technology is repeatable for almost any district heating system and requires little land. Because the borehole field is underground, the land above it can be used as parkland, grazing land, open space, or for the solar collectors needed to provide heat stored in the field.

Transition highlight
Another borehole thermal-energy storage facility serving a district heating system resides in Braedstrup, Denmark. This system, installed in 2012, serves 1,500 customers. The borehole field is honeycomb shaped, with 48 boreholes, each 45 meters deep, 3 meters apart, and containing U-shaped pipes. The boreholes, along with two additional water tanks (total volume of 7,700 cubic meters), are fed by 18,600 square meters of solar collectors. The storage field, which has a volume of 19,000 cubic meters for heating and a storage capacity of 400 megawatt-hours of heat, warms to

55–60 degrees Celsius during summer, and cools, after heat is removed from it, to 12–15 degrees Celsius during winter. An electric heat pump, powered by onsite WWS electricity, is used to raise the field temperature to 80 degrees Celsius for distribution to buildings year-round. The round-trip efficiency of the storage (energy sent to heat pumps divided by energy collected) is about 63 percent, slightly higher than that of Drake Landing. In other words, of all the heat that enters storage, 63 percent is sent to heat pumps for distribution to consumers and the rest is lost. The investment was $0.77 per kilowatt-hour of heat produced.[110]

5.1.3.2 PIT THERMAL-ENERGY STORAGE

Pit thermal-energy storage consists of a lined pit dug into the ground and filled with a storage material, such as water or a mix of gravel and water, then covered with insulation and soil. The pit can be small or large. The material in the pit is supplied with heat primarily during the summer. The heat is extracted mostly during the winter and sent to a boiler, where it is distributed via district heating to buildings for air and water heating.

The largest pit storage worldwide is the water-pit facility in Vojens, Denmark, completed in 2015. It is 13 meters deep, 610 meters in circumference, and filled with 203,000 cubic meters of water. The pit is lined underneath and on its sides with welded plastic. Soil, a few meters thick outside the lining, becomes warm, insulating the pit. The surface of the water is also covered with a floating plastic lid topped with 60 centimeters of insulating expanding clay and a draining system to remove rainwater. The water temperature during summer can reach 95 degrees Celsius but is maintained at a maximum of 80 degrees Celsius to extend the life of the liner.[111] Almost 5,500 solar thermal panels heat the water. The pit is connected to the district heating system of Vojens, which has 2,000 customers. The solar collectors plus pit storage facility provide 45–50 percent of the city's annual heat demand.[112] The cost of heat from the Vojens pit storage system is competitive, without subsidy, with the cost of heat from a gas boiler. In fact, heating bills decreased by 10–15 percent with the implementation of this seasonal heat storage system.[111]

> **Transition highlight**
> In 2015, a water-pit storage plant was built to supply the district heating system serving 99 percent of the 2,500 residents in Gram, Denmark. The storage pit is 15 meters deep, 10 meters of which is below ground. The water volume is 122,000 cubic meters. The facility has 44,800 square meters of solar collectors, which directly and through pit storage supply 61 percent of the town's heat demand.[113]

5.1 DISTRICT HEATING AND COOLING

Five other pit storage systems coupled with district heating exist in Marstal, Dronninglund, Toftlund, and Hoje-Taastrup in Denmark, and in Tibet, China.

The Marstal system, completed in 2012, serves 2,200 residents. It includes 33,365 square meters of solar collectors, tank water storage, and 75,000 cubic meters of water-pit storage that stores 6 gigawatt-hours of heat. Marstal's peak water temperature is 88 degrees Celsius, round-trip efficiency is 52 percent, and the cost is $0.52 per kilowatt-hour.[110]

The Dronninglund system, which serves 3,300 residents, consists of 37,573 square meters of solar collectors plus 62,000 cubic meters of water-pit storage. It first operated in March 2014. It has a heat storage capacity of 5.4 gigawatt-hours, a peak temperature of 89 degrees Celsius, 81 percent round-trip efficiency, and costs $0.51 per kilowatt-hour.[110]

The Toftlund system, which serves 3,250 residents, consists of 27,000 square meters of solar collectors and 85,000 cubic meters of water-pit storage. Heat is supplied from the solar collectors or from a heat pump, where the heat can be obtained from an electric boiler. The storage system began operating during June 2017. The system has a heat storage capacity of 6.9 gigawatt-hours, round-trip efficiency of 72 percent, and costs $0.70 per kilowatt-hour.[114]

The Hoje-Taastrup system, which began commercial operation in 2023, stores 3,300 megawatt-hours of heat in 70,000 cubic meters of water. In this system, water can be heated to 90 degrees Celsius, which is higher than in previous systems, due to a new polypropylene membrane that permits higher water temperatures.

Finally, a 15,000-cubic-meter water-pit system that stores 700 megawatt-hours of heat was completed in the town of Langkazi, Tibet, China, in 2018. This system is fed by 22,275 square meters of solar-thermal collectors. Langzi is at an elevation of 4,200 meters. At that height, the boiling temperature of water is 84 degrees Celsius. As such, the maximum water temperature is kept at 65 degrees Celsius. The system is used for short-term heat storage rather than for seasonal heat storage, because heat losses are too high at the altitude of the town, but the system still supplies 90 percent of the town's heat year-round due to the high year-round solar insolation at the site.

5.1.3.3 AQUIFER THERMAL-ENERGY STORAGE

Aquifer thermal-energy storage is similar to pit storage, except that the water used with aquifer storage for storing heat is naturally occurring water in underground layers of groundwater, called aquifers. Aquifers can also store cold water. Aquifers contain permeable sand, gravel, sandstone, or limestone. Aquifers can be used for underground thermal-energy storage if they are encapsulated by impervious layers above and below them, and if natural groundwater flow is slow.

With aquifer storage, one or several pairs of wells are drilled into an aquifer. During summer, cold water is extracted from one well in each pair, heated with building heat (thereby cooling the building), then returned to the other well, where it mixes with and heats the water in that well. During winter, warm water from the warm well is extracted, the heat is removed to heat the building, and the cold water is returned to the cold well.

Because aquifers cannot be insulated, heat storage at temperatures greater than 50 degrees Celsius is efficient only for large, deep aquifers with volumes greater than 50,000 cubic meters and a low surface-to-volume ratio.[115] As such, the maximum temperature in a shallow aquifer for heating is generally limited to 20 degrees Celsius. For cooling, shallow aquifers are generally used.

By 2018, about 2,800 aquifer storage systems existed worldwide, storing about 2.5 terawatt-hours of heat each year.[116] Of these, about 85 percent were located in the Netherlands, and 10 percent in Sweden, Denmark, and Belgium combined.[116] The rest are located in the United Kingdom, Germany, Japan, Türkiye, and China. Some notable aquifer systems built to date include the following.

The **Eindhoven aquifer system** is a 1.7-million-cubic-meter heat and cold storage reservoir under Eindhoven University of Technology, in the Netherlands. Completed in 2001, it supplies both direct cooling during summer and low-temperature heating during winter to heat pumps to maintain comfort in 20 university buildings. Sixteen pairs of wells were constructed. The aquifer lies between 28 and 80 meters below ground level. Its natural temperature of 11.8 degrees Celsius varies during the year from 6 to 16 degrees Celsius due to the extraction or addition of heat. The maximum heating and cooling energy delivered is 15–30 gigawatt-hours per year. The maximum charge and discharge rate of the reservoir for both heating and cooling is 17 megawatts.[115]

The **Stockholm airport aquifer system**, completed in 2011, consists of a one-million-cubic-meter storage reservoir under the Arlanda Airport in Stockholm. The system is used for heating and cooling the airport. During winter, heat stored in the aquifer from summer is used to preheat air used for ventilation in the terminal buildings and to melt snow at the gates. The extraction of heat during winter cools the aquifer. As such, by summer, the aquifer water is cold. The cold water is extracted during summer and used

5.1 DISTRICT HEATING AND COOLING

to air-condition the terminals. The aquifer well lies 15–30 meters below ground. The maximum heating and cooling energy delivered is 20 gigawatt-hours per year.[115] The temperature in the aquifer varies from 2 to 25 degrees Celsius during the year due to the extraction or addition of heat.

Transition highlight
The **Riverlight Project aquifer system** consists of a 180,000-cubic-meter storage reservoir under the Riverlight Project in London, along the River Thames. Completed in 2014, it serves 806 residential apartments in addition to several commercial businesses. The water pipes are linked to heat pumps, which provide cooling during the summer and heating during the winter. The aquifer system consists of four warm wells and four cold wells drilled into the London chalk aquifer. The well depths are about 100 meters each. The thickness of the aquifer in this layer is about 25 meters. The maximum heating and cooling energy delivered is 1.4 gigawatt-hours per year.[115] The temperature in the aquifer varies from 8 to 24 degrees Celsius during the year due to the extraction or addition of heat.

5.1.4 STANFORD UNIVERSITY 100 PERCENT RENEWABLE ELECTRICITY, HEAT, AND COLD ENERGY SYSTEM

Stanford University, in California, developed a district heating and cooling system powered by solar PV. The PV also powers other electrical needs of the university. In fact, sufficient PV was purchased to supply 100 percent of the campus' annual electricity demand and almost 100 percent of the heating and cooling demands with WWS by 2022. In 2015, the electricity, heat, and cold demands were about 208, 158, and 225 gigawatt-hours per year, respectively. For an average California home that uses 7 megawatt-hours per year of electricity, this corresponds to the demand of roughly 32,000 homes.

The **Stanford Energy System Innovations** project is a fourth-generation district heating and cooling system. It was first conceived in 2009 and became operational in March 2015, replacing a fossil-gas cogeneration plant that supplied 80 percent of the electricity and heat for the university campus. The system consists of a 2.3-million-gallon insulated steel hot-water tank, two 4.75-million-gallon insulated steel cold-water tanks, three waste-heat-driven 2,270-tonne heat pumps (heat recovery chillers) that raise the temperature of the hot-water tank and expel cold to the cold-water tanks, three hot-water generators (boilers) to provide extra heat to the hot-water tank, four 2,720-tonne cold-water generators (chillers) to provide extra cold to the cold-water tanks, cooling towers (13,200 tonnes total) to cool water further for the cold-water tanks and provide additional waste heat for the heat pumps, a 35.4-kilometer hot-water-pipeline system, a 42.2-kilometer cold-water-pipeline system, a heat recovery and distribution system that takes advantage of the fact that different parts of the university have both heating and cooling needs at the same time, a planning model for optimizing the distribution of heating and cooling around campus every 15 minutes, and a high voltage substation that receives electricity from the grid when necessary.[105]

The heat recovery system collects waste heat from buildings and moves it to a hot-water loop, where it is fed to the heat pumps. The recycling of waste heat eliminates the discharge of waste heat to the air. The round-trip efficiency of the heating portion of this system (ratio of the energy returned as heating after storage to the energy in the electricity used to heat the water for storage) is around 83 percent. That of the cooling portion is around 84.7 percent.[106]

The project requires electricity, as do all buildings on campus. To provide this electricity, the university first built 5 megawatts of solar PV on university rooftops, then signed a power purchase agreement over 25 years to generate 67 megawatts from a new solar PV farm in Rosamond, California, dedicated to the campus. The farm came online during December 2016. The university then signed another power purchase agreement for dedicated power from an 88-megawatts solar PV farm near Lemoore, California. However, a 2019 wildfire damaged the Rosamond farm. Repairs were completed in early 2022. The combination of the district heating and cooling project and the 160 megawatts of solar PV mean that 100 percent of the Stanford University campus' annual electricity demand and nearly 100 percent of the university's heating and cooling demand were supplied by clean, renewable energy by 2022.

5.2 INDIVIDUAL HEATING AND COOLING UNITS IN BUILDINGS

Buildings not connected to district heating and cooling systems need a heating and cooling source. In this section, rooftop solar water heaters, heat-pump air and water heaters, and passive heating and cooling are discussed.

5.2.1 ROOFTOP SOLAR WATER HEATERS

Rooftop solar water heaters (or domestic hot-water systems) have been used for over 130 years. In 1891, inventor **Clarence Kemp** of Baltimore, Maryland,

5.2 INDIVIDUAL HEATING AND COOLING UNITS IN BUILDINGS

patented the first commercial solar water heater. The technology was adopted regionally right away. By 1897, solar water heaters had been installed on the rooftops of 30 percent of the homes in Pasadena, California.[117]

In 1909, **William Bailey** redesigned the rooftop solar water heater by separating the solar collector from the storage tank. Cold water from the storage tank in a house or garage would now flow through a copper pipe that was coiled as it passed through a shallow, insulated box under glass that was placed on the roof. The coils maximized the area of the pipe exposed to the sun, which heated the water in the pipe. The hot water then flowed to the top of the tank. The name given for this new device was the Day and Night Solar Heater.[117] Whereas 1,000 units were sold in 1920 in Los Angeles, the number declined to 40 in 1930, not only due to the Great Depression, but also because fossil gas was being piped into people's homes and solar heating could not compete in price. However, in southern Florida, 100,000 solar water heaters were sold before World War II. In the mid 1950s, Frank Bridgers designed the Bridgers & Paxton Solar Building, the world's first commercial office building to use solar water heating plus passive heating.

Today, all solar hot-water systems include a solar collector and a storage tank, but there are two primary types of systems, active and passive. Active systems use a pump to circulate water and use either direct or indirect circulation. An **active direct-circulation system** has a pump that circulates water through a rooftop solar collector. The water heated by the collector goes to a storage tank, where the water sits or is used immediately. Direct systems work well so long as the temperature of water in the collector exceeds the freezing point of water. An **active indirect-circulation system** has a pump that circulates a nonfreezing heat-transfer fluid through the rooftop solar collector, where the fluid is heated during the day. The heat from the fluid is transferred to water by means of a heat exchanger. The water is then stored in the storage tank or used immediately. **Passive systems** use city water pressure to push water from a city water source to a rooftop solar collector. Sunlight heats the water in the collector. As water is used in the building, the water in the collector moves to the household storage tank, where a heat-pump water heater warms the water further. Thus, the solar collector preheats the water, reducing the energy required by the heat pump.

Whereas solar hot-water systems are useful, they are often more costly and less efficient than using a heat-pump water heater powered by rooftop solar PV. Having only rooftop PV, but with more PV panels to provide electricity for a heat pump to heat water, is more cost-effective than having both PV (but less of it) and a solar water heater on a roof. The reason is that hiring two different contractors is almost always more expensive than hiring one to do a larger job.

5.2.2 HEAT PUMPS

Heat pumps are efficient devices that extract heat or cold from the air, the ground, water, or a hot or cold waste stream, and move the heat or cold to where it is needed. A heat pump that extracts heat or cold from the ground is a ground-source heat pump; one that extracts heat or cold from the air or from a waste stream of air is an air-source heat pump. One that extracts heat or cold from water, such as from a swimming pool or lake, is a water-source heat pump.

The advantage of a ground- or water-source heat pump over an air-source heat pump is that temperatures under the ground or in the water are relatively stable, even when the air outside is very cold or very hot; thus, ground- and water-source heat pumps are more efficient under extremely cold or hot air conditions than air-source heat pumps. The advantage of an air-source heat pump is that it is easier and less expensive to install and maintain because none of its parts is buried underground or in the water.

Traditionally, ground-source heat pumps extracted heat and cold from the ground through coils buried in a shallow backyard pit spread over a large area and covered with soil. The development of a technology to drill two narrow but deep holes vertically has reduced the cost of installing ground-source heat pumps significantly.[118]

When heat pumps are used for cooling, they operate just like air-conditioners or refrigerators, which move heat from a room or refrigerator, respectively, to the outside of the house or refrigerator. The difference between a heat pump and an air-conditioner or refrigerator is that the heat pump provides both heating and cooling. An air-conditioner and refrigerator provide only cooling.

One type of heat pump is the **ductless mini-split heat pump**. As its name implies, it does not require ducts to move hot or cold air around a building. Instead, an indoor air-handling unit is placed in each of several rooms or zones of the building. Multiple indoor units are each connected to a single outdoor compressor/condenser unit by thin pipes containing a power cable, suction tubing, refrigerant, and a condensate drain line. Refrigerants have low boiling points so evaporate at temperatures much lower than the boiling point of water. Evaporation requires energy, so when air transfers sufficient heat to a liquid refrigerant to cause the refrigerant to boil (thus,

5.2 INDIVIDUAL HEATING AND COOLING UNITS IN BUILDINGS

evaporating the liquid), the air cools. Similarly, when cool air passes by a gaseous refrigerant, heat is transferred from the refrigerant to the air, causing the refrigerant to condense and release heat to the air.

For air-conditioning (cooling of warm indoor air), a fan in the indoor heat pump unit blows warm air from a room over evaporator coils. Simultaneously, a thin pipe brings liquid refrigerant to the evaporator coils. Inside the evaporator, the liquid pressure is dropped, cooling the refrigerant. Heat is transferred by conduction from the room air to the cold refrigerant, bringing the refrigerant to its boiling point, evaporating the refrigerant. The room heat is therefore converted to latent heat stored in the refrigerant molecules. The room air that passed over the coils is now cold and flows back to the room, cooling the room. The gaseous refrigerant containing the heat then travels by tubing to the outside unit, where it is heated further by compression. In a condenser, the hot gas then transfers some of its heat by conduction to the cooler (relative to the hot refrigerant) air outside, causing the refrigerant's temperature to drop below its boiling point and condense back into a liquid. The condensation releases the stored latent heat, which also passes to the outside air. In sum, a ductless heat pump moves heat from a room to the outside air, cooling the room, using only electricity as a power source.

To heat cold indoor air, the heat pump runs in reverse. Gaseous refrigerant in the outdoor unit is compressed, heating the refrigerant. The hot gas is then piped to the evaporator coils. Simultaneously, cold air from the room is blown over the coils. At the coils, the cold room air causes the gaseous refrigerant to condense and release its latent heat into the air from the room. This warm air then flows back into the room. The refrigerant is now liquid and cold and flows back to the outside unit. Dropping the pressure cools the liquid refrigerant even more. At the outside unit, heat from the outside air is transferred by conduction to the cold, liquid refrigerant, raising its temperature above its boiling point, evaporating the refrigerant while cooling the outside air. The cycle then starts again. In sum, heat from outside air is transferred to the room to warm it.

Air-source heat pumps become less efficient for heating as outdoor air becomes very cold, and less efficient for cooling when outdoor air becomes very hot. However, even at −20 degrees Celsius, air-source heat pumps are almost twice as efficient as electric-resistance heaters.[119] To guard against poor performance, a resistance heating element is sometimes added to back up an air-source heat pump to ensure a room stays warm at very low outdoor temperatures. Ground- and water-source heat pumps largely solve the low-temperature problem because ground and water temperatures change only slowly. Thus, ground- and water-source heat pumps are often recommended for locations that experience extreme seasonal heat or cold or both. Air-source heat pumps are recommended for mild climates.

Whereas heat-pump air heaters and air-conditioners move heat between a building and the outside, **heat-pump water heaters** often extract heat from air within the room they reside in. This extraction reduces the air temperature in the room by 1–3 degrees Celsius. In the summer, such air cooling is advantageous, because the door to the room can be opened, providing additional cooling to the house, reducing energy requirements for air heating. In the winter, the extra cooling slightly increases the air heating requirement.

An air-source heat-pump water heater works as follows. First, liquid refrigerant inside the water heater is brought to an evaporator. At the evaporator, the refrigerant's pressure is dropped, cooling the refrigerant. A fan inhales warm (relative to the cool refrigerant) room air and blows it over the evaporator's coils. Heat from the room air then moves by conduction to the refrigerant, heating it to its boiling point, evaporating the refrigerant. During evaporation, heat from the room is stored as latent heat in the gas molecules of the refrigerant. Thus, the inhaled room air is now cooler, and it is expelled back to the room. The gaseous coolant is heated further by compression in a compressor. The gas then moves to a condenser, where it condenses back to a liquid, releasing its latent heat and compressional heat to water in the tank, heating the water. The net result is hot water and a slightly cooler room that the water heat sits in.

Because heat pumps move – rather than create – heat and cold, they reduce the energy needed to heat or cool air or water in a building by a factor of three to five compared with conventional gas heaters or electric-resistance heaters. The efficiency of a heat pump is defined by its **coefficient of performance**, which is the ratio of useful heating or cooling required to the work (electricity input) required. The coefficient of performance of a heat pump exceeds one, which means that the heat pump moves more than one unit of heat or cold for every unit of electricity required to run the heat pump.

Air-source heat pumps have coefficients of performance of 3.8 (2.5–5.5) when the outdoor temperature is 10 degrees Celsius, declining to 2–4 at 0 degrees Celsius outdoor temperature and to 1.8–2 at −20 degrees Celsius.[119] Ground-source heat pumps, on the other hand, have coefficients of performance at all outdoor temperatures of 4.2–5.2,[120] because their source of heat is deep soil, whose temperature is independent of outdoor air temperature. In comparison, **electric-resistance heaters** have a coefficient of performance of about 0.97, and fossil-fuel-powered boilers have a coefficient of performance of about 0.8. Since only 1 joule of electricity is needed to move 2.5–5.5 joules of hot or cold air with an air-source heat pump, such heat pumps reduce power demand compared with fossil-gas boilers by 70–85 percent. The use of heat pumps for all air and water heating worldwide is estimated to reduce world all-purpose end-use power demand compared with a business-as-usual case by about 13.3 percent.[5]

Heat pumps are now used not only for normal air and water heating and air-conditioning but also for dishwashing, clothes drying, and hot tubs. For example, the AdoraDish V6000 dishwasher uses a heat pump, reducing electricity use by 30 percent compared with a conventional dishwasher.[121] Northern Lights sells hot tubs running on a heat pump.

> **Transition highlight**
> In 1938, J. Ross Moore of North Dakota commercialized the first electric clothes dryer after tinkering with the technology since 1915. The first **electric heat-pump clothes dryer** was commercialized in Europe in 1997. By 2005, heat-pump clothes dryers were being manufactured in Japan as well. By 2009, about 4 percent of European clothes dryers were heat-pump dryers.[122] In 2005, the first heat-pump washer–dryer combo was commercialized, in Japan. Today, heat-pump dryers and washer–dryer combos are available worldwide. Heat-pump dryers require about half the electricity of conventional electric-resistance clothes dryers. However, heat-pump dryers themselves cost more upfront. This upfront-cost difference can be erased for new-construction homes if a ventless heat-pump dryer is purchased because the construction cost of vents is usually more than the cost of the dryer itself. As such, a ventless electric heat-pump dryer in a new-construction home will almost always have a lower lifecycle cost than a conventional electric dryer due to both the energy cost savings and the construction cost savings, despite the higher upfront cost of the heat-pump dryer. A heat-pump dryer will also have a lower lifecycle cost if it is installed in an existing building and used regularly.[22]

5.2.3 PASSIVE HEATING AND COOLING IN BUILDINGS

The energy required to heat and cool buildings can be reduced significantly with **passive heating and cooling techniques**. Passive techniques include installing insulation; thermal-mass materials; ventilated facades; window blinds, awnings, and films; and night ventilation.[123] Each of these is discussed, in turn.

5.2.3.1 INSULATION

The most basic passive energy-saving technique is to insulate a building thoroughly to minimize heat loss during winter and cold loss during summer. This means insulating walls, ceilings, floors, and pipes; tightly sealing cracks in doors and windows; and using thick (up to triple-pane) windows. Insulation reduces energy use during all seasons.

5.2.3.2. THERMAL-MASS MATERIALS

Thermal-mass materials are building materials used to absorb, store, and release heat in a building in such a way as to keep building temperature relatively constant during the day and the night. The two main types of thermal-mass materials are sensible heat storage materials and latent heat storage materials.

Sensible heat storage materials are materials that do not change temperature much during the day or night, thus helping to keep buildings at a near constant temperature. Good sensible heat storage materials have both a high heat capacity and moderate thermal conductivity.

Heat capacity is the energy required to increase the temperature of 1 cubic meter of a substance by 1 degree Celsius. Substances with high heat capacities absorb lots of sunlight during the day without increasing their temperature much and release lots of heat at night without decreasing their temperature much. Substances with low heat capacities warm rapidly during the day upon absorbing sunlight and cool rapidly at night upon releasing heat. The heat capacity of liquid water is 3.5 times that of sand. Thus, the same sunlight heats a given volume of sand 3.5 times as much as it does water. This is why ocean beach water feels cool during the day, whereas sand feels hot. Conversely, the loss of the same heat at night cools sand 3.5 times as much as it cools water.

Conduction is the transfer of energy from molecule to molecule. The thermal conductivity of a substance is a measure of its ability to conduct heat in the presence of a temperature difference between one end of the material and the other. Steel is very conductive, but pinewood is not. A good thermal-mass material should have a moderate thermal conductivity. If the thermal conductivity is too high, such as with steel, the material will transfer its heat (or cold) too fast to the air or to another material. If the conductivity is too low, such as with wood, the material will not conduct its heat (or cold) to the air quickly enough.

Effective thermal-mass materials have high heat capacities and moderate thermal conductivities. Such materials include water, concrete,

5.2 INDIVIDUAL HEATING AND COOLING UNITS IN BUILDINGS

marble, granite, and brick and are usually built into walls or floors. Water, for example, can be placed in tubes inside a wall or under a floor, or sit in plastic containers ("**water bricks**") in a room. Thermal-mass materials modulate the temperature of a building relative to using wood or steel. In other words, they store heat during the day without raising the temperature of the building and slowly release the stored heat to the air in the building at night, keeping the building warm at night. A disadvantage of sensible heat storage is it requires a large volume of storage material to have an impact on building temperature if the building is large.

Latent heat storage materials are materials that modulate building temperature by changing phase near room temperature. As such, they are also called **phase-change materials**. Such materials include paraffin wax, fatty acids, and salt hydrates. Much less mass of a phase-change material is needed than of a thermal-mass material to modulate building temperature. When a phase-change material absorbs heat, its temperature quickly reaches its melting point, at which point the material absorbs more heat in order to melt without raising its temperature further.

For example, if paraffin wax is distributed within south-facing walls of a northern-hemisphere building, the wax absorbs heat and increases its temperature until the melting point is reached. Melting points can be 37–65 degrees Celsius, depending on which wax is used. During melting, the wax still absorbs heat, but that heat melts the wax rather than raising its temperature. By absorbing heat without increasing in temperature, the melting wax keeps the building cool. After the sun goes down, the wax releases heat to the air, warming the air and cooling the wax. Once it cools to its freezing point (similar to its melting point), the melted wax freezes (solidifies), releasing latent heat to the air, warming the air more without dropping the temperature of the wax further. Thus, phase-change materials keep the temperature of a wall relatively constant during the day and night. A phase-change material can be encapsulated within a wallboard, mixed into concrete, or combined with insulation material.

5.2.3.3 VENTILATED FACADES

A **ventilated facade** is a weather-protective wall, separated from the main wall of a building by air. Air in the space between the facade and wall flows and mixes with outdoor air, providing ventilation. The air space also benefits the building acoustically. The facade can be made of a thermal-mass material to reduce temperature swings.

5.2.3.4 WINDOW BLINDS, AWNINGS, AND FILMS

Window blinds control the sunlight entering buildings. Lowering a window blind reduces the amount of sunlight – including damaging ultraviolet light – entering a building during a sunny day. Raising the blind increases the sunlight entering the building during a cloudy day. A **window awning** is a roof-cover that extends horizontally out above a window to protect it from sunlight or rain. A retractable awning that is open during summer and closed during winter shades a window during summer days and allows sunlight through a window during winter days, modulating building temperature. Similarly, putting a transparent or translucent **window film** on a window reduces the penetration of sunlight into a building by increasing the window's solar reflectivity. Window films also reduce the outgoing heat from the building. An advantage of a window film is it reduces ultraviolet radiation entering a building, reducing damage to furniture and human skin and eyes. During winter, though, the reduction in sunlight entering a room due to a window film may exceed the reduced heat loss from the room to the outside, cooling the room more than desired. However, in the annual average, a window film usually conserves energy and always reduces ultraviolet radiation.

5.2.3.5 NIGHT VENTILATION

Night ventilation is the use of natural ventilation to remove heat from a building during the night. During the day, thermal-mass materials in a building heat up due to their absorption of solar and heat radiation. During the night, thermal-mass materials release their heat to the building, keeping the building warm. However, in hot climates, building heat can accumulate during day and night over periods of days to weeks, even with thermal-mass materials present. To shed some of this heat, night ventilation is useful. Night ventilation takes advantage of either nighttime winds or heat-generated pressure to push indoor heat through a stack on the roof to the air outside. Since the indoor air is now cooler, the thermal-mass material in the building cools more as well. The next day, the thermal-mass material and building heats to a lower temperature than it does with no nighttime ventilation.

5.3 WWS ELECTRIC APPLIANCES AND MACHINES

A key step in electrifying all energy sectors is to ensure electric appliances and technologies that replace fossil-fuel ones perform at least as well and at a similar or lower cost. Here, some electric replacement technologies are described.

5.3 WWS ELECTRIC APPLIANCES AND MACHINES

5.3.1 ELECTRIC-INDUCTION COOKERS

In a 100 percent WWS world, **electric-induction cookers** (induction burners) will replace cookers that run on fossil gas, propane, wood, waste, dung, or coal. Induction cookers are also better than electric-resistance cookers. An induction cooker consists of a ceramic plate with a coiled wire beneath it. When the cooker is turned on, an electric current runs through the coil, generating a fluctuating magnetic field, but not heat, in the cooker itself. This is why the cooker does not feel hot when it is touched.

Induction cookers work only with pots and pans that have bases made of iron or stainless steel, both of which are highly resistive to electricity transfer and stick to magnets. Highly conductive and nonmagnetic materials, such as aluminum, copper, glass, and ceramics, do not work well. Once an iron or stainless-steel pot or pan is placed on a cooker, the magnetic field in the cooker induces many smaller electric currents in the pot or pan's metal base. Because iron does not conduct well, much of the electricity in the small currents turns to heat by resistance, heating the metal. The heat is produced in the metal base of the pot or pan, but not in the cooker. As such, the cooker itself feels only warm due to conduction of heat from the base of the pot or pan to it.

Whereas a fossil-gas cooker converts about 32 percent of fossil-gas energy to heat used to heat food, and an electric-resistance cooker converts about 75–80 percent of the electricity it uses to such heat, an electric-induction cooker converts about 85 percent of electricity to such heat.[124] As such, an induction cooker uses 68 percent less energy than a fossil-gas cooker does to provide the same heat. Most heat from fossil-gas heating dissipates to the air. Due to the efficiency difference, water boils with an induction cooker in about half the time of a fossil-gas cooker. Induction cookers also cause less food scorching because they have fewer hot spots within a pot or pan than fossil-gas cookers.

Transition highlight
Electric-induction cookers have the potential to eliminate enormous loss of life. Millions of children and adults, primarily in developing countries, die each year, due to breathing air pollutants from the indoor burning of wood, waste, dung, and coal for home cooking and heating. Induction cookers can eliminate the need to burn fuels for cooking. Induction cookers can also save time for the hundreds of millions of people who walk many kilometers daily to collect wood or waste for cooking. Single induction cookers cost $40–$100. All that is needed, aside from a burner, is an

electricity source and an iron or stainless-steel pot or pan. While bringing electricity to remote villages in developing countries is challenging, the advent of low-cost PV and batteries now permits the widespread adoption of clean cooking in remote communities, saving many lives.

5.3.2 ELECTRIC FIREPLACES

Wood and fossil-gas fires in fireplaces are fire risks and sources of health-damaging carbon monoxide, nitrogen oxide, organic gas, and particles inside and outside the home. Such fires are cozy and warm and give a home character, but they also produce indoor and outdoor air pollution. Wood fires can also cause smoke damage and unpredictable impacts of their embers that burst from a fireplace into a living room. Fortunately, warm and visually appealing electric fireplaces are now available to replace wood and gas fireplaces. These eliminate the air pollution and smell of wood or gas burning while providing the same warmth and coziness.

5.3.3 ELECTRIC LEAF BLOWERS

Possibly the most annoying fossil-fuel machine today is the gasoline **leaf blower**. It not only smells and creates noxious air pollution, but it is also noisy, since it has no muffler or other means of noise suppression. Both corded and battery-powered electric leaf blowers are available. They perform the same task as a gasoline leaf blower but do not smell and are quiet. Some battery-powered leaf blowers run for 1–5 hours at low speed or 10–50 minutes at full speed and take 30–240 minutes to charge, depending on battery size. Most models allow the battery to be swapped, so an electric leaf blower can be used continuously if multiple charged batteries are available for swapping.

5.3.4 ELECTRIC LAWNMOWERS

Another noisy machine that creates a smell and unhealthy air pollution is the gasoline **lawnmower**. Cost-competitive (over the lifetime of the lawnmower) battery-powered lawnmowers are readily available to replace gasoline lawnmowers.[125] Not only do electric lawnmowers eliminate the air pollutants (unburned hydrocarbons, oxides of nitrogen, carbon monoxide, and particles) that gasoline lawnmowers emit, but they also start instantly, are quiet, and require less maintenance (since they have no engine and fewer parts) than gasoline mowers. Although the upfront cost of an electric lawnmower may be more than that of a gasoline mower, electric mowers have no gasoline cost and need less equivalent energy (in the form of electricity) than gasoline mowers because the ratio of work output to energy input of electricity powering a motor is about four times that of gasoline powering an engine. As such, the energy and overall cost to run an electric mower over 10 years is less than that to run a gasoline mower, even if the upfront cost of the electric mower is higher. Runtimes of electric lawnmowers are currently 1–2 hours before recharging or battery swapping is needed. Battery charging may take 3–4 hours.

5.3.5 OTHER APPLIANCES AND TECHNOLOGIES

All other home, business, and industrial technologies that run on fossil fuels or bioenergy fuels have an electric counterpart that is needed as part of a 100 percent clean, renewable energy economy. One example technology is the chain saw. Andreas Stihl of Germany invented the first **electric chain saw** for woodcutting in 1926. Stihl invented a gasoline-powered chain saw three years later. Both electric and gasoline chain saws are commercial today. Advantages of electric over gasoline chain saws are that the former produce minimal noise, emit no fumes, require no mixing of oil and gasoline, require minimal maintenance, are lighter, and cost less. Cordless versions have traditionally lasted about an hour before needing a recharge and have been used for smaller jobs than gasoline chainsaws. However, with the improved storage capacity of batteries and with battery swapping capabilities, both limits are now less of a concern.

5.4 INCREASING ENERGY EFFICIENCY AND REDUCING ENERGY USE

Two additional goals of a 100 percent WWS system are **reducing energy use** and **increasing the efficiencies** of buildings and electric appliances, machines, and tools. Measures taken to implement these goals are **demand-side energy conservation** measures. Demand-side conservation reduces the need for energy and the cost of electricity.

Reducing energy use involves changing behavior. Some examples of such behavior change include

- using public transit or telecommuting instead of driving,
- carpooling and minimizing the number of driving trips required,
- teleconferencing instead of flying on an airplane to a conference,

- eating foods that require less energy to produce,
- eating locally sourced food instead of food transported over a long distance,
- ensuring electronic equipment is shut off when it is not in use.

Increasing energy efficiency involves relatively low-cost investments that can result in large energy savings. Some examples of energy-efficiency measures are

- replacing incandescent and compact fluorescent light bulbs with light-emitting diode (LED) light bulbs;
- controlling lighting with a sensor or timer;
- replacing electric-resistance air and water heaters and clothes dryers with electric heat-pump versions;
- **weatherizing** (sealing) cracks and gaps in windows, doors, fireplaces, and skylights;
- improving wall, floor, ceiling, and pipe insulation;
- implementing green building standards in new-building construction to minimize building energy use;
- adding advanced lighting controls;
- washing clothes in cold water;
- cleaning and/or replacing filters in heat pumps and air filtration systems;
- using ductless heating and air-conditioning to eliminate heat and cold losses from pipes;
- using triple-glazed windows to reduce heat loss from windows;
- installing thermal-mass materials in walls and floors to modulate temperature changes in a building;
- installing a ventilated facade to shield a building from extreme heat or cold;
- installing window blinds and awnings to control sunlight into buildings;
- applying window films to reduce heat loss and sunlight intake;
- installing night ventilation cooling in a building;
- improving air-flow management in a building;
- installing passive solar heating in a home;
- using more energy-efficient appliances;
- performing building energy audits to identify where energy is wasted;
- improving data center design;
- improving the efficiencies of solar cells, wind turbines, batteries, and electric cars;
- designing cities to facilitate the use of public transit and bicycles and to improve traffic flow; and
- building high-rise apartments instead of single-family homes to reduce construction materials and heat loss.

For example, weatherizing a home with caulk stops leaks around the edges of windows, doors, and fireplaces. Whereas weatherizing costs only a few to tens of dollars, it can save hundreds to thousands of dollars over several years by reducing continuous leaks of hot air during winter and cold air during summer. Similarly, replacing an old appliance with an energy-efficient one can save hundreds to thousands of dollars over the life of the new appliance.

Increasing energy efficiency and reducing energy help to avoid the use of fossil fuels and their resulting pollution. As such, these measures accomplish the same goal as replacing fossil fuels with WWS, but usually at lower cost. Because energy-efficiency measures and reducing energy use are low cost, they should be prioritized in strategies to address global warming, air pollution, and energy security. However, the measures cannot solve the problems on their own. A full solution also requires a transition to 100 percent WWS and the elimination of nonenergy emissions.

5.5 ALL-ELECTRIC HOME

Part of transitioning society to 100 percent clean, renewable energy and storage is to build new buildings, including residences, that produce their own energy and emit no pollution. To illustrate the components and benefits of a clean, renewable residence, I provide an example of a 100 percent WWS new-construction home, my own home.

The home, completed in 2017, has two storeys with approximately 278 square meters (3,000 square feet) of home floor space and 46.5 square meters (500 square feet) of garage space. The structure itself is made of prefabricated steel, 80 percent of it recycled. Advantages of a prefabricated steel structure are that it eliminates wood waste during assembly of the structure, it eliminates mold and termites associated with the structure itself (the exterior of the house and floors are still wood), and it produces walls that are perfectly flat, with corners that are exactly at 90-degree angles. The precise construction reduces the risk of air leaks and mistakes in constructing the rest of the building.

The home uses double-glazed windows that include a krypton-gas-filled cavity in lieu of a third glazing to suppress conduction and convection between glazes and to reduce 99.5 percent of incoming ultraviolet radiation. The low-conductivity fiberglass window frames have insulating foam within them and triple weather stripping around them. The windows are insulated to a similar degree as triple-glazed windows.

No fossil-gas pipes are connected to the property. Instead, the house and two battery-electric vehicles are powered by forty-three 320-watt roof-mounted solar PV panels (total nameplate capacity of 13.76 kilowatts). Since the expected output from each panel is 298.7 watts, the real peak total output is 12.84 kilowatts. The panels are mounted at a 15-degree tilt, facing south-by-southwest, in four rows, with at least 0.914 meters between rows. The home is also connected to the electricity grid. Although the rooftop PV panels produce more annual electricity than the building and its electric vehicles consume, enough solar PV electricity is not always available at the exact times it is needed to meet instantaneous electricity demand.

In order to improve the matching of power demand with supply over time, the home was built with wall-mounted batteries with a total peak discharge rate of 13.2 kilowatts and a total storage capacity of 25.6 kilowatt-hours. Thus, if all batteries were filled to capacity, they could discharge electricity at a rate of 13.2 kilowatts for 1.94 hours or 1 kilowatt for 25.6 hours. The batteries and PV are connected to inverters.

Because the home has no fossil gas, all heating, cooling, and cooking is done with electric appliances. For air heating and air-conditioning, ductless mini-split heat-pump air heaters and air-conditioners are used. Nine indoor units are placed in different rooms or zones of the house, and two compressor/condensers are placed outside either to release heat to or extract heat from the outside air. The indoor units are connected to the outdoor units by tubes containing a refrigerant, thus the system is ductless. For water heating, an air-source heat-pump water heater is used. The source of the heat is the air in the mechanical room in which the water heater resides. During heat-pump water-heater use, heat from the air is transferred to the water heater, and the room slightly cools. Thus, during summer, the mechanical room door can be opened to cool the rest of the home. Hot water is circulated around the home through pipes with a circulation pump on a timer. The timer is set so that the circulation pump runs only during times of low electricity price. For cooking, an electric-induction cooktop is used. Induction cookers use 70 percent less energy than fossil-gas cookers and boil water in half the time yet don't feel hot when they are touched. Other major electric appliances include a clothes washer, clothes dryer, refrigerator, microwave oven, convection oven, toaster, garbage disposal, light-emitting diode (LED) lights, televisions, computers, garage door, and air-filtration system. Energy-efficient versions of these were purchased.

The overall strategy for powering the home with rooftop solar PV and wall-mounted batteries is as follows. During the night, the batteries supply electricity to the home. Thus, by morning, the battery level is usually at its lowest limit. During the morning, the PV begins generating electricity, which is consumed first by household appliances. Excess electricity after that is used to charge the batteries. Excess electricity after that is sold to the grid. Once the batteries become fully charged, usually by mid-morning, they sit idle until near sunset. At that point, the batteries provide household power until they are depleted, at which point, grid electricity is used. If the batteries do not drain during the night, the house runs 24 hours per day on PV and batteries.

Electric cars at the home are charged at night for cost reasons, as discussed shortly. The two electric cars charged during the week both draw power quickly (one draws up to 20 kilowatts and the other, up to 8 kilowatts) and consume a lot of energy (up to 85 kilowatt-hours and 53 kilowatt-hours, respectively). The batteries (peak discharge rates of 13.2 kilowatts and peak energy storage of 25.6 kilowatt-hours) cannot supply the peak charge rate of one car (20 kilowatts) or the total energy requirement of either car. Thus, the peak charge rate of the 20-kilowatt car from the batteries is limited to 13.2 kilowatts and the peak energy supplied by the batteries to either car is 25.6 kilowatt-hours. Rooftop PV or grid electricity simultaneously supplies the remaining electricity for both cars. More batteries are needed to eliminate the use of grid electricity.

For the first seven years of household electricity use, including during winter, the PV system produced about 120 percent of household plus vehicle electricity use. In other words, there were no electric bills (aside from a $10 per month hookup fee), no fossil-gas heating bills (because no fossil gas was connected to the property), and no gasoline bills (because the cars are battery-electric). In fact, excess electricity was sent back to the grid, and until 2023, when utility compensation changed, the utility paid an average of $800 per year for it.

In California and several other states, many **community-choice-aggregation (CCA)** utilities have arisen. These utilities procure clean, renewable electricity by signing power purchase agreements with wind, solar, geothermal, and hydroelectricity farms. Some CCAs provide the electricity at competitive prices to customers who sign up with them. The CCAs also purchase electricity from homeowners who have roof PV or a small backyard wind turbine. In some cases, the CCA pays for the PV electricity at the same rate a ratepayer would pay for electricity at the same time of day. Thus, if a ratepayer pays 25 cents per kilowatt-hour for electricity generation at 4 p.m., a homeowner who sends electricity back to the grid at that time also receives 25 cents per kilowatt-hour as payment for that electricity. The transmission and distribution portion of a ratepayer's bill is handled by a different utility, but transmission and distribution costs are often applied only when more electricity is drawn from the grid than is sent back to the grid.

Paying different electricity rates at different times of the day is called **time-of-use pricing**. Often, there is a period during weekdays (peak period), such as from 2 p.m. to 9 p.m., during which electricity price is highest, another period (off-peak period), such as from 11 p.m. to 7 a.m., during which the price is lowest, and periods of intermediate-price (**partial-peak**) electricity price (all other times). Time-of-use pricing may differ between weekends and weekdays. One weekend distribution may be to charge peak prices from 3 p.m. to 7 p.m. and off-peak prices during all other hours.

An alternative to time-of-use pricing is to use constant pricing all day and to charge increasingly higher rates per kilowatt-hour of electricity used with the more electricity that is used. The advantage of time-of-use pricing over this method is that the former is a built-in method of **demand-response management** of the grid. In other words, time-of-use pricing helps utility operators balance supply of electricity with demand for it.

With one method of demand response, utility operators give individuals and businesses financial incentives not to use electricity at certain times of the day when electricity demand is high. This is done by increasing electricity prices or by paying customers not to use electricity during those times. Higher prices between 2 p.m. and 9 p.m., for example, shift electricity use from that period to another period, such as between 11 p.m. and 7 a.m., when electricity prices are lower. Thus, nighttime car charging is encouraged by low nighttime electricity prices.

The construction of a super-efficient all-electric home, with heat pumps, electric appliances, and electric cars, that is powered by solar PV and batteries not only eliminates bills for electricity, fossil gas, and gasoline and results in payments for the excess electricity generated, but it also eliminates installation costs related to fossil gas. For example, it eliminates the cost of a gas hookup fee and gas pipes.

Transition highlight

Eliminating fossil gas and electrifying a new home can save $6,000–$23,000 in upfront costs and $3,000–$10,000 per year in electricity and fuel bills. Upfront costs avoided include those for a fossil-gas hookup fee ($3,000–$8,000) and fossil-gas pipes ($3,000–$15,000). In some cases, homeowners can receive up to $1,000 per year by selling excess electricity to their utility. These savings offset part of the capital cost of installing solar and battery systems that provide all household and vehicle needs. Such systems have a total upfront capital cost ranging from $30,000–$115,000. The capital and maintenance cost of heat pumps for air and water heating and air-conditioning and other electric appliances and vehicles are similar to those of fossil-fuel equivalent appliances and vehicles. As such, a solar-plus-battery system powering the home can pay for itself in 4.5–4.7 years with government incentives (which have been about 35 percent of the capital cost in the United States) and in 8–9.2 years without subsidies. This payback time is decreasing yearly, as solar and battery costs are declining year by year.

5.6 MICROGRIDS

A **microgrid** is a group of distributed electricity- and/or heat-generating resources, connected to energy demands, that acts as a single controllable entity. A microgrid may be isolated from or connected to a main grid. If it is connected, it can be disconnected to run in island mode.[126] Microgrids can serve remote communities, hospitals, islands, military forward operating bases, emergency-relief zones, and data centers. A remote community with a dozen buildings, for example, may obtain all of its electricity from rooftop PV, store its electricity in batteries, and use PV and battery output to power lights, appliances, electric-induction cooktops, heat pumps for air and water heating and air-conditioning, food production, and water purification.[127]

A WWS microgrid may use a wind turbine, a PV array, a run-of-the-river hydropower plant, or an enhanced geothermal plant as an electricity source. Aside from meeting building demand, electricity from a microgrid may also charge battery-electric vehicles, extract water from air, purify water, and produce food in a greenhouse or container. A more advanced microgrid may use electricity to produce hydrogen that is stored, then use the stored hydrogen in a fuel cell to reproduce electricity and heat as needed. Thermal-mass materials, such as water bricks, may also be placed in buildings to keep the buildings from getting too hot during the day and too cold during the night. Microgrid control algorithms are important for ensuring that all components of a microgrid work together in tandem.[126]

A **mobile microgrid** is a microgrid in which all equipment can be moved rapidly from one place to another. Mobile microgrids are important for disaster relief zones and military forward operating bases. For example, after a hurricane that displaces people from their homes and/or causes widespread power outages, a mobile microgrid can be moved into the disaster zone to provide electricity for medical care, lighting, appliances, heating, and/or cooling. Similarly, military bases in combat zones need to be mobile and so require mobile electricity generation and storage as well.

6

WWS SOLUTIONS FOR INDUSTRY

The industrial sector creates products made of metal, plastic, rubber, concrete, glass, and ceramics, among other materials. Energy is needed in industry for heating, cooling, drying, curing, and melting, and for electric appliances and machines. Industrial heat ranges from low- to high-temperature heat. About half of industrial heat is high-temperature heat (above 400 degrees Celsius) and the other half, low- (30–200 degrees Celsius) and medium- (200–400 degrees Celsius) temperature heat.[128] High-temperature heat is used for plastics and rubber manufacturing, casting, steel production, other metal production, glass production, lime calcining in cement manufacturing, metal heat-treating and reheating, ironmaking, and silicon extraction from sand. Low- and medium-temperature heat are used for drying and washing during food production, chemical manufacturing, distilling, cracking, pulp and paper manufacturing, and petroleum refining, among other processes. This chapter first discusses the current sources of energy used in industry and then discusses WWS alternatives for these sources. The chapter also includes methods of eliminating chemical emissions from steel, concrete, silicon, and ammonia manufacturing.

6.1 CURRENT ENERGY SOURCES FOR INDUSTRY

Energy for industrial heat is currently obtained from combustion fuels, electricity, and steam.

Combustion fuels used for producing industrial heat include fossil gas, coal, fuel oils, liquefied gases, and biomass. These fuels are burned in ovens, fired heaters, kilns, and melters. The resulting heat is transferred either directly or indirectly to the material being melted. With **direct (convection) heating**, combusted gases come into direct contact with the material. With **indirect (radiant) heating**, the hot gases pass through radiant burner tubes or panels separated from the material, and the heat is transferred to the material radiantly.

Electricity-based heating technologies today include electric-arc furnaces; electric-induction furnaces; electric-resistance furnaces, kilns, and boilers; electric crackers; electron-beam heaters; dielectric heaters (including radio-frequency dryers and microwave processors); and electric heat pumps. All except for the dielectric heaters and electric heat pumps produce high-temperature heat (greater than 400 degrees Celsius). The remaining technologies produce primarily medium- and low-temperature heat. All technologies use either direct or indirect electric heating. With direct electric heating, an electric current is sent directly to a material, heating the material by resistance heating. With indirect electric heating, an electric current in one material induces a current in a second material that is highly resistive, causing the second material to dissipate the electricity to heat.

Steam-based heating technologies heat materials directly with steam or indirectly through a heat-transfer mechanism. Almost all steam heating is for low-temperature (less than 200 degrees Celsius) processes, such as pulp and paper manufacturing, chemical manufacturing, and petroleum refining. Currently, most of the steam for these processes is generated simultaneously (cogenerated) with electricity produced by the burning of a fossil fuel or biomass.

Worldwide, about 17 percent of all carbon dioxide emissions from anthropogenic sources comes from burning fossil fuels to produce low-to-high-temperature heat for industrial processes.[129] Another 8.4 percent of carbon dioxide emissions result from chemical reactions arising during industrial processes.[129] These emissions, which occur during steel manufacturing, concrete production, silicon purification, and ammonia production, are called **process emissions**.

Of all industrial heat produced worldwide in 2022, only 4 percent was produced from electricity.[130] In a 100 percent WWS world, all industrial-sector low-to-high-temperature heat will be provided by electricity. In addition to removing emissions related to industrial heat, it is necessary to eliminate nonenergy carbon dioxide process emissions from industry. Industrial-sector electrification and alternative methods of producing steel, concrete, silicon, and ammonia to eliminate their process emissions are discussed next.

6.2 ARC FURNACES

An **electric-arc furnace** is a spherical-shaped furnace for melting metal. It has a retractable roof and contains three graphite electric rods. The floor of the furnace is coated with a heat-resistant material that is used to collect the molten metal. Scrap steel or iron is fed into the furnace, the roof is closed, and the electric rods (cathodes) are lowered onto the metal. An electric current that passes between each rod and an anode mounted under it (at the bottom of the furnace) creates an arc (extremely hot, bright light). The current that passes from each negatively charged cathode rod to the positively charged anode melts the metal, as does the radiant heat emitted by the arc. When the metal is melted, the metal alloy, slag (stony waste matter separated from metal during smelting), and oxygen formed by the process are removed through side doors.

The arc furnace derives from the carbon arc lamp. In 1800, Sir Humphrey Davy (1778–1829) invented the **carbon arc lamp**, which consists of two carbon rods with an electric current running through them. When the rods first contact each other, a spark is ignited. The rods are then pulled apart slowly. The current forms an extremely bright arc of light across the air gap. The bright arc forms because the hot (3,600–6,300 degrees Celsius) tips of the carbon rods vaporize and ionize, creating a plasma made of positive carbon ions and free electrons at high temperature. The electrons turn the gas into a good conductor that can be maintained at the high temperature. When the current strikes the ionized carbon vapor, a bright light forms. The carbon rods slowly burn away as they gasify, requiring the distance between them to be adjusted as well. Carbon arc lamps were the major form of street lighting and industrial indoor lighting worldwide from 1801 to 1901. Their disadvantages are that they produce short ultraviolet wavelengths of radiation, which are dangerous to humans; they create a buzzing sound; and they produce flickering light and sparks, which can cause fires.

In 1878, **Carl W. Siemens** (1829–1906) extended the idea of the arc lamp to build and patent an electric-arc furnace. James B. Readman (c. 1850–1927) subsequently invented (1888) and patented (1889) an arc furnace in Edinburgh, Scotland, and applied it to produce phosphorus. Paul Heroult (1863–1914) of France developed a commercial arc furnace for steel production in 1900. In 1905, he went to the US to work with several steel companies, including the Sanderson Brothers Steel Company in Syracuse, New York, which installed the first commercial electric-arc furnace worldwide in 1906.

Arc furnaces are used in foundries (workshops in which metals are melted and cast into different shapes), steel mills, and silicon extraction facilities. In steel mills, they are used to produce steel from recycled scrap metal, reducing the energy needed versus making steel from raw ores. Because arc furnaces require a lot of electricity, they are often used when electricity prices are low, thus their use responds well to time-varying electricity pricing. As such, arc furnaces can help manage the grid by using electricity primarily when excess is available or its price is low.

In a 100 percent clean, renewable energy world, arc furnaces are one technology that will replace fossil fuels and biomass to produce high-temperature heat.

6.3 INDUCTION FURNACES

Another method of melting metals such as iron, steel, copper, aluminum, and precious metals with electricity is with an **electric-induction furnace**. Induction furnaces can melt from 1 kilogram to 100 tonnes of metal at a time. An induction furnace consists of a nonconductive crucible surrounded by a large coil of copper wire. Metal is placed inside the crucible to be melted. A strong alternating electric current is sent through the wire coil, creating a rapidly reversing magnetic field that induces circular electric currents (called eddy currents) inside the metal. The metal heats by resistance heating as the eddy currents pass through it and dissipate. Because the heating of the targeted material results from electric currents that are induced by a magnetic field created from electricity passing through a coiled wire, the heating is by electromagnetic induction.

Because the metal is heated by induction, the temperature of the metal rises to no higher than the temperature required to melt the metal. This prevents the loss of some alloying elements and results in high efficiencies. Thus, an induction furnace differs from an arc furnace, where the temperature rises above that required to melt the metal. The heating efficiency (ratio of heat output to electricity input) of an electric-induction furnace approaches 100 percent, but the overall system efficiency is a bit less due to losses in the power electronics.[131]

In a 100 percent WWS world, induction furnaces are another electric-powered technology that will replace fossil fuels and biomass for producing high-temperature heat.

6.4 RESISTANCE FURNACES, KILNS, AND BOILERS

A third electricity-based method of obtaining high-temperature heat is with an electric-resistance furnace, boiler, or kiln. Whereas a **furnace** is a general heating chamber, usually used for melting a material, a **kiln** is a special type of furnace for heating a ceramic material, such as clay, pottery, or glass. A **boiler** boils water to produce high-temperature steam. In all cases, direct current (DC) electricity passes from a negative electrode (cathode), through a conducting material, to a positive electrode (anode). The conducting material heats up due to resistance within the material. In **direct**

resistance heating, the material targeted for heating or melting is itself the conductor of electricity. In **indirect resistance heating**, a separate heating element is heated and transfers its heat by a combination of conduction, convection, and radiation to the material targeted for heating or melting.

Some applications of direct resistance heating are the heating of long metal rods, the heating of iron-containing metals prior to forging, and the continuous annealing (heating followed by slow cooling) of wire.

Indirect resistance heating is used to heat solids, liquids, and gases in many industries. It is used in the heating and metals industries for melting, hot working, plasma heating, stress relieving, and preheating. It is also used for heating in the food, paper, print, textile, rubber, plastic, glass, and ceramic industries.

6.5 ELECTRIC CRACKERS

Crackers are pieces of industrial equipment used to break down complex chemicals into simpler products using heat and pressure. Most crackers today break down large organic chemicals, such as naphtha, into simpler ones, such as ethene, propene, and butadiene, which are then used to make plastics. The source of most naphtha is distillation from crude oil. With 100 percent WWS, the use of all oil, fossil gas, and coal for energy will be eliminated, but the use of oil products for plastics will remain until a replacement for oil-based plastics is found. Today, about 90 percent of oil is used for energy, and 10 percent is used for nonenergy purposes, such as plastics and medicine (Section 8.1.4). No fossil gas or coal will be needed for nonenergy purposes with 100 percent WWS. During naphtha cracking, the naphtha is first mixed with 600-degree-Celsius steam produced under high pressure. The mixture is then briefly (for milliseconds) heated in a tube up to 750–950 degrees Celsius in the absence of oxygen to break the naphtha down. This process is called steam cracking. In a traditional cracker, fossil gas is burned to provide the heat needed. With an **electric steam cracker**, the fossil-gas heating is replaced by electric-resistance heating, where the heating elements are placed either around the tube or in the walls of the heating chamber.[132]

6.6 ELECTRON-BEAM HEATERS

Electron-beam heating is an electricity-based method of melting metals or modifying materials with a fine beam of electrons. When the high concentration of electrons hits a solid material, the kinetic energy of the electrons is converted to heat, which melts the material. The electrons are produced with an **electron gun**, which ejects a narrow stream of electrons from a heated cathode and accelerates them using high voltage. The beam is obtained by using electric and magnetic fields to force free electrons in a vacuum into a straight line. The power per unit area of electron-beam heating is one thousand times that of peak sunlight reaching the surface of the Earth. Around 50–80 percent of the energy in the electron beam is transferred to the material.

Electron-beam heating takes place in a large vacuum furnace and is applied to mass-produce steel and purify metals, such as titanium, vanadium, tantalum, molybdenum, tungsten, zirconium, niobium, and hafnium. The electronics industry uses many of these metals. A **vacuum furnace** is a furnace in which the material being operated on is surrounded by a vacuum, or absence of air. Temperatures in a vacuum furnace usually range from 1,000 to 3,000 degrees Celsius. Electron-beam heating is also used in vacuum chambers to weld and precisely cut materials to make machines, evaporate and deposit thin layers of metal on solar cells, and modify surface layers of metals.

6.7 DIELECTRIC HEATERS

A fourth method of electric heating is **dielectric heating**, which is what microwave ovens use to cook food. Dielectric heating is used for low-temperature applications. Dielectric heating, also known as electronic heating, uses electromagnetic radiation in the frequency range covering radio frequencies or microwave frequencies. The former type of dielectric heating is referred to as **radio-frequency heating**, and the latter as **microwave heating**.

In a dielectric heater, radio waves or microwaves are used to heat a **dielectric material**, which is a poor conductor of electric current. Radio-frequency heating is often applied to heat large materials because the heating is more uniform than with microwave heating. As such, radio-frequency heating is used for most dielectric industrial heating applications, including gluing, welding, plastic production, preheating, bread baking, textile drying, adhesive and paper drying, microwave preheating, and vulcanizing rubber. Microwave heating is used primarily for the tempering of meat and other food-processing applications.[133]

6.8 ELECTRIC HEAT PUMPS AND SOLAR HEATERS

Heat is also needed for many low- and medium-temperature processes in a 100 percent WWS world. Such heat is currently obtained as a coproduct of fossil-fuel combustion or biomass combustion for electricity production. These sources will be replaced, in part, by electric heat pumps and solar-thermal heaters. A heat pump can produce heat up to 160 degrees Celsius.[134] This heat can be used to boil water to produce steam of a similar temperature.

Common solar-thermal heaters that provide low- to moderate-temperature heat for industry include parabolic-trough collectors with and without thermal storage, flat-plate solar collectors with hot-water storage, and linear Fresnel direct steam-generation collectors without storage.[135,136] Parabolic-trough collectors produce heat at 60–350 degrees Celsius. Flat-plate solar collectors produce heat below the boiling point of water (30–70 degrees Celsius). Linear-Fresnel collectors generally produce heat above 300 degrees Celsius. Electricity-generating parabolic-dish concentrated solar power (CSP) plants produce 100–500 degrees Celsius heat. Similarly, electricity-generating central-tower receiver CSP plants produce heat of up to 600 degrees Celsius.[137] However, a manufacturing plant that uses such heat needs to be near the solar heat production, otherwise, heat losses from ducts or pipes transporting the heat long distance will be significant.

An advantage of solar-to-heat over electricity-to-heat technologies is that solar-to-heat conversion efficiencies are higher than wind- or solar-to-electricity-to-heat conversion efficiencies. On the other hand, solar heat is available during the day only, whereas WWS electricity may be available day or night.

6.9 FIREBRICKS FOR STORING AND PROVIDING INDUSTRIAL HEAT

One way to almost eliminate all emissions arising from the production of low-to-high-temperature heat for industry is to produce all process heat from electricity, as just described, where the electricity comes from WWS sources. However, due to the variability of wind and solar, for example, electricity or heat storage is also needed to ensure that the industrial heat is available continuously, 24 hours per day. Whereas many electricity storage options are available (Chapter 3), using variable WWS electricity immediately to produce high-temperature heat and storing that heat is a better option, as discussed next.

Refractory bricks are bricks that can withstand high temperatures without damage to their structures. They have been used to insulate kilns, furnaces, fireplaces, fireboxes, and ovens for thousands of years. **Firebricks** are refractory bricks that can, with one composition, store heat, and with another, insulate the firebricks that store the heat.[138,139,140] Heat-storing

6.9 FIREBRICKS FOR STORING AND PROVIDING INDUSTRIAL HEAT

firebricks have high specific heats and densities so that they can absorb a lot of energy with little temperature increase, and they have high melting points. They are surrounded either by another type of firebrick that is more insulating, and then by steel to reduce heat loss further,[139,141] or simply by a thick steel container.[142]

Refractory bricks were used to line primitive kilns dug into the ground during the early Bronze Age (4000–3000 BCE), ironmaking furnaces during the Iron Age (1500–500 BCE), crucibles for molten glass since the early 1600s, and steelmaking furnaces since the mid 1850s.[143] Firebricks, like refractory bricks, are usually made of ceramic material containing a combination of alumina (Al_2O_3), silica (SiO_2), magnesia (MgO), and chromia (Cr_2O_3). The portions of each constituent depend on the desired peak temperature, insulating properties, mechanical properties, and resistance to corrosion of the firebrick. Alumina is used to raise a firebrick's melting point, and silica is used for its insulating properties. A high melting point is needed so that a firebrick can withstand high temperatures; the insulating property is needed so that it does not lose heat to the outside rapidly. In a heat storage enclosure, some firebricks are used for heat storage, and others, for insulation. Those used for storage should have a high specific heat, high density, and high melting point.

Another firebrick option is pure low-grade solid carbon (graphite), which can be heated to 2,400 degrees Celsius.[142] This technology has several challenges associated with keeping its cost low, including the fact that graphite slowly vaporizes and its use of radiant heating limits its ability to transfer heat for many applications without additional heat-transfer technology.

Heat-storing firebricks are heated with electricity in one of two ways. The first is to connect a metallic alloy or ceramic electric-resistance heater to the firebrick.[139] However, such a heater cannot easily deliver heat to the center of a firebrick array. In addition, whereas such a heater is well suited for 1,100 degrees Celsius, it fails relatively quickly at 1,500 degrees Celsius because oxygen diffuses through its outer protective coating at such temperatures.

Direct resistance heating (DRH) is another way to heat a firebrick.[139] With DRH, an electric current fed to an electrically conductive firebrick dissipates to heat, permitting the firebrick's temperature to rise to 1,800 degrees Celsius.[144,145] An electrically conductive firebrick contains a conductive metal oxide, such as chromia, that is doped with, for example, 2–5 percent nickel oxide or magnesium. The dopant allows the firebrick to reach a desired temperature of up to 1,800 degrees Celsius. The doped chromia itself may be molded into a firebrick or molded together with alumni, silica, and/or magnesia into a firebrick.[144,145] Only a fraction of all firebricks used in this heating solution are electrically conductive; the rest are insulating bricks. Aside from reaching higher temperatures, another advantage of DRH over external heaters is that DRH results in no temperature drop between the heating element and the firebrick because the brick itself is the heating element. Finally, DRH is insensitive to voltage, current, or frequency and thus does not require expensive power electronics, or any electronics if connected directly to a photovoltaic array.

Firebricks may be organized in a pattern that allows air to flow through channels in them. The process heat may be drawn from the firebricks on demand by passing ambient or recycled air through channels in the bricks, yielding low- to- high-temperature air,[139] or it may be obtained from the emission of infrared radiation directly from the red-hot bricks.[142]

The temperature of a group of firebricks is not the same as the temperature achieved in the material being heated. The temperature of the material being heated depends on the specific heats and masses of both the firebricks and the material being heated and on the heat loss between the two. For example, graphite firebricks supplying 1,500 degrees Celsius heat to a material may need the graphite heated to 1,800–2,000 degrees Celsius to account for both the material's properties and heat loss.

Because firebricks are made from common materials, the cost per kilowatt-hour-thermal of a firebrick storage system is less than one-tenth of the cost per kilowatt-hour-electricity of a battery system. Firebricks themselves are low cost because they can be made of inexpensive heat storage materials and because no heat exchanger is needed. Firebricks avoid additional cost by avoiding the need for furnaces, kilns, boilers, and crackers.

Computer simulations across 149 countries indicate firebricks are a remarkable tool in reducing the cost of moving the world to clean, renewable energy.[43] Relative to a base case with no firebricks, using firebricks was calculated to reduce, among all 149 countries, 2050 battery capacity by 14.5 percent; yearly hydrogen production for grid electricity by 31 percent; underground low-temperature heat storage capacity by 27.3 percent; onshore-wind nameplate capacity by 1.2 percent; land needs by 0.4 percent; and annual energy cost by 1.8 percent. In sum, the recent commercialization of firebricks suggests a large-scale solution to addressing industrial process heat emissions is possible.

6.10 STEEL MANUFACTURING

Steel manufacturing requires significant electricity and heat, and results in air pollution and carbon dioxide emissions. With steel manufacturing,

6.10 STEEL MANUFACTURING

carbon dioxide emissions arise not only from burning a fossil fuel or biomass for high-temperature heat but also from chemical reaction during the extraction of pure iron from iron ore. Carbon dioxide emissions arising from chemical reaction are called **process emissions**.

Steel can be produced from raw iron ore or recycled metal. Steel produced from iron ore is produced in two stages. The first is called ironmaking, and the second, steelmaking.

In the **ironmaking** step, molten pure iron metal is extracted from solid iron ore (iron oxide). A **blast furnace** is filled with iron ore, some relatively pure solid carbon in the form of coke (coal heated in the absence of air), and **limestone** (calcium carbonate). Hot air containing oxygen is then forced through the bottom of the blast furnace. It reacts with the coke to form carbon monoxide and heat. The carbon monoxide gas then reacts with the iron ore and the coke to produce molten pure iron metal and carbon dioxide. Due to the heat, the limestone simultaneously decomposes to calcium oxide and carbon dioxide. The calcium oxide then reacts with and removes sandy remnants of the iron ore to form a waste product, called slag. Slag is less dense than molten iron, so it floats above the pure iron. Slag is then cooled and removed for use in roads, leaving pure iron behind. Thus, traditional steelmaking releases carbon dioxide not only through fossil-fuel and biofuel burning to produce high temperatures but also from chemical reactions during the production of pure iron.

Steelmaking is the second step in steel production. In this step, impurities are removed from the raw iron. Then, carbon and other alloying elements are added to make crude steel. Impurities removed include phosphorus, sulfur, and excess carbon. Alloying elements added include chromium, nickel, vanadium, and manganese.

Steelmaking is performed in one of two ways. **Primary steelmaking** involves the use of new iron from ironmaking. **Secondary steelmaking** involves the melting of recycled scrap steel in an electric-arc furnace to produce new steel.

The main method of primary steelmaking is the **basic oxygen steelmaking** method. With this method, the molten iron and impurities from the blast furnace are mixed with scrap steel and placed in a **basic oxygen furnace**. Oxygen is then blown through the furnace and reacts with carbon in the molten mix to form carbon dioxide, which is released to the air. Calcium oxide in the molten mix also reacts with phosphorous and sulfur, the products of which rise to the top as slag and are removed. Finally, alloys are mixed in, and the molten steel is poured into preshaped molds, where it cools and hardens.

With secondary steelmaking, scrap metal is melted in an arc furnace. Oxygen is blown through the metal to help remove the carbon and speed the meltdown of the metal by increasing combustion. Calcium, phosphorous, and sulfur are removed, and alloys are mixed in, in a manner similar to the process in the basic oxygen furnace.

In sum, the sources of carbon dioxide during the two-step steel formation process are (a) its emissions during fossil-fuel combustion to produce heat in the blast furnace and in the basic oxygen furnace, (b) its chemical release during reaction of carbon with iron ore, (c) its release upon the thermal decomposition of limestone during ironmaking, and (d) its release during the chemical reaction of carbon with oxygen during steelmaking. In addition, in an arc furnace, there is a small amount of carbon dioxide released due to the vaporization of graphite and its reaction with oxygen. The overall carbon emissions during the ironmaking plus steelmaking process using a blast furnace and basic oxygen furnace are about 1,870 kilograms of carbon dioxide per tonne of steel.[146] Of this, ironmaking produces about 70–80 percent of the carbon dioxide emissions.

6.10.1 REDUCING CARBON EMISSIONS WITH HYDROGEN DIRECT REDUCTION

An alternative to extracting molten iron from iron ore with coke during ironmaking is to extract the pure iron with hydrogen gas (H_2), where the hydrogen is produced with 100 percent WWS electricity.[146] Such hydrogen is **green hydrogen**. This process is called the **hydrogen-direct-reduction** process. The main extraction reaction involves mixing iron ore with hydrogen gas to produce pure molten iron plus steam. This reaction occurs optimally at a temperature of around 800 degrees Celsius, which is lower than the temperature needed in a blast furnace.[146] The reaction eliminates the process emissions of carbon dioxide produced from the purification of iron ore to iron during ironmaking. However, an injection of carbon into the molten iron during the steelmaking process is still needed to create an iron-carbon alloy (0.002–2.14 percent carbon) to strengthen the steel. In addition, some carbon is still emitted from the thermal decomposition of limestone and subsequent emission of carbon dioxide.

If the heat required for the hydrogen-direct-reduction ironmaking process is obtained with an electric-resistance furnace instead of with fossil fuels, if the hydrogen for the process is produced by electrolysis (passing of electricity through water), if an electric-arc furnace is used for the steelmaking process, and if all electricity is from 100 percent WWS, the

6.10 STEEL MANUFACTURING

hydrogen-direct-reduction process emits only 53 kilograms of carbon dioxide per tonne of steel, or only 2.8 percent of the emissions of the blast furnace/basic oxygen furnace process (1,870 kilograms of carbon dioxide per tonne of steel).[146] The only carbon dioxide emissions during the hydrogen-direct-reduction process are from oxidation of injected carbon in the arc furnace, the thermal decomposition of limestone, and oxidation of the vaporized carbon in the arc furnace electrodes.

The system just described is a 100 percent clean, renewable energy system, but still results in a residual of 53 kilograms of carbon dioxide per tonne of steel (2.8 percent of the original emissions). The remainder will likely be released to the air. Capturing the remaining carbon dioxide requires electricity, and even if the electricity is from WWS, it is better to use that WWS electricity to displace a fossil-fuel electricity source than to capture carbon dioxide, since displacing a fossil-fuel source eliminates not only carbon dioxide from the source but also air pollution and upstream mining and emissions. Carbon capture does not reduce mining or air pollution.

Transition highlight
During June 2021, a steel manufacturing plant in Lulea, Sweden, created pure metallic iron (sponge iron) from hydrogen for the first time. The hydrogen was produced by electrolysis, where the electricity came from wind, thus the hydrogen was green hydrogen. The plant subsequently produced nearly carbon-free steel commercially from the purified iron in July 2021. As of 2024, the plant had produced 5,000 tonnes of green steel. Based on the results of this plant, the Swedish steel industry plans to transition all steel factories to green steel ones.

To that end, a green-steel plant being constructed in Boden, Sweden, is expected to open in 2025. This plant will produce five million tonnes of green steel per year by 2030 from the hydrogen-direct-reduction process. The hydrogen will come from an 800-megawatt electrolyzer. An electric-arc furnace will be used for steelmaking. Green-steel plants are also being built in Finland, Spain, France, Germany, the United States, and China, among other countries.[147]

6.10.2 REDUCING CARBON EMISSIONS WITH LASERS

Another way to separate pure iron from iron ore is to heat the iron ore directly with lasers.[148] With this process, an array of small lasers is pointed toward the iron ore, forming a spotlight, melting the iron ore in seconds. Lasers require only 70 percent of the energy needed for hydrogen direct reduction and result in the same emission reduction, so long as the electricity used for the lasers comes from WWS.[148] Lasers can also be applied to iron ore with low (below 67 percent) iron content, whereas hydrogen direct reduction is applied primarily to iron ore with higher iron content.

6.10.3 REDUCING CARBON EMISSIONS WITH MOLTEN OXIDE ELECTROLYSIS

Yet another method of extracting molten iron from iron ore during ironmaking is with **molten oxide electrolysis**.[149] With this technique, iron ore is first heated in a molten electrolyte soup above 1,961 degrees Celsius, where it decomposes to produce a different form of iron oxide, called magnetite, and oxygen. The molten electrolyte soup contains silicon, aluminum, and magnesium and helps electricity flow. The soup is heated further, past the melting point of pure iron, which is 2,084 degrees Celsius. Above this temperature, electricity that passes through the soup, turns magnetite into pure molten iron, which sinks to the bottom of the cauldron, where it is drained. As such, the molten oxide electrolysis process produces pure iron without emitting chemically produced carbon dioxide.

6.11 CONCRETE MANUFACTURING

Concrete is a mixture of **aggregate** (sand, gravel, and crushed stone) and paste (water and **ordinary Portland cement**). The paste binds the aggregate together, making a hard surface. Concrete is used for roads, foundations, buildings, runways, sidewalks, driveways, and a variety of other purposes.

Joseph Aspdin (1778–1855) of Leeds, UK, invented ordinary Portland cement in the early nineteenth century. He formed it by burning powdered limestone and clay on his kitchen stove. Today, cement contains limestone, shells, or chalk, which all contain calcium carbonate, mixed with clay, shale, slate, blast furnace slag, silica sand, or iron ore. After being heated to 1,500 degrees Celsius, these ingredients form a hard substance that is ground into a fine, powdery cement.

The concrete industry produces about 8 percent of the world's fossil-fuel carbon dioxide emissions,[150] or about 6.4 percent of all anthropogenic (fossil-fuel plus permanent-deforestation) carbon dioxide emissions. These emissions are equivalent to about 1.18 tonnes of carbon dioxide per tonne of cement produced.[151] Of the total, 1.6 percent is from the quarrying of raw materials, 42.4 percent is from electricity and heat production during the cement manufacturing process, 46.3 percent is from chemical

6.11 CONCRETE MANUFACTURING

reaction (process emissions) during cement manufacturing, and 9.7 percent is from the production of concrete from cement and the transport of concrete.[108] As such, almost half of carbon dioxide emissions from concrete production are process emissions, and the rest are emissions related to energy (electricity production, heat production, and transport).

The process emissions during cement manufacturing arise due to the chemical reaction of calcium carbonate (limestone) with clay at a high temperature to produce clinker (a mix of oxides of silicon, iron, aluminum, and calcium) and carbon dioxide. The clinker is then mixed with gypsum (plaster of Paris) to form cement. The cement is subsequently mixed with water to form a paste, which is combined with the aggregate to form concrete.

Five ways of reducing process emissions and/or energy emissions from concrete manufacturing are to (1) use basalt instead of limestone to react with clay to produce clinker, (2) use geopolymer concrete instead of concrete derived from ordinary Portland cement, (3) use Ferrock instead of concrete, (4) recycle concrete, and (5) make concrete that traps carbon dioxide. In all cases, energy must be supplied by WWS to maximize carbon dioxide reductions.

6.11.1 BASALT-BASED CONCRETE

All process emissions of carbon dioxide during concrete production are due to the reaction of limestone rock with clay. Limestone contains both calcium and carbonate. The calcium is needed to produce clinker. The carbonate in limestone is released as carbon dioxide gas upon the reaction of limestone with clay at high temperature. Basalt is a rock more common than limestone that contains calcium and other chemicals, but no carbonate. Basalt and other similar calcium silicate rocks with no carbonate can be ground, in the place of limestone, during cement production, eliminating process carbon dioxide emissions entirely while still producing the exact same ordinary Portland cement.[152,153] On March 14, 2024, the first basalt-based concrete in history was poured in Seattle, Washington State, for a foundation, wall, and ramp.[153]

6.11.2 GEOPOLYMER CONCRETE

Geopolymer concrete was named and developed in the 1970s by Joseph Davidovits (b. 1935), a French material scientist. It is a hardened mixture of geopolymer cement, aggregate, and water. **Geopolymer cement**

consists of any natural or industrial waste material containing aluminosilicate minerals mixed with an alkali solution.[154] Waste materials include fly ash, granulated blast-furnace slag, rice-husk ash, or metakaolin. **Fly ash** is abundant in most countries since it is a waste product of 100 years of coal burning for electricity generation. **Slag** is a waste product of steel production. **Rice-husk ash** is an abundant waste product of rice milling. **Metakaolin** is a form of the clay mineral, kaolinite. Alkali solution options include sodium hydroxide, potassium hydroxide, and/or sodium silicate mixed with water. The cement is cured at a temperature of 100 degrees Celsius to provide strength. The cement is then mixed with aggregate and water to form concrete.

Slag-based geopolymer cement consists of fly ash, ground-granulated blast-furnace slag, and an alkali solution. Rock-based geopolymer cement consists of volcanic rock, fly ash, slag, and an alkali solution.

Because geopolymer concrete does not use calcium carbonate and because it does not need energy for high-temperature kilns, producing it results in about 80 percent lower carbon dioxide emissions than producing concrete from ordinary Portland cement.[154] Most of the carbon dioxide emission reduction is due to eliminating the off-gassing of carbon dioxide from calcium carbonate chemical reaction. The remaining reduction is due to the fact that geopolymer concrete does not require the use of extreme high-temperature kilns, thus it reduces energy consumption by about 50 percent versus concrete derived from ordinary Portland cement. If the remaining energy is provided by WWS, geopolymer concrete emissions of carbon dioxide are almost eliminated. Other benefits of geopolymer concrete are that it is more resistant to freezing and thawing cycles and to corrosion by acid rain than is concrete from ordinary Portland cement. The costs of the two types of concrete are similar.

> **Transition highlight**
> Geopolymer concrete has been used in many projects to date. For example, 70,000 tonnes of it were used to construct the Toowoomba Wellcamp airport in Queensland, Australia.

6.11.3 FERROCK

Another commercialized alternative to concrete is **Ferrock**, or iron carbonate ($FeCO_3$).[155] Ferrock is derived by first mixing waste steel dust containing iron oxide with crushed glass containing silicon dioxide, limestone, kaolinite or another clay, stabilizers, promoters, and a catalyst into a mixer at room temperature. The mixture is then poured into a mold containing

6.11 CONCRETE MANUFACTURING

seawater. The filled mold is put into a curing chamber, where carbon dioxide from a furnace is injected. The iron, carbon dioxide, and salt water react together to form Ferrock and hydrogen gas. When the final product dries, it is about five times harder and more flexible than ordinary Portland cement. The production of Ferrock not only avoids the chemical carbon dioxide emissions and most energy emissions from concrete production, but it also traps carbon dioxide and produces hydrogen, which can be used for other applications.

6.11.4 CONCRETE RECYCLING

Another method of reducing carbon dioxide emissions from concrete manufacturing is with **concrete recycling**. Concrete structures or roads are often demolished. Historically, such concrete has been sent to a landfill. However, if the concrete is uncontaminated (free of trash, wood, and paper), it can be recycled. Rebar (steel reinforcement) in concrete can also be recycled, as magnets can remove it. The rebar can then be melted and used for other purposes. The broken concrete is crushed. Crushed concrete is often used as gravel in new construction projects or as aggregate in new concrete.

6.11.5 SEQUESTERING CARBON DIOXIDE IN CONCRETE

Trapping carbon dioxide in a material, as done in Ferrock, is a method of offsetting emissions of chemically produced carbon dioxide emissions from the concrete production process. Trapping carbon dioxide within concrete itself is another option.[156] The clinker in ordinary Portland cement contains calcium oxide. If carbon dioxide from any source is mixed with the clinker, it will react with the calcium oxide to form calcium carbonate within the cement. Upon drying, the solid calcium carbonate strengthens the cement. Even if the cement breaks, the carbon dioxide, trapped in the calcium carbonate, will not break free because calcium carbonate is a solid bound to the cement.

Remaining carbon dioxide emitted chemically during the cement formation process can theoretically be captured upon emissions. However, capturing carbon dioxide requires equipment, electricity, and pipelines. Even if the electricity is from WWS, it is far better to use that WWS electricity and the money for the equipment to displace coal or fossil-gas electricity than to capture carbon dioxide. This is because displacing a fossil-fuel electricity source with WWS electricity (a) eliminates more carbon dioxide than does using the WWS electricity to run carbon capture equipment and (b) eliminates the air pollution, upstream mining and emissions, and fossil-fuel infrastructure, which carbon capture equipment does none of. In addition, carbon capture equipment extracts only 10–80 percent of carbon dioxide from a source in the annual average. Further, of the captured carbon dioxide, 82 percent worldwide is used for enhanced oil recovery. During that process, 47–109 percent of the captured carbon is released back to the air (Section 8.4). Thus, carbon capture attached to cement is an opportunity cost that increases carbon dioxide, air pollution, mining, and fossil infrastructure and is not recommended for reducing process emissions during cement manufacturing.

6.12 SILICON PURIFICATION

Pure silicon is a common chemical in alloys and semiconductors. Alloys, including aluminum–silicon and iron–silicon alloys, are the major end use of pure silicon. Such alloys are used to make transformer plates, engine blocks, cylinder heads, and machine tools. Semiconductors are used in computers, microelectronics, and solar photovoltaic cells.

Pure silicon (Si) is extracted from silicon dioxide (SiO_2), also called silica and quartz. The source of silicon dioxide is either sand or a mine. The most common process of silicon extraction is **carbothermic reduction of silica**. It involves heating silicon dioxide together with pure graphitic carbon up to a temperature of 2,200 degrees Celsius in an electric-arc furnace. The reaction produces pure silicon and carbon dioxide gas. The silicon is extracted, and the carbon dioxide is released to the air as process emissions. If electricity for the arc furnace is derived from a fossil-fuel energy source, additional carbon dioxide is emitted. For some applications, such as photovoltaics, the silicon is still not pure enough and more purification is needed.

Several alternative methods of purifying silicon from silica exist that do not involve releasing carbon dioxide. These methods involve either dissolving silica in a mixture of organic alcohol and a base, reacting silica with aluminum or magnesium at moderate temperature (450–650 degrees Celsius), or placing silica in a molten salt electrolyte solution and running electricity through the solution.[157] Alternatively, pure silicon can be obtained from calcium silicate instead of from silica. The calcium silicate is dissolved in a mixture of three salts – calcium chloride, magnesium chloride, and sodium chloride – at a warm temperature. In solution, the calcium silicate dissociates into pure silicon.[158] All materials for this process are relatively inexpensive. Developing one of these alternative methods and using WWS electricity to provide the heat needed eliminates energy and process carbon dioxide emissions from silicon production.

6.13 AMMONIA MANUFACTURING

About 80 percent of ammonia (NH_3) manufactured worldwide is used to make fertilizer. The rest is used for cleaning; purifying water; refrigerating; and manufacturing plastics, explosives, textiles, pesticides, dyes, and chemicals.[96] The most common way to produce ammonia is the Haber–Bosch process, which involves combining one molecular nitrogen (N_2) gas molecule from the air with three molecular hydrogen (H_2) gas molecules produced chemically to produce two ammonia molecules. In 2020, about 146.85 million tonnes per year of nitrogen in ammonia were produced among 145 countries. This necessitated about 31.7 million tonnes per year of hydrogen. Almost all hydrogen for ammonia today is from steam methane reforming, where the methane originates from fossil gas. Such hydrogen is referred to as gray hydrogen. Gray hydrogen results in two sets of CO2 emissions: from the use of fossil gas for energy and the chemical source of hydrogen.

If hydrogen is instead produced from WWS electricity through electrolysis (green hydrogen), all carbon dioxide and air-pollution emissions during ammonia production are eliminated. Using green instead of gray hydrogen does not affect the process of manufacturing ammonia. It affects only how the hydrogen is produced. Creating and compressing hydrogen for ammonia production using WWS electricity is estimated to require about 170.5 gigawatts of annually averaged power in 2050, an 18.8 percent decrease versus producing hydrogen from steam methane reforming.[96]

Transition highlight
In September 2024, three Danish companies (Topsoe, Skovgaard Energy, and Vestas) joined forces in Ramme, Denmark, to begin manufacturing 5,000 tonnes of ammonia per year directly from WWS electricity (12 megawatts of wind and 50 megawatts of solar PV). This was the first time a hybrid wind and solar farm had been connected to both an electrolyzer to produce hydrogen and a plant that manufactures ammonia from hydrogen. This plant avoids the production of hydrogen for ammonia from fossil gas.

7

SOLUTIONS FOR NONENERGY EMISSIONS

WWS technologies eliminate energy-related emissions of air pollutants that affect health and climate. Energy-related emissions are responsible for about 90 percent of anthropogenic air-pollution health problems worldwide and 75–80 percent of anthropogenic greenhouse gas emissions. As such, some emissions that affect human health and climate do not come from energy sources, but they must still be reduced or eliminated in order to help solve the air-pollution and climate problems the world faces. Such nonenergy emissions include gases and particles from open biomass burning; methane from agriculture and landfill waste; halogens from leaks and their reckless disposal; and nitrous oxide from fertilizers, industry, and wastewater treatment. This chapter discusses these sources of emissions and methods of controlling them.

7.1 OPEN BIOMASS BURNING AND WASTE BURNING

Open biomass burning is the burning of evergreen forests, deciduous forests, woodland, grassland, and agricultural land, either to clear land for other use, to stimulate grass growth, to manage forest growth, or to satisfy a ritual. Agricultural fields are often set on fire after a harvest to remove straw, thus clearing land for a new crop the next spring. Sugarcane fields are often burned before harvest to remove the outer leaves around the sugarcane stalks to facilitate sugarcane extraction. **Waste burning** is the burning of trash, such as in a landfill, open pit, garbage can, or backyard incinerator. Such waste burning is illegal in many countries but still occurs in others. **Wildfires** are fires triggered accidentally by a campfire, debris burning, an electrical spark from a transmission line, or a cigarette; or intentionally, by arson. About 17 percent of all global carbon dioxide emissions worldwide are from fires (open biomass burning, waste burning, and wildfires). Humans cause 100 percent of open biomass burning and waste-burning fires and about 93 percent of all fires. Nature causes the rest.[6]

7.1 OPEN BIOMASS BURNING AND WASTE BURNING

All fires produce not only gases that warm the climate (carbon dioxide, methane, ozone precursors, nitrous oxide, carbon monoxide, and water vapor) but also climate-warming particles (black and brown carbon) and direct heat. Black and brown carbon, along with other particles emitted during open biomass burning (ash, organic carbon aside from brown carbon, and sulfate) cause substantial health impacts to people and animals who breathe them in. In addition, the oxides of nitrogen and organic gases from open biomass burning result in elevated levels of ozone, formaldehyde, and other gases that affect human health. Waste burning emits the same chemicals as biomass burning but also toxic chemicals from burning plastics, paints, varnishes, pesticides, medical waste, and chemical byproducts.

While some argue that open biomass burning followed by regrowth of vegetation produces no net increase in carbon dioxide to the air, that contention is incorrect. Although the carbon dioxide released upon burning is offset by carbon dioxide taken from the air during photosynthesis to regrow the vegetation burned, the time lag between burning and regrowth (1–10 years for savannah and 80 years for a forest, for example) always increases carbon dioxide in the air.[159] The contention is also misleading because biomass burning emits black and brown carbon, water vapor, methane, ozone precursors, nitrous oxide, carbon monoxide, and heat, all of which increase global temperatures. These are not recycled like carbon dioxide is. As such, biomass burning always causes net global warming.[6]

The only solution to open biomass burning is to stop it. No technology can control its emissions. An alternative to burning agricultural waste straw is to till it into the soil. An alternative to sugarcane burning is to cut away the leafy parts of the sugarcane before harvest and mix them into the soil. Since humans cause 100 percent of open biomass burning and waste burning, both types of fires are largely preventable through government policies restricting burning and discouraging the conversion of forest land to agricultural land or another land-use type.

Similarly, the only method of reducing the impacts of waste burning is to stop it. If waste is burned in an incinerator, many of the resulting emissions can theoretically be controlled with emission-control technologies; however, no technology eliminates all emissions, including of carbon dioxide. Thus, even incinerators with emission controls produce substantial pollution. The best control of waste burning is to instead recycle the waste or put it in a landfill. Waste should not be dumped into the oceans since plastics do not degrade for centuries. The accumulation of plastics in the oceans has caused an environmental catastrophe, resulting in the **Great Pacific Garbage Patch**. This is a plastic wasteland that consists of two distinct garbage patches in the North Pacific Ocean, one near Japan (Western Garbage Patch) and the other between California and Hawaii (Eastern Garbage Patch). Collectively, the two are three times the area of France.

7.2 METHANE FROM AGRICULTURE AND WASTE

Methane is a long-lived greenhouse gas that selectively absorbs specific long wavelengths of heat radiation, trapping some of that radiation near the surface of the Earth. Methane not only increases global warming, but it also increases background levels of ozone, another greenhouse gas, through chemical reaction.

Anthropogenic sources of methane include open biomass and waste burning, fossil-gas mining leaks and pipeline venting, fossil-fuel combustion, and biological sources enabled by human activity. Biological sources include bacterial production of methane in landfills, rice paddies, the stomachs and manure of farm animals, and sewage treatment plants. Policies controlling biomass and waste burning are needed to reduce methane from those sources. A 100 percent WWS energy system eliminates the need for methane for, thus emissions of methane from, energy systems. This section discusses controlling methane from human-enabled biological sources.

The root biological source of most methane is **methanogenic (methane-producing) bacteria**, which live in environments lacking oxygen. Methanogenic bacteria consume organic material and excrete methane. Ripe oxygen-depleted environments include the digestive tracts of cattle, sheep, and termites; manure from cows, sheep, pigs, and chickens; rice paddies; landfills; and wetlands. All of these sources aside from termites and wetlands are human-enabled.

The main method of reducing methane from bacteria in the digestive tracks and manure of animals is to reduce human consumption of meat and poultry. This requires people changing their diets, which can have additional health benefits. Another method is to change the diet of animals that release methane. For example, studies indicate that feeding cattle red or brown seaweed can reduce methane emissions from them substantially.[160]

Methane from manure can also be captured with a methane digester. A **methane digester**, or manure digester, is an airtight tank in which manure is processed after water is separated from it. The manure is heated and stirred to simulate the inside of a cow's stomach. Methane emitted by bacteria within the manure rises from the manure to the top of the tank, where it is captured in a bag or piped out of the digester, into a storage tank.

7.2 METHANE FROM AGRICULTURE AND WASTE

Methane gas from landfills, like from a digester, can be captured directly. **Landfill gas** is often a mixture of nearly 50 percent methane, nearly 50 percent carbon dioxide, and trace amounts of nonmethane organic gases. Landfill gas is extracted by drilling multiple half-meter-wide boreholes up to 30 meters deep into a section of the landfill. Trash is then removed so that a well, which consists of a perforated or slotted siding with a cap on the bottom, can be installed. Most of the top of the well is sealed to prevent gas escape. A wellhead is installed at the top through which gas is piped to its end destination. In some cases, a network of multiple vertical and horizontal pipes can capture the landfill gas. Once captured, the landfill gas is usually filtered to separate out the methane from the other gases.

> **Transition highlight**
> **Rice paddies** release a significant amount of methane globally, both through the leaves of rice plants themselves and in the oxygen-depleted environment of the flooded soil in which rice plants usually grow. On average, rice-paddy soil is flooded for four months of the year. Direct seeding of rice plants, instead of transplanting rice plants into already-flooded paddies, can reduce the time needed for flooding down to one month. This reduces methane emissions from rice paddies by 15–90 percent. A system of pipes can also be used to capture rice-paddy methane just as it captures methane from landfill gas.

Once biological methane from a digester, landfill, or rice paddy is captured, what should be done with it? In a 100 percent WWS world, the best use of captured biological methane may be to produce hydrogen through methane pyrolysis. The hydrogen should then be used in a hydrogen-fuel-cell-electric vehicle for long-distance transport, in a hydrogen fuel cell to produce grid or microgrid electricity, or for steel or ammonia manufacturing.

With **methane pyrolysis**, methane (CH_4) is heated to above 1,000 degrees Celsius in the absence of oxygen until the bonds between the carbon atom and its four attached hydrogen atoms are broken.[161] If a metal catalyst, such as nickel or iron is present, the temperature needed can be lowered to about 600 or 800 degrees Celsius, respectively.[161] Adding a small amount of cobalt can further lower the temperature needed. In all cases, the carbon atoms from multiple methane molecules coalesce into carbon nanotubes, solid carbon, graphite or carbon black,[161] all of which are solids that can be used for industry or buried. The free hydrogen atoms combine into molecular hydrogen (H_2) gas molecules.

If the methane comes from a biological source and the heat comes from a WWS source, no carbon dioxide is emitted during methane pyrolysis. However, standard methane pyrolysis converts only about 80–95 percent of the input methane into solid carbon and hydrogen.[161] The rest is released to the air. The process also produces some air pollutants due to incomplete decomposition of some methane during heating. Tests suggest that methane conversion can approach 100 percent efficiency when microwave heating instead of conventional heating is used during pyrolysis.[161]

In sum, using methane captured from a digester, landfill, or rice paddy and heat from a WWS source to produce hydrogen through methane pyrolysis reduces climate-relevant emissions more than does allowing the methane to be released to the air, while slightly increasing air pollution. However, using methane from fossil gas to produce hydrogen by methane pyrolysis is not good for either the climate, air quality, or the environment. This is because the mining, transporting, and processing of fossil gas results in substantial land damage and emissions of gaseous air pollutants, black carbon particles, methane, and carbon dioxide beyond the emissions associated with methane pyrolysis itself.[162]

7.3 HALOGENS

Anthropogenically emitted halogens are responsible for about 9 percent of global warming (see Figure 1.1). Halogens are still used today as refrigerants, solvents, blowing agents, fire extinguishants, and fumigants. They enter the atmosphere primarily upon evaporation when they leak or when the appliances containing them are drained. Their persistence in the atmosphere and their strong warming per molecule make them potent greenhouse gases.

The main methods of reducing halogen emissions and their impacts on climate are (a) replacing halogens used in equipment and appliances that have high global-warming potentials with halogens that have low warming potentials, (b) requiring tougher standards for sealing halogens in the equipment and appliance, and (c) requiring tougher standards for disposing of halogens at the end of the life of the equipment or appliance they are contained in. For these suggestions to be effective, they need to be implemented and enforced worldwide.

7.4 NITROUS OXIDE

Nitrous oxide is a potent greenhouse gas with a long lifetime. It is produced largely by bacteria and contributes to about 4.3 percent of global warming. In fact, 67–80 percent of anthropogenic nitrous oxide originates

7.4 NITROUS OXIDE

from agriculture.[163] In particular, bacteria in nitrogen-containing fertilizers expel substantial amounts of nitrous oxide. In addition, the cultivation of legumes (plants in the pea family) results in the conversion, by bacteria, of atmospheric nitrogen gas to nitrous oxide, which is released to the air. A third agricultural source of nitrous oxide is the bacterial conversion of solid waste of domesticated animals.

Some methods of reducing nitrous oxide emissions from fertilizers are (a) using less nitrogen-based fertilizer, (b) cultivating leguminous crops that don't require fertilizer in the crop rotation, and (c) reducing tillage to reduce the breakdown of organic fertilizer, thereby reducing reaction and release of chemicals.

The remaining anthropogenic sources of nitrous oxide are fossil-fuel combustion, open biomass and waste burning, industrial processes, and wastewater treatment. Transitioning fossil energy to 100 percent WWS will eliminate fossil combustion sources of nitrous oxide. Reducing biomass and waste burning will reduce nitrous oxide from these sources.

The two main industrial sources of nitrous oxide are the production of nitric acid for use in fertilizers and the production of adipic acid for use in the production of nylon fibers and plastics. To date, nitrous oxide emissions from adipic acid production have been reduced effectively with emission-control technologies in several plants, so the expansion of such technologies to adipic acid plants worldwide will help reduce nitrous oxide emissions. Similarly, nitrous oxide emission-control technologies for nitric acid production plants are available and could be implemented worldwide with stringent policies.[164]

The source of nitrous oxide in wastewater is organic material from human or animal waste. Modulating the dissolved oxygen content in the wastewater treatment process can control the nitrous oxide content of wastewater.[165,166]

8

WHAT DOESN'T HELP

Although coal, fossil gas, oil, and bioenergy have provided energy that has helped societies until today, such energy has come at a cost, namely, increased air pollution, ground and water contamination, land despoilment, global warming, and energy insecurity. While such a cost was accepted in the past, much of the public can no longer stomach the continuous death and illness caused by air pollution from these fuels. Scientists and policymakers similarly no longer believe society can maintain a safe climate and function securely if these fuels continue to be used.

Realizing that it needs to adapt if it wants to stay in business, the fossil-fuel industry first proposed replacing coal with fossil gas as a "bridge fuel" to renewables. It then supported the development of four technologies that it argued would help reduce carbon emissions: carbon capture, synthetic direct air carbon capture, blue hydrogen, and carbon-based electro-fuels. If adopted, these technologies would permit the industry to continue producing fossil fuels in perpetuity. Similarly, the agriculture industry argued that, because biomass, biofuels, and biogas were all renewable fuels, the industry should be allowed to continue producing energy, especially if carbon capture were added to some of their facilities. Lastly, the nuclear industry, which has suffered slowing development, rising costs, a major meltdown, and unresolved waste issues since 2000, made a concerted effort to be part of the solution by proposing the use of new small and large nuclear reactors. Because the fossil-fuel, agriculture, and nuclear lobbies are strong in many countries, the solutions proposed by these lobbies were incorporated, along with WWS technologies, into "all-of-the-above" policies to address climate change. "**All-of-the-above**" policies involve promoting nearly all technologies, regardless of their air-pollution impacts, carbon emissions, security risk, cost, effectiveness, or length of time between planning and operation, that allege to address climate. An additional technological fix added to the mix was geoengineering, which consists of a variety of measures largely to increase the reflectivity (albedo) of the Earth to cool the Earth's surface.

However, all of these non-WWS "miracle" technologies result in greater global warming, air pollution, land degradation, or all three compared with WWS alone. Some of these technologies increase energy insecurity and/ or take up to 23 years between planning and operation while costing substantially more than WWS technologies. As such, they are opportunity costs compared with investing in WWS so are not recommended. This chapter discusses these non-WWS technologies and delineates the reasons why they are not needed or helpful for solving global warming, air pollution, and/or energy-security problems.

8.1 BRIEF HISTORY OF FOSSIL FUELS

Prior to the Industrial Revolution of the mid-1700s, the world relied primarily on biomass but also on some coal for its energy, in the form of heat. Discoveries and inventions from the 1650s onward, however, led ultimately to biomass and coal, and later fossil gas and oil, being used to produce mechanical energy for industry and transportation and, eventually, to produce electricity.

8.1.1 WOOD AND COAL BURNING BEFORE THE INDUSTRIAL REVOLUTION

In ancient Greece and Rome, wood was burned for heating, cooking, and to produce high temperatures for industrial processes. One major industry was the smelting of silver ore, which contained lead, to produce silver coins. Smelting of lead itself arose during the Roman Empire because the Romans used lead in cookware, pipes, face powders, rouges, and paints. Lead emissions from smelting between 500 BCE and 300 CE were so great that lead concentrations measured deep in Greenland ice cores, far away from Greece and Rome, were four times the levels in earlier samples.[167] The Romans also burned wood to smelt ores containing copper, zinc, and mercury.[168] Ice-core data suggest that the smelting of copper to produce coins not only by the Romans but also in China during the Song Dynasty (960–1279 CE), increased atmospheric copper concentrations appreciably.[169]

As noted by the poet Horace, thousands of wood-burning fires in Rome blackened buildings.[170] Air-pollution events in Rome were called "heavy heavens."[171]

During the fall of the Roman Empire, wood burning for smelters declined significantly. However, around 1000 CE, following the discovery of lead and silver mines in Germany, Austria, Hungary, and other parts of central and eastern Europe, smelting increased again.[167] Wood burning for heating and cooking continued throughout the decline of the Roman Empire in all population centers of the world.

Sea coal was introduced to London by 1228 and gradually replaced the use of wood as a fuel in lime kilns and forges. Wood shortages may have led to a surge in sea coal use by the mid 1200s. Lime kilns were used to produce quicklime, a building material. The pollution in London due to the burning of sea coal in lime kilns became sufficiently severe that a commission was ordered by King Edward I in 1285 to study and remedy the situation. The commission met for several years. Finally, in 1306, the king banned the use of coal in lime kilns. The punishment was "grievous ransom," which may have meant fines and furnace confiscation.[172] However, by 1329, the ban had been either lifted or lost its effect.

Between the thirteenth and eighteenth centuries, the use of sea coal and charcoal increased in England. Coal was used not only in lime kilns and forges but also in glass furnaces, brick furnaces, breweries, and home heating. One of the early writers on air pollution was **John Evelyn** (1620–1706), who wrote *Fumifugium or The Inconveniencie of the Aer and the Smoake of London Dissipated* in 1661. He explained how smoke in London was responsible for the fouling of churches, palaces, clothes, furnishings, paintings, rain, dew, water, and plants. He blamed "Brewers, Diers, Limeburners, Salt and Sopeboylers" for the problems.

8.1.2 THE INDUSTRIAL REVOLUTION AND THE GROWTH OF COAL

In 1679 the French-born English physicist **Denis Papin** (1647–1712) invented the pressure cooker while working with the English chemist and physicist **Robert Boyle** (1627–1691). In this device, water was boiled under a closed lid. The addition of steam to the air in the cooker increased the total air pressure exerted on the cooker's lid. Papin noticed that the high pressure pushed the lid up. The phenomenon gave him the idea that steam could be used to push a piston up in a cylinder, and the movement of the cylinder could be used to do work. Although he designed a model of such a cylinder-and-piston steam engine in 1690, Papin never built one. Nevertheless, the pressure cooker is the first device used to create mechanical energy from heat produced from burning either biomass or coal.

In 1698, **Thomas Savery** (1650–1715), an English engineer, used the concept of the pressure cooker to patent the first practical steam engine, which is a machine that converts heat to mechanical energy. The first steam engine was used to pump water out of coal mines, replacing the use of horses for this task. The engine worked when water was boiled to produce steam that was transferred to a steam chamber. A closed pipe between the steam chamber and the water to be removed in the mine was then opened. Liquid water drops were then sprayed into the steam chamber to recondense the steam to liquid, creating a vacuum that sucked the water to be removed through the pipe, from the mine to the steam chamber. The pipe from the chamber to the mine was then closed, and another pipe from the chamber to outside the mine was opened. Finally, the boiler was fired up again to produce more steam to force the liquid water through the second pipe from the steam chamber to outside the mine.

In 1712, **Thomas Newcomen** (1663–1729), also an English engineer, developed a more efficient steam engine than Savery by introducing a piston. Boiling water sent steam into a steam chamber under the piston, creating pressure that pushed the piston and an attached lever up, producing mechanical energy. Liquid water drops were then sprayed into the steam chamber to recondense the steam, creating a vacuum that pulled the piston and lever down. Newcomen's engine was used to pump water out of mines and to power waterwheels. Steam engines in the early eighteenth century were inefficient, capturing only 1 percent of their input energy.

In 1763, Scottish engineer and inventor **James Watt** (1736–1819) was given a Newcomen steam engine to repair. He quickly found that allowing evaporation and condensation to occur in the same chamber was inefficient. The spraying of cold water to condense steam, thereby creating a vacuum to pull the piston down, resulted in more heat required to produce steam to push the piston back up if both operations occurred in the same chamber rather than in different chambers. Watt overcame this shortcoming by developing two chambers: one in which condensation occurred and the second in which evaporation occurred. The condensation chamber stayed cold due to the spraying of cold water into it and the evaporation chamber stayed warm due to the boiling of water under it. In 1769, Watt patented this revised steam engine, which had twice the efficiency of the previous one.

Watt made further modifications until 1800, including an engine in which the steam was supplied to both sides of the piston and an engine in which motions were circular instead of up and down. Watt's engines were used not only to pump water out of mines but also to provide energy for paper, iron, flour, cotton, and steel mills; distilleries; canals; waterworks; and locomotives. For many of these uses, steam engines were located in urban areas. Because they required the burning of large quantities of coal, they increased urban air pollution. Pollution became particularly severe because, although Watt had improved the steam engine, it still captured only 5 percent of the energy it used by 1800.[173] Because the steam engine was a large, centralized source of energy, it was responsible for the shift from the artisan shop to the factory system of industrial production during the **Industrial Revolution** of 1760–1880.[174]

The first utility-scale electricity generation from coal in the world occurred on September 4, 1882, with the opening of the Pearl Street Power Station in New York City, owned by Thomas Edison. After that, coal use for heat, mechanical energy, and electricity soared, reaching 8 billion tonnes in 2013. Due to a slow transition away from coal, consumption declined to 7.4 billion tonnes in 2021. However, by 2023, coal consumption had rebounded to a record 8.3 billion tonnes, with China consuming about 48.2 percent of the total. Given that there are still 1.05 trillion tonnes of known coal reserves remaining in the world, it would take 126 years to deplete all coal at the 2023 rate of consumption.

8.1.3 THE HISTORY OF FOSSIL GAS

Fossil gas is a colorless, flammable gas containing about 88.5 percent methane by mass plus smaller amounts of ethane, propane, butane, pentane, hexane, nitrogen, carbon dioxide, and oxygen.[175] It is often found near petroleum deposits. It originates from the compressional heating, over millions of years, of ancient algae, phytoplankton, and sea life that fell to the ocean floor.

Its first recorded discovery was in the Greek Temple of Apollo in Delphi on Mount Parnassus, Greece. Starting between 800 BCE and 1400 BCE, and ending in 392 CE when the Roman Emperor Theodosius forbade it, Greek and Roman rulers traveled to Delphi to listen to the high priestess of the temple, called the Pythia, or the **Oracle of Delphi**. The Pythia would enter into a trance inside a small chamber before channeling prophecies from the god Apollo. The Greek author, Plutarch, wrote that before entering the trance, the Pythia would inhale sweet-smelling, perfume-like vapors from crevices in the floor that arose from a fissure or spring below the temple. Recent studies indicate that, underlying the Temple of Apollo are two earthquake faults that cross each other. Seismic activity is thought to have released fossil gas, which contains methane and ethene, a sweet-smelling gas that can cause hallucinations. This theory is supported by the fact that water from a nearby spring also contains methane and ethene.[176]

8.1 BRIEF HISTORY OF FOSSIL FUELS

Around 500 BCE in Szechuan, China, a gas was found seeping from the ground. Bamboo was used to pipe the gas to a nearby village, where it was burned to evaporate water from salty brine water that had been mined from deep underground, in order to isolate the salt.

In 1609, the scientist **Jan Baptist van Helmont** (1580–1644), born in modern-day Belgium, introduced the term **gas** into the chemical vocabulary. He produced what he called *gas silvestre* ("gas that is wild and dwells in out-of-the-way places") by fermenting alcoholic liquor, burning charcoal, and acidifying marble and chalk. The gas he discovered in all three cases, but did not know at the time, was **carbon dioxide**. It wasn't until 1756 that **Joseph Black** (1728–1799), a Scottish physician and chemist, isolated carbon dioxide and identified its properties. Black, though, called carbon dioxide "fixed air" because of its ability to attach or "fix" to compounds exposed to it. Fixed air was renamed to carbon dioxide in 1781 by the French chemist **Antoine Laurent de Lavoisier** (1743–1794). During his experiments, van Helmont also discovered that heating coal or wood produced a vapor, which is known now to be methane, the main component of fossil gas and biogas.

In 1626, French explorers observed Native Americans igniting a gas that seeped from the ground near Lake Erie in North America. The gas, which likely contained methane, was burned to produce light and heat.

In November 1776, the Italian physicist and chemist **Alessandro Volta** (1745–1827) observed bubbles rising to the surface in a shallow, marshy portion of Lake Maggiore, Italy. In 1778, he analyzed the gas, which formed from the microbial decay of plants in the marsh in the absence of oxygen, thus was biogas, and discovered that biogas contained a new, combustible chemical, methane.

The first recorded use of a gas for lighting occurred in 1783, when Professor **Jan Pieter Minckeleers** (1748–1824) of the University of Louvain, in modern-day Belgium, produced a gas, which contained mostly methane by heating coal. This gas was not fossil gas or biogas. It was, instead, **manufactured gas**, because it was produced synthetically. He used the gas to light his lecture hall at the university.

In 1794, the French civil engineer **Philippe LeBon** (1767–1804) patented a technique to obtain manufactured gas by distilling sawdust or coal. The first commercial plant producing manufactured gas obtained from coal was built by the London and Westminster Gas Light and Coke Company in 1812. They installed wooden pipes to the Westminster Bridge and lit the bridge using the gas on New Year's Eve in 1813. In the United States, the first commercial manufactured gas plant was the Gas Light Company of Baltimore, established in 1816.

Fossil gas was not used, except incidentally, anywhere in the world until after **William Hart** (1797–1865) dug the first fossil-gas well in Fredonia, New York, in 1821. Hart saw bubbles percolating up through creek water and dug 8.2 meters down with a shovel to create a well that allowed more gas to flow. He then used hollowed-out logs connected by tar and rags to create a pipeline for the gas. The gas was ultimately distributed by America's first fossil-gas company, the Fredonia Gas Light Company, established in 1858.

Manufactured gas, fossil gas, and biogas in the 1800s were used primarily for lighting, However, in the late 1800s, the invention of the electric light bulb by Thomas Edison resulted in the eventual replacement of gas lighting with electric lighting. In addition, in 1885, **Robert Bunsen** (1811–1899) invented the Bunsen burner at the University of Heidelberg, Germany. This device burns gas to produce a continuous flame for heating, sterilization, or combustion. After this invention, the use of gas for heating and cooking accelerated. With the discovery of more fossil-gas wells, manufactured gas was replaced more and more by fossil gas. Manufactured gas eventually disappeared entirely in 1966 in North America and in the 1980s in Europe.

Today, fossil gas is used for home air heating and cooking, water heating, clothes drying, high-temperature industrial heating, and electricity production. The first electricity was produced from fossil gas in 1937, when Sun Oil used a gas turbine to generate electricity at a chemical plant in Philadelphia. The first public electricity generation with fossil gas occurred in 1939, in a plant in Neuchatel, Switzerland, where a 4-megawatt turbine was powered by fossil gas.

8.1.4 DISCOVERY OF OIL

On August 27, 1859, **Colonel Edwin Drake** (1819–1880) discovered oil 21 meters below ground near Titusville, Pennsylvania. The well also produced fossil gas. The oil and gas were piped almost 9 kilometers to Titusville, cementing the ability of oil and fossil gas to be piped long distances. Oil was used primarily to produce transportation fuels but also to produce heat, chemicals, asphalt, lubricants, waxes, plastics, clothing, cosmetics, and gum. Today, in the clothing industry, oil is used to produce acrylic, rayon, vegan leather, polyester, nylon, and spandex. In the sporting goods industry, it is used in basketballs, golf balls, football helmets, surfboards, fishing rods, skis, and tennis rackets. Oil products are also used in perfume, cosmetics, hand lotion, toothpaste, soap, shaving cream, combs, shampoo, contact lenses, and eyeglasses. Even in medicine, oil has deep roots. Its products are used in artificial limbs, hearing aids, heart valves, antihistamines, aspirin, and IV bags. In buildings, oil is used in roofing material, insulation, linoleum flooring, furniture, pillows, rugs, house paint, dishes, nonstick pans, and detergents. Today, about 10 percent of all oil is used for materials; the rest is used for energy, mostly for transportation, but also for heat and electricity.

Now that the origins of the use of coal, oil, and fossil gas for energy have been explored, it is time to look at why they should no longer be used and are unnecessary in a clean, renewable energy future.

8.2 COMPARISON OF ENERGY TECHNOLOGIES

When evaluating different proposed solutions to air pollution, global warming, and energy insecurity, agnostic metrics are needed to compare the damage caused by each technology performing the same task.

One such metric, which compares the potential climate impact of one energy technology over another, is **CO_2-equivalent emissions**. These are the emissions of carbon dioxide from a source plus the emissions of all other gas and particle pollutants from the source, each multiplied by the pollutants' ability to warm the planet over 20 or 100 years relative to carbon dioxide's ability to warm the planet over the same period. Different time frames are considered because carbon dioxide has a long life in the atmosphere. So, if another chemical, such as methane, which has a much shorter life but a more powerful warming potential per molecule during that short life, is emitted in sufficient quantities, the CO_2-equivalent emissions of carbon dioxide plus methane will be much higher over 20 years than over 100 years. Given that damaging climate impacts are occurring today over periods of only a few years, and short-lived powerful global warming agents such as black carbon and methane cause much more damage over the short term than over the long term, it is important to compare lifecycle emissions of different technologies over short periods, such as 20 years.[177] However, impacts over 100 years are also important because long-lived warming chemicals emitted today will impact the climate of future generations in addition to that of the current generation.

Energy technologies result in several types of CO_2-equivalent emissions. These include lifecycle emissions, opportunity-cost emissions, anthropogenic heat emissions, anthropogenic water-vapor emissions, emissions risk due to the leakage of captured carbon dioxide, and emissions due to the covering or clearing of land for energy development, among others. These emissions are discussed, in turn.

8 WHAT DOESN'T HELP

8.2.1 LIFECYCLE EMISSIONS

Lifecycle emissions are all CO_2-equivalent emissions that arise during the construction, operation, and decommissioning of an electricity-generating plant per unit of electricity generated, averaged over 20 or 100 years. For a fossil-fuel or nuclear plant, operating a plant includes mining, transporting, and processing the plant's fuel; running the plant's equipment; repairing the plant; and disposing of waste during the plant's life. WWS requires no fuel mining and has no fuel waste. Figure 8.1 provides lifecycle emissions of several electricity-generating technologies.

8.2.2 OPPORTUNITY-COST EMISSIONS

Opportunity-cost emissions are emissions from the background electric-power grid due to the longer time lag between planning and operation of one energy technology relative to another and the longer downtime needed to refurbish one technology at the end of its life relative to the other.[178]

For example, if Plant A takes 4 years and Plant B takes 10 years between planning and operation, the background grid will emit pollution for 6 more years with Plant B than with Plant A. The emissions during those additional 6 years, when averaged over the lifetime energy production of the technology, are opportunity-cost emissions. Similarly, if Plant A has a useful life of 20 years and requires 2 years of refurbishing to last another 20 years, and Plant B has a useful life of 30 years but takes only 1 year of refurbishing, then Plant A is down an additional 5.9 years out of every 100 years for refurbishing compared with Plant B. During those additional years that Plant A is down, the background grid emits pollution that must be accounted for.

The time lag between planning and operation of a plant includes planning time and construction time. The planning time includes the time required to identify a site, obtain a site permit, purchase or lease the land, obtain a construction permit, obtain financing and insurance for construction, install transmission lines, negotiate a power purchase agreement, and obtain an operating permit. The construction time includes the time required to build the plant, connect it to the transmission infrastructure, and obtain a final operating license.

The planning-to-operation time of a wind or solar farm is generally two to five years, which includes a planning time of one to three years and a construction period of one to two years.[178] The construction time applies to both onshore and offshore wind. For example, the 407-megawatt

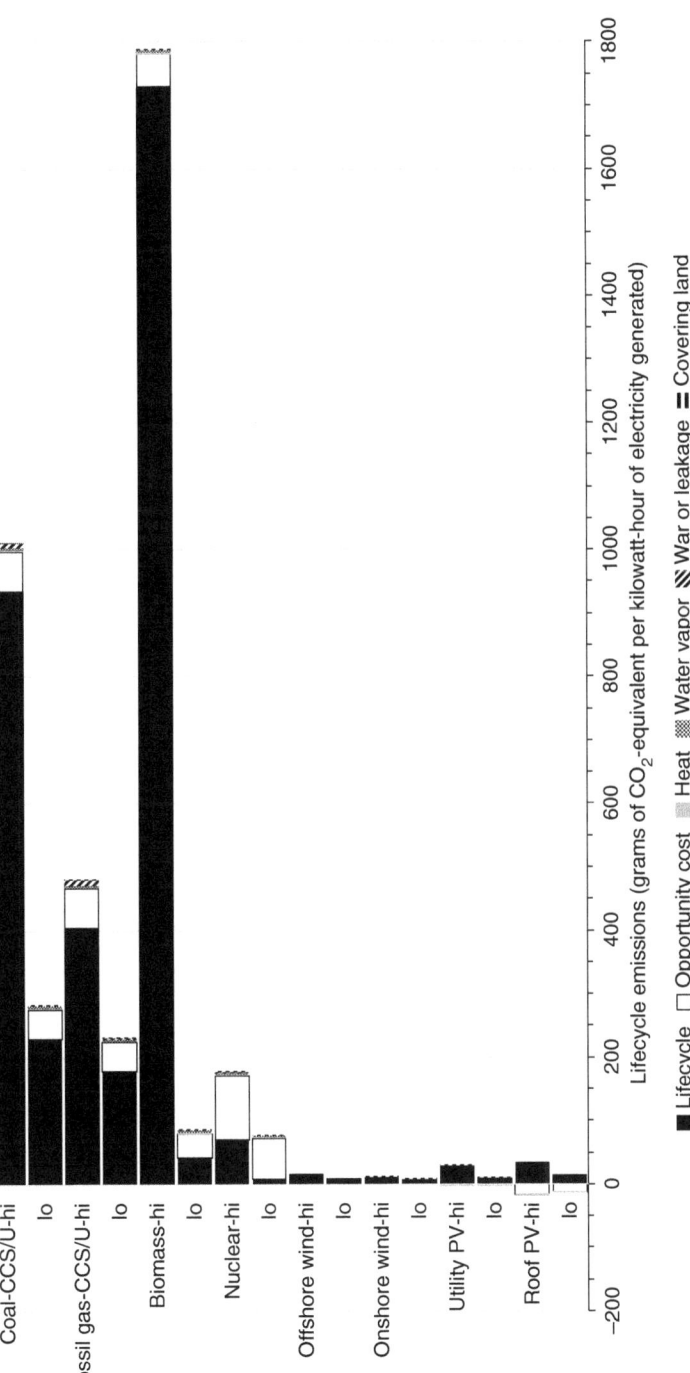

Figure 8.1 Low and high estimates of total 100-year-averaged CO_2-equivalent emissions from several electricity-producing technologies. The total for each bar is the sum of lifecycle emissions ("Lifecycle"), opportunity-cost emissions ("Opportunity cost"), anthropogenic heat emissions ("Heat"), anthropogenic water-vapor emissions ("Water vapor"), emissions from nuclear-weapons proliferation or carbon-capture leakage ("War or leakage"), and emissions from loss of carbon storage by covering land and vegetation with building materials ("Covering land"). CCS/U is carbon capture and storage or use. Opportunity-cost emissions are relative to onshore wind, so negative opportunity-cost emissions (for rooftop PV) are due to the shorter time lag between planning and operation of roof PV versus onshore wind. All wind and solar have negative heat emissions; wind has negative water-vapor emissions (see text). Source of calculations.[10]

(49 turbine) Horns Rev 3 offshore wind farm, in the North Sea off the west coast of Denmark, required 1.8 years to build.[179] The 903-megawatt Shanwei Jiazi I and II offshore wind farms in China began construction in 2021 and were completed by December 2022 after less than two years.

Wind turbines often last 30 years before refurbishing, and the refurbishing time is three months to one year. For rooftop PV, the time between planning and operation can be as little as six months. For geothermal electricity, the planning-to-operation time is 3–6 years; for hydroelectric, it is 8–16 years.

The planning phase of a coal-fired power plant without carbon-capture equipment is generally 1–3 years, and the construction phase is another 5–8 years, for a total of 6–11 years between planning and operation. The planning-to-operation time of a coal plant with carbon capture is similar to one without, since the carbon-capture equipment can be installed during the long period of coal-plant construction. The planning-to-operation time of a fossil-gas power plant without carbon capture is less than that of a coal plant. But adding carbon capture is likely to extend the planning-to-operation time of a fossil-gas plant to that of a coal plant (6–11 years). The planning-to-operation time of almost all nuclear plants worldwide today is 12–23 years, up from 10–19 years in the 2010s.

For technologies that take a long time before operation, such as nuclear and hydroelectric power, the opportunity-cost emissions can be greater than the lifecycle emissions (Figure 8.1).

8.2.3 ANTHROPOGENIC HEAT EMISSIONS

In a coal, oil, fossil-gas, and biomass electric-power plant, fuel is burned to generate electricity. For coal and oil, about 65–67 percent of the energy in the fuel is released as waste heat, and the rest is converted to electricity. For fossil gas, 40–60 percent of the energy is released as waste heat. For biomass, about 74 percent is released as waste heat. A nuclear reactor also releases about 65 percent of the energy in uranium used to generate electricity as waste heat.[6] In all cases, most of the waste heat is absorbed by liquid water that is released to a river, a lake, or the ocean. The rest of the waste heat is released directly to the air.

The heat released due to combustion or nuclear reaction is one source of **anthropogenic heat**. Other sources are the dissipation of electricity to heat, the burning of fuels to move vehicles, and the burning of biomass. For example, when electricity is used to run an appliance, the electricity is converted to heat. This is most obvious in an electric-resistance heater,

8.2 COMPARISON OF ENERGY TECHNOLOGIES

where the electricity is converted directly to heat to warm a room. However, any electrical device or appliance, such as a light bulb or a dishwasher, converts all electricity consumed to heat. Energy must be conserved, so even if electricity is first used to produce light in a light bulb, the wavelengths of visible light emitted by the bulb are absorbed by building materials, causing the temperature of the building to rise.

Similarly, when gasoline is used to move a car, only 17–20 percent of the energy in the gasoline is used to move the car. The rest is waste heat released to the air. Even the energy used to move the car is converted to heat due to the resistance caused by friction between the car's tires and the road, and between the car and air molecules in its path. If no friction occurred, the car would be a perpetual-motion machine, running for an infinite distance with no heat production once moving. Even with an electric vehicle, where 80–90 percent of the electricity is converted to energy used to move the car, that electricity is also dissipated to heat by friction.

Whereas fossil-fuel and nuclear electricity generators produce anthropogenic heat, solar and wind generators decrease such heat. Solar PV and CSP convert sunlight to electricity, thereby reducing the amount of sunlight reaching the surface below the PV panel or CSP mirrors, cooling the surface. Wind turbines similarly cool the air by extracting kinetic energy from the air and converting it to electricity. **Kinetic energy** is the energy embodied in air due to its motion. For each kilowatt-hour of electricity produced by a wind turbine, 1 kilowatt-hour of kinetic energy is extracted from the wind. Since energy in the wind is ultimately converted to heat when the wind blows against objects or the ground, taking energy out of the wind cools the air.

All electricity produced for the grid eventually dissipates to heat as well. However, for purposes of comparing the heat released by different electricity-generating technologies, the heat from electricity consumption is assumed to be the same regardless of the electricity-generating technology, so is not included in anthropogenic heat calculations. Figure 8.1 provides anthropogenic heat fluxes for different energy generators, converted to CO_2-equivalent emissions.

8.2.4 ANTHROPOGENIC WATER-VAPOR EMISSIONS

Two major sources of anthropogenic water-vapor emissions to the air are combustion and evaporation. Burning fossil fuels, bioenergy, and vegetation releases water vapor due to the chemical reaction between the hydrogen in the fuel and oxygen in the air. Electricity-generating plants running on fossil fuels, bioenergy fuels, and uranium use liquid water for cooling.

Some of this heated water evaporates from a cooling tower or as it flows back to the river, lake, or ocean that it came from. Many CSP plants also use water cooling, although some use air cooling. Similarly, whereas non-binary geothermal plants and some binary plants use water cooling, thus emit water vapor, binary plants that use air cooling do not emit water vapor. Water also evaporates from reservoirs behind hydropower dams. Anthropogenic water vapor worldwide may cause about 0.23 percent of gross global warming.[6]

On the other hand, wind turbines reduce water evaporation, since they reduce wind speeds downwind of them, and lower wind speeds mean less evaporation from ground water and surface water.[180,181]

In sum, while electricity-generating plants that run on fuel combustion add water vapor to the air, wind turbines reduce water vapor to the air. Since water vapor is a greenhouse gas, electricity generators that burn fuels slightly increase global warming, whereas wind turbines slightly decrease it. The small changes in warming and cooling due to changes in water vapor for each generator are reflected in changes in CO_2-equivalent emissions in Figure 8.1.[10]

8.2.5 LEAKS OF CARBON DIOXIDE SEQUESTERED UNDERGROUND

The storage of carbon dioxide underground due to carbon capture and storage runs the risk of carbon dioxide leaking back to the air through existing fractured rock or overly-porous soil or new fractures caused by an earthquake. The ability of a storage site to sequester carbon dioxide for decades to centuries varies with location and tectonic activity. Estimated leakage rates have ranged from 0.1–10 percent per 1,000 years.[182] However, because liquefied carbon dioxide injected underground is under high pressure, it may leak more readily than expected through horizontal and vertical fractures in rocks. Because carbon dioxide is an acid, it will also erode rocks over time. If a leak occurs, it may or may not be detected. If detected, a leak may or may not be sealed. Figure 8.1 provides an estimate of the CO_2-equivalent emissions arising from leaks of carbon dioxide stored underground.[10]

8.2.6 EMISSIONS FROM COVERING LAND OR CLEARING VEGETATION

Another source of carbon dioxide associated with electricity generation is from covering land or clearing vegetation. Covering land with impervious construction material, such as asphalt or concrete, eliminates vegetation

8.3 WHY NOT FOSSIL GAS AS A BRIDGE FUEL?

and its carbon above the soil and the ability of soil to accumulate carbon due to the decay of vegetation. Similarly, clearing land of vegetation reduces the carbon stored in the vegetation and its roots. Normally, when grass dies, the dead grass contributes to organic carbon in soil. The grass then regrows, removing carbon from the air by photosynthesis. If the soil is instead covered with concrete, the grass no longer exists and can no longer remove carbon or store it in the soil. However, existing carbon stored underground remains.

The carbon emissions due to covering land for an energy facility can be estimated simplistically by first summing the land areas covered by the facility; by the mine (if any) where the fuel is extracted from; by the roads, railways, or pipelines needed to transport the fuel; and by the waste disposal site associated with the facility. This summed area is then multiplied by the organic carbon stored in vegetation and in the soil under the vegetation per unit of area that is lost. The loss of carbon is then converted to a loss of carbon per unit of electricity produced by the energy facility over a period, such as 100 years. Figure 8.1 shows the resulting CO_2-equivalent emissions due to this process for each type of energy facility.

8.2.7 TOTAL CO_2-EQUIVALENT EMISSIONS

A summary of the emissions just discussed (lifecycle emissions plus several types of emissions not included in standard comparisons) indicates WWS technologies have lower overall CO_2-equivalent emissions than nuclear, biomass, fossil gas with carbon capture, or coal with carbon capture (Figure 8.1). The lower CO_2-equivalent emission rate of WWS is one reason WWS is proposed. Other reasons are that WWS also eliminates air pollution, whereas fossil-fuel and biomass combustion, even with carbon capture, do not. WWS also eliminates continuous fuel mining, thus land and ecosystem damage, and energy security compared with other technologies. Finally, WWS reduces meltdown risk, weapons-proliferation risk, uranium-mining lung-cancer risk, and waste-storage risks associated with nuclear. These issues are discussed next with respect to individual non-WWS technologies.

8.3 WHY NOT FOSSIL GAS AS A BRIDGE FUEL?

Most fossil gas today is used for electricity generation, building heat, or industrial heat. Because fossil gas is not very dense, it can be stored only in a large container or an underground storage facility. As such, fossil gas is

8 WHAT DOESN'T HELP

often compressed or liquefied for transport and storage. **Compressed natural gas (CNG)** is fossil gas compressed to less than 1 percent of its gas volume at room temperature. **Liquefied natural gas (LNG)** is fossil gas cooled to −162 degrees Celsius, the temperature at which it condenses to a liquid at outdoor air pressure. LNG has a volume that is one six-hundredth of the volume of the original gas.

Both CNG and LNG can be sent through pipelines, although different pipelines are needed for each. Both can also be stored and used in automobiles designed to run on fossil gas. Both can be transported by truck or bus with a special fuel tank and stored at a power plant when a pipeline is not available.

For overseas transport, CNG from a pipeline is often converted to LNG at a marine export terminal, put on a tanker ship with super-cooled storage tanks, then shipped overseas. At the import terminal, it is regasified and piped to its final destination – either a power plant, industrial company, or company that transmits and distributes the fossil gas to buildings for heating or other purposes.

Fossil gas is mined from underground conventional or unconventional wells. Conventional wells are shallow and small in area and produce oil, fossil gas, or both. Unconventional wells are deep and large in area and include shale gas and coal-bed methane wells. **Hydraulic fracturing (fracking)** is used in both well types, but mostly in unconventional shale gas wells. **Shale** is sedimentary rock composed of a muddy mix of clay mineral flakes and small fragments of quartz and calcite. Large shale formations containing fossil gas can be found in many areas of North America and worldwide, some close to population centers.

In the United States in 2022, at least 79 percent of all fossil gas mined and 95 percent of new fossil gas mined came from shale rock, thus was extracted by fracking.[183] Fracking requires large volumes of water, laced with chemicals, forced under pressure to fracture and refracture the shale rock to increase the flow of fossil gas. As the water returns to the surface over days to weeks, it is accompanied by methane that escapes to the air. As such, more methane leaks during the fracking of unconventional shale wells than from conventional gas and oil wells. About 5 percent of fracked wells leak methane and the other components of fossil gas immediately, 50 percent leak after 15 years, and 60 percent leak after 30 years.[183] Methane, which has a strong atmospheric warming agent,[184] also leaks during the transmission, distribution, and processing of fossil gas.[185,186,187]

For electricity production, fossil gas is usually used in either an **open-cycle gas turbine** or a **combined-cycle gas turbine**. In an open-cycle

8.3 WHY NOT FOSSIL GAS AS A BRIDGE FUEL?

turbine, air is sent to a compressor, and the compressed air and fossil gas are both sent to a combustion chamber, where the mixture is burned. The hot gas expands quickly, flowing through a turbine to perform work by spinning the turbine's blades. The rotating blades turn a shaft connected to a generator, which converts a portion of the rotating mechanical energy into electricity.

The main disadvantage of an open-cycle turbine is its exhaust contains waste heat that could otherwise be used to generate more electricity. A combined-cycle turbine routes that heat to a heat-recovery steam generator, which boils water with the heat to create steam. The steam is sent to a steam turbine connected to the generator to generate 50 percent more electricity than with the open-cycle turbine alone. Thus, a combined-cycle turbine produces about 150 percent of the electricity of an open-cycle turbine with the same input fossil gas and the same carbon dioxide emissions.

On the other hand, the ramp rate (rate at which a turbine can produce its maximum electricity from a cold start) is 20 percent per minute for an open-cycle turbine. This is two to four times that of a combined-cycle turbine (5–10 percent per minute). In other words, the less efficient open-cycle turbine, which emits 50 percent more carbon dioxide per unit of electricity generated, is more useful for filling in gaps in grid electricity supply than a combined-cycle turbine.

It has long been suggested that fossil gas could be used as a **bridge fuel** between coal and renewables.[188] The two main arguments for this suggestion are (1) fossil gas emits less CO_2-equivalent emissions per unit of energy produced than coal and (2) fossil-gas electric-power plants are better suited to be used with intermittent renewables than is coal. However, these justifications are incorrect and insufficient. Fossil gas is not recommended for use together with WWS technologies for multiple reasons, as discussed in the following sections.

8.3.1 CLIMATE IMPACTS OF FOSSIL GAS VERSUS COAL

When used in an electric-power plant, fossil gas substantially increases, rather than decreases, global warming compared with coal over a 20-year time frame. Over a 100-year time frame, coal causes more warming than fossil gas, but the difference is small. Also, whereas coal causes more air-pollution damage than fossil gas, both coal and fossil gas cause far more air-pollution damage than WWS. As such, spending money on either fossil gas or coal represents an opportunity cost relative to spending the same money on WWS.

Considering only lifecycle emissions, generating 1 kilowatt-hour of electricity with fossil gas using a combined-cycle gas turbine or an open-cycle gas turbine causes 2.3 and 2.8 times, respectively, the warming over a 20-year time frame as generating 1 kilowatt-hour with coal. Over a 100-year time frame, using fossil gas, with the same turbines, may cause only 36 and 19 percent less warming, respectively, than using coal.[10] The fact that fossil gas causes far more warming than coal over a 20-year time frame is a concern because of the severe damage global warming is already causing. More fossil-gas emissions in the short term may trigger difficult-to-reverse impacts, such as the complete melting of the Arctic sea ice.

The reasons fossil gas causes more warming than coal over a 20-year time frame and only slightly less warming than coal over a 100-year time frame are as follows. First, although fossil gas burned in an open-cycle or combined-cycle turbine produces less carbon dioxide per unit of electricity than burning coal, more methane (a greenhouse gas) generally leaks and is vented to the air during the mining, transport, and processing of fossil gas than of coal. Because methane is a potent short-term greenhouse gas, the emission of more methane contributes to more short-term warming by fossil gas than coal. Second, and more important, burning coal emits more nitrogen and sulfur oxides per kilowatt-hour of electricity generated than does burning fossil gas. Nitrogen and sulfur oxides produce cooling aerosol particles, which offset or mask much of coal's global warming. The cooling impacts of these particles are through their reflection of sunlight to space and their enhancement of cloud thickness. Thicker clouds reflect more sunlight, cooling the ground. As such, coal's emissions of nitrogen and sulfur oxides, which are both short-lived chemicals, cause short-term cooling, offsetting much of the short-term warming caused by carbon dioxide from coal.

Regardless, neither fossil gas nor coal is recommended in a 100 percent WWS world. Considering lifecycle emissions alone and before considering carbon capture, fossil gas causes 56–72 times the warming, per unit of electricity generated, of wind, averaged over 100 years.[10] Coal causes about 88 times as much warming as wind. With carbon capture and considering all emissions, fossil gas and coal also cause much more warming than does wind (Figure 8.1).

8.3.2 AIR-POLLUTION IMPACTS OF FOSSIL GAS VERSUS COAL AND RENEWABLES

Whereas fossil gas for electricity causes more warming than coal over 20 years and slightly less warming than coal over 100 years, coal emits more

8.3 WHY NOT FOSSIL GAS AS A BRIDGE FUEL?

health-damaging air pollutants than does fossil gas. Nevertheless, both fossil gas and coal are much worse for human health than are WWS technologies, which emit zero air pollutants during their operation. WWS emits pollutants only during the manufacture and decommissioning of WWS equipment. Such WWS-related emissions will disappear to zero as all energy is transitioned to 100 percent WWS, since at that point, even mining and manufacturing will be powered with WWS.

More carbon monoxide, volatile organic carbon, methane, and ammonia are emitted during fossil-gas production and use than during coal production and use in the US. Coal production and use emit more nitrogen oxides, sulfur dioxide, and particulate matter.[189] Emissions from the mining, transport, processing, and consumption of fossil gas may cause 5,000–10,000 deaths each year in the US from air pollution.[190] Coal-related emissions may cause 20,000–50,000 deaths per year. Thus, both fuels cause air-pollution deaths, but coal causes more. The reason is that nitrogen oxides and sulfur dioxide from coal convert to particles in the air, and those particles plus the particles emitted directly from coal plants cause severe health damage.

In sum, coal causes more deaths than does fossil gas, but both coal and fossil gas cause far more deaths than do WWS technologies. The combination of the much higher CO_2-equivalent emissions and air-pollution deaths due to fossil gas compared with WWS renders fossil gas neither a bridge fuel nor a safe chemical.

8.3.3 FOSSIL GAS IS NOT NEEDED FOR PEAKING OR LOAD FOLLOWING

Another argument for using fossil gas as a bridge fuel is that it is useful for load following and peaking. Thus, it prevents blackouts on the electricity grid. The argument continues that WWS technologies are intermittent and need fossil-gas-powered load-following or peaking plants to back them up when wind and solar outputs are low.

Whereas fossil-gas plants do help with peaking and load following, they are not needed. Other types of WWS electric-power storage options available include hydroelectric-power reservoir storage, pumped hydropower storage, stationary batteries, grid-hydrogen storage, CSP with storage, flywheels, compressed-air energy storage, and gravitational storage with solid masses.

Transition highlight
By 2024, the costs of wind-battery and solar-battery electricity-generating systems were already much lower than those of fossil-gas peaker plants.[191]
In fact, in just four years, from 2020 to 2024, California and Texas each

installed over 8 gigawatts of batteries (the peak power of about eight 1-gigawatt nuclear reactors) on their main grids, starting from virtually nothing, in addition to installing a lot of solar photovoltaics and, in Texas, wind turbines.

8.3.4 LAND REQUIRED FOR FOSSIL-GAS INFRASTRUCTURE

The continuous use of fossil gas for electricity and heat results in the continuous and cumulative degradation of land for as long as fossil-gas use continues. Wells must be dug, and pipes must be laid every year to supply a world thirsty for fossil gas. When gas wells become depleted after 20–30 years, new wells must be drilled. Fifty thousand new fossil-gas wells are drilled each year in North America alone to satisfy fossil-gas demand.[192] The land area required for the well pads, roads, and storage facilities for these new wells amounts to 2,500 square kilometers of additional land consumed per year. Once a gas well is depleted, it is sealed and abandoned, and a portion of the abandoned land cannot be used for any other purpose. The United States alone has 1.3 million active oil-and-gas wells and 3.2 million abandoned ones. Worldwide, an estimated 29 million wells are abandoned. Two-thirds of these are estimated to leak methane and other hydrocarbons.[193] The fossil-gas infrastructure also requires land for underground and above-ground pipes, power plants, fueling stations, and underground storage facilities.

The flammability of fossil gas further results in explosions with fatal consequences in homes and urban areas. For example, on September 9, 2010, a fossil-gas explosion in a San Bruno, California, neighborhood destroyed 38 homes, killed 8 people, and injured 58 others.

Fossil-gas blowouts are also a danger to the local and regional community. For example, from October 23, 2015, to February 11, 2016, a major breach in the Aliso Canyon, California, underground fossil-gas storage facility caused 97,100 tonnes of methane, 7,300 tonnes of ethane, and a host of other hydrocarbons to spew into the air. The result was not only damage to the health of nearby residents but also the possible death from air pollution of up to a few dozen people across California.[194]

The fossil-fuel infrastructure currently occupies about 1.3 percent of the land area of the United States.[10] This is due not only to active and abandoned oil-and-gas wells but also to coal mines, oil refineries, millions of kilometers of pipelines and the clear-cut areas around them, power plants, fueling stations, fossil-gas-processing stations, and fossil-gas storage facilities. Whereas all fossil fuels contribute to this land area degradation, fossil-gas' share is growing due to the replacement of coal by fossil gas, particularly by hydraulically fracked gas. The damage due to fracking includes damage not only to the landscape but also to groundwater, into which fracked fossil gas often leaks. Additional damage occurs to roads, which must carry heavy trucks associated with fossil-gas development. Gas flaring is another form of damage, as flaring emits soot (particles containing black carbon), which causes health damage, warms the air, evaporates clouds, and melts snow.

8.4 WHY NOT CARBON CAPTURE WITH FOSSIL GAS OR COAL?

A proposal to help solve the climate problem that does little more than keep the fossil-fuel and bioenergy industries in business is to capture carbon dioxide emitted from stationary fossil-fuel, bioenergy, and chemical emission sources before the gas escapes from the exhaust stack of the source. The carbon dioxide is then piped and either stored underground or used by industry. The carbon dioxide is captured with equipment added to the plant.

This solution does not help for four reasons. First, carbon capture always increases carbon dioxide, because it always requires equipment and electricity to run the equipment. Even in the best case, where renewable electricity powers the capture equipment, that same amount of renewable electricity will always reduce more carbon dioxide simply by replacing any combustion source of carbon dioxide. Thus, using renewable electricity for carbon capture is an opportunity cost that always increases carbon dioxide. Second, carbon capture always increases emissions of health-affecting gases and particles aside from carbon dioxide because of the additional electricity carbon capture requires. That extra electricity, if dirty, produces more emissions than no capture and, if clean, is no longer able to replace a dirty electricity source, thereby allowing the current plant to continue polluting, as well as preventing the elimination of pollution from another plant. Third, again due to its additional need for electricity, carbon capture may increase the land degradation from the mining of fossil fuels compared with no capture. Fourth, carbon capture permits fossil-fuel infrastructure to continue and, in fact, may increase it due to the additional energy requirements. In sum, carbon capture diverts funds from lower-cost renewables that reduce even more carbon dioxide and air pollutants and far more effectively than does carbon capture. These issues are discussed in detail next.

Carbon capture and storage (CCS) is the separation of carbon dioxide from other exhaust gases following fossil-fuel or biofuel combustion or chemical release, such as during cement or steel manufacturing. The captured carbon dioxide is then pressurized and transported, via pipeline, to an underground geological formation (such as a saline aquifer), a depleted oil-and-gas field, or an unminable coal seam. The exhaust gases not captured are emitted to the air or filtered further. Geological formations worldwide may theoretically store up to two trillion metric tonnes of carbon dioxide, which compares with a 2022 carbon dioxide emission rate from fossil-fuel combustion and industrial-process chemical reaction of about 38.5 billion metric tonnes per year.[129]

Another proposed CCS method is to inject the carbon dioxide into the deep ocean. Because carbon dioxide is an acid, this method acidifies the ocean. Since the Industrial Revolution, a high fraction of carbon dioxide emitted to the air worldwide has dissolved in ocean water, similarly increasing ocean acidity. Some carbon dioxide injected into the deep ocean will also eventually work its way back to the surface and evaporate to the air. A third proposed sequestration method is to mix captured carbon dioxide with concrete, trapping the gas inside the concrete.

Carbon capture and use (**CCU**) is the same as CCS, except the carbon dioxide isolated during carbon capture is sold to industry to pay back the cost of the carbon-capture equipment and its energy consumption. To date, the major application of CCU has been **enhanced oil recovery**. In fact, by the end of 2023, about 82.1 percent of all carbon dioxide captured worldwide for storage or use had been used for enhanced oil recovery.[195] With this process, captured carbon dioxide is piped to an oil field, where it is pumped underground. There, it binds with the oil, reducing the oil's density and allowing it to rise to the surface faster. Once the oil is extracted, some of the carbon dioxide is separated from it and sent back into the field. However, during transport and separation, and due to the energy needed to inject carbon dioxide into the well, 30–40 percent of all carbon dioxide originally captured is released back to the air.[196,197] Enhanced oil recovery also results in about two additional barrels of oil for every tonne of carbon dioxide injected into the oil field. When this is burned, it produces even more air pollution and carbon dioxide emissions.

> **Transition highlight**
> Burning two barrels of oil releases 0.86 tonnes of carbon dioxide to the air. Because one tonne of carbon dioxide added to an oil field for enhanced oil recovery produces two barrels of oil, 86 percent of that carbon dioxide is released back to the air upon burning the oil. However, only 20–80 percent of the oil produced from enhanced oil recovery is new oil that does not replace oil that would otherwise be produced. Thus, only the emissions

8.4 WHY NOT CARBON CAPTURE WITH FOSSIL GAS OR COAL?

from burning 20–80 percent of the oil due to enhanced oil recovery can be considered new emissions. Burning such oil releases 17–69 percent of the carbon dioxide originally captured. Adding that to the 30–40 percent released during enhanced oil recovery gives a total release rate of originally captured carbon dioxide of 47–109 percent.

Another proposed use of carbon dioxide has been to create carbon-based fuels (**electro-fuels**) to replace gasoline, diesel, methanol, and jet fuel. Carbon-based electro-fuels suffer multiple issues, as discussed in Section 8.7.

8.4.1 AIR POLLUTION AND CLIMATE IMPACTS OF FOSSIL SOURCES WITH CARBON CAPTURE

Proponents of carbon capture claim that capture equipment removes 90 percent or more of carbon dioxide from a fossil-fuel exhaust stream. However, real data show that the annually averaged carbon dioxide capture efficiency is only 10–80 percent.[198] This is because the carbon-capture equipment runs less efficiently over time; the carbon-capture equipment is down, due to unplanned or planned maintenance; and/or the facility taking the carbon dioxide for use (e.g., enhanced oil recovery) or storage is temporarily not taking the carbon dioxide.

The capture efficiency accounts only for carbon dioxide captured from a smokestack. But running carbon-capture equipment requires electricity. The amount of such electricity is embodied in the energy penalty. The **energy penalty** of carbon capture is the percentage of fuel burned in an electricity-producing plant that must be dedicated to carbon capture for a fixed quantity of work output by the plant. It has a theoretical range of 11–40 percent[199] and a practical range of 13–44 percent.[200] The electricity is used to separate the carbon dioxide from the gas mixture, compress the carbon dioxide for pipe transport, move the carbon dioxide through the pipeline, and store or use the carbon dioxide.

The electricity required for the energy penalty usually comes from fossil fuels. Mining, transporting, and using such fuels produces emissions of carbon dioxide, methane, and air pollutants. If the fossil fuels used to run the capture equipment are from a different power plant from the plant housing the equipment, then none of the emissions associated with the extra fuel used to run the equipment are captured. Otherwise, only 10–80 percent are captured. In addition, none of the carbon dioxide, methane, or air pollutants emitted during the mining of the fossil fuels used at the power plant where the capture equipment is housed are ever captured.

8 WHAT DOESN'T HELP

When all carbon dioxide and methane emissions are accounted for from mining, transporting, and combusting fossil fuels for normal electricity production in a fossil-fuel power plant, and the energy penalty associated with carbon capture equipment is also accounted for, the overall capture efficiency of the plant with carbon capture declines to between zero and 25 percent. Even in the best case of using renewable electricity to run the carbon-capture equipment, that renewable electricity would reduce more carbon dioxide by replacing the fossil fuel in the first place. As such, even in the best case of using renewable electricity to run capture equipment, carbon capture always increases carbon dioxide relative to using that renewable electricity to replace a fossil source.

Further, capture equipment captures only carbon dioxide, not health-affecting air pollutants. Also, no health-affecting pollutants are captured during the mining of fossil fuels for normal power-plant operations or the additional fuel needed for the energy penalty. Because more fossil fuels are needed to run the capture equipment, carbon capture increases air pollution relative to no capture. In fact, most equipment capturing carbon dioxide from exhaust gas uses a nitrogen-based solvent to bind to the carbon dioxide. The chemical reaction results in the production of some ammonia gas, which is released to the air. Because ammonia is soluble, it dissolves in aerosol particles containing water, producing the ammonium ion, increasing aerosol-particle mass. Aerosol particles cause 90 percent of air-pollution deaths. Thus, a lot of carbon capture equipment produces a harmful air pollutant released to the local area.

To summarize,

1. Carbon-capture equipment captures 75–90 percent of the carbon dioxide in a concentrated exhaust stream when operating at its fullest and best.
2. However, the carbon-capture equipment may be shut down regularly, either due to a lack of demand for the captured carbon dioxide or due to scheduled or unplanned maintenance of the capture equipment. These factors result in the yearly averaged capture rate of the carbon dioxide exhaust stream declining to 10–80 percent.[198]
3. Carbon-capture equipment does not capture the upstream carbon dioxide or other greenhouse gas emissions resulting from mining, transporting, or processing the fossil fuel used in the plant emitting carbon dioxide. Accounting for these emissions further reduces the overall carbon dioxide capture rate.

8.4 WHY NOT CARBON CAPTURE WITH FOSSIL GAS OR COAL?

4. Carbon-capture equipment does not capture any of the health-affecting air pollutants from the fossil-fuel exhaust stream or from the upstream mining, transporting, or processing of the fossil fuel. Such pollutants include carbon monoxide, nitrogen oxides, sulfur dioxide, organic gases, mercury, toxins, black carbon, brown carbon, and fly ash, all of which affect health.
5. A fossil plant with carbon capture needs 13–44 percent more electricity to run the carbon-capture equipment than is needed by a plant without the equipment.[200] If that energy comes from a fossil fuel, the plant then emits even more uncaptured carbon dioxide. If that energy comes from a renewable source, the use of the renewable energy for carbon capture prevents the energy from displacing fossil-fuel energy. In both cases, fossil-fuel mining and infrastructure, and air pollution increase 13–44 percent relative to no capture.
6. Most carbon-capture equipment releases ammonia, a harmful pollutant, to the air.
7. If the captured carbon dioxide is used for enhanced oil recovery or to produce synthetic fuels, 47–109 percent of the captured carbon dioxide is released back to the air.
8. If the captured carbon dioxide is sequestered underground, some of it leaks back into the air over time.
9. Even in the best case of using renewable electricity to power carbon-capture equipment, carbon capture increases carbon dioxide, air pollution, fossil-fuel mining, and fossil-fuel infrastructure because that renewable electricity could otherwise reduce more carbon dioxide and eliminate all air pollution, mining, and infrastructure from a fossil-generating source simply by replacing the source.
10. Thus, the use of renewable electricity for carbon capture is an opportunity cost that increases carbon dioxide, air pollution, fossil-fuel mining, and fossil-fuel infrastructure.[68,201,202]

8.4.2 CARBON CAPTURE PROJECTS

By the end of 2023, carbon dioxide was actively being captured in 41 projects worldwide: 15 fossil-gas-processing facilities, 3 coal-fired power plants, 2 oil refineries, 4 ethanol refineries, 6 chemical plants, 1 iron-and-steel manufacturing plant, 7 facilities for producing hydrogen from fossil gas for ammonia production, 2 multisource carbon dioxide pipelines, and 1 direct-air-capture plant.[195] Of all the carbon dioxide captured among these projects, 82.1 percent was used for enhanced oil recovery.[195] The rest was sequestered underground.

8 WHAT DOESN'T HELP

The maximum possible capture capacity among these 41 projects was 64.8 million tonnes of carbon dioxide per year,[195] which is only 0.17 percent of the 38.5 billion metric tonnes per year emitted in 2022 due to fossil-fuel combustion and chemical reactions.[129] However, of the carbon dioxide going to enhanced oil recovery, 47–109 percent is released back to the air (Section 8.4). This leaves the maximum capture capacity at 0.1 percent of world emissions. Further, the actual capture rate ranges from 10–80 percent of the maximum capture capacity, reducing the capture rate to an average of 0.045 percent of world emissions.

As of 2024, only two fossil-fuel electric-power-generating plants with carbon-capture equipment had sufficient public data to analyze. The carbon-capture equipment of one of the plants was shut off in 2019. In both cases, the separated carbon dioxide was used for enhanced oil recovery. These projects, plus a third, the largest carbon capture project in the world at the time, in which carbon dioxide from a liquefied natural gas-processing facility was being captured and stored, are discussed. All three captured little carbon and increased air pollution and fossil-fuel mining.

8.4.2.1 BOUNDARY DAM PROJECT

The first electric-power plant with CCU equipment was the Boundary Dam power station in Estevan, Saskatchewan, Canada. This plant has been operating with CCU equipment on one coal boiler connected to a steam turbine since October 2014. The cost of the retrofit project was $1.47 billion, which included a $240 million subsidy from the Canadian government. The remaining $1.23 billion was paid for by coal-plant electricity customers as an additional charge added to their normal bills. From 2015 through the end of 2023, the average carbon dioxide capture efficiency from the exhaust stream of this plant was 57 percent.[203] Whereas, during some years, the capture efficiency was as high as 66 percent, in others, it was as low as 36 percent. The captured carbon dioxide has all been sold for enhanced oil recovery. Of all carbon dioxide used for enhanced oil recovery, 30–40 percent is lost during carbon dioxide transport to and processing within the oil field.[196,197] Another 20–80 percent is emitted due to the additional oil produced, depending on whether it is replacing other oil or is new oil. That means that only some or none of the original coal-plant emissions are stored underground, on average, and the rest are released to the air. Even that capture rate ignores the fact that the Boundary Dam project required the coal plant to produce 25 percent more electricity to run the carbon-capture equipment, and this resulted in 25 percent more coal mining, combustion,

8.4 WHY NOT CARBON CAPTURE WITH FOSSIL GAS OR COAL?

and emissions. In addition, none of the carbon dioxide or air pollution from the mining of the coal was ever captured.

8.4.2.2 PETRA NOVA PROJECT

The second coal plant with carbon-capture was the W. A. Parish coal power plant near Thompsons, Texas. The plant was retrofitted with carbon-capture equipment as part of the Petra Nova project and began using the equipment during January 2017. The carbon-capture equipment received 36.7 percent of the emissions from a 654-megawatt boiler at the coal plant. The equipment required almost 0.5 kilowatt-hours of electricity to run per kilowatt-hour of electricity produced by the coal plant. An entire fossil-gas turbine with a heat-recovery boiler was built to provide this electricity. A cooling tower and water treatment facility were also added. The retrofit cost $1 billion ($4,200 per kilowatt) beyond the coal-plant cost.[201] Due to poor economics, carbon capture was halted at the plant in December 2019.

During operation, captured carbon dioxide was compressed and piped to an oil field, where it was used to enhance oil recovery. Carbon dioxide from the fossil-gas turbine was not captured. The mining and transport of fossil gas also emitted uncaptured carbon dioxide, methane, and air pollutants. Upstream carbon dioxide, methane, and air pollutants from the coal plant itself were also uncaptured. Finally, most of the captured carbon dioxide was released back to the air during enhanced oil recovery.

More specifically, during 2017, only 55.4 percent of the carbon dioxide from the coal-combustion exhaust was captured. When uncaptured carbon dioxide from fossil-gas combustion and uncaptured carbon dioxide and methane from mining and processing the coal and fossil gas used are also accounted for, the overall capture rate of CO_2-equivalent emissions was only 10.8 percent over a 20-year time frame and 20 percent over a 100-year time frame.[201] Of the carbon dioxide captured, 47–109 percent was then emitted back to the air during enhanced oil recovery (Section 8.4). During the entire three-year operation of the plant, the capture efficiency from the coal exhaust was a bit higher, 65 percent.[198] However, this does not change the conclusion that this project resulted in virtually no reduction in CO_2-equivalent emissions, while increasing fossil-gas mining and air pollution versus no capture.

Using wind instead of fossil gas to power the carbon-capture equipment could have reduced CO_2-equivalent emissions before enhanced oil recovery by 37.4 percent over 20 years and 44.2 percent over 100 years, versus no capture, which is a greater reduction than with gas. The decrease is greater because wind results in no fossil-gas mining or combustion emissions.

However, using the wind electricity that powered the carbon-capture equipment instead to replace coal electricity directly would reduce CO_2-equivalent emissions over both 20 and 100 years by 49.7 percent, more than in the other cases. Thus, it is better to use new WWS electricity to eliminate a carbon dioxide source, and thus prevent carbon dioxide from being emitted in the first place, than to use the same WWS electricity to extract carbon dioxide from the source with carbon capture.

The climate benefit of using WWS electricity to replace fossil-fuel electricity instead of to power carbon-capture equipment is only part of the story. Carbon capture does not reduce any air pollution from the coal plant or coal mine. Instead, it increases air pollution when a fossil fuel is used to provide power for the capture equipment. Even when wind powers the capture equipment, air pollution continues from the coal plant and coal mine. Only when wind replaces the coal plant itself does air pollution from both the coal plant and coal mine decrease.

In sum, the social cost (equipment plus health plus climate cost) of coal-CCU powered by fossil gas is over twice that of wind replacing coal directly and even higher than that of doing nothing. The social cost of using wind to power CCU equipment is also much higher than using the same wind to replace a coal plant. In other words, the best strategy is to use WWS to replace fossil fuels.[201] This conclusion is independent of the fate of the carbon dioxide after it leaves the carbon-capture equipment, and it applies to CCS or CCU with bioenergy or cement manufacturing as well.

When the fate of captured carbon dioxide is considered, the problem deepens. If carbon dioxide is sealed underground, few additional emissions may occur. However, if the captured carbon dioxide is used to enhance oil recovery, its current major application, 47–109 percent of the captured carbon dioxide is released back to the air. If the captured carbon dioxide is used to a create carbon-based fuel to replace gasoline and diesel, energy is still needed to produce the fuel, the fuel is still burned in vehicles (creating pollution), and little net carbon dioxide is captured. A fourth application is to use the carbon dioxide to produce carbonated drinks. However, most carbon dioxide in carbonated drinks is released back to the air during consumption.

8.4.2.3 GORGON PROJECT

A third project is the world's largest CCS plant, which is attached to the Gorgon liquefied natural gas (LNG) facility on Barrow Island, 130 kilometers off the northwest coast of Australia. On March 21, 2016, the facility started converting a portion of the extracted gas to LNG for export and using the rest for local consumption. The raw gas from the Gorgon field near Barrow Island contains about 15 percent carbon dioxide. The Australian government permitted the LNG facility on the condition that 80 percent of the carbon dioxide from the gas be captured and injected two kilometers below Barrow Island, starting the day LNG exports commenced. The CCS equipment cost $3 billion.

Delays prevented capture and injection until August 2019. Inefficiencies and problems limited capture rates thereafter. As a result, over the five years between 2016 and 2021, only 32 percent of the carbon dioxide emitted from the LNG facility exhaust stream was captured.[198] The rest was released to the air. Also, a portion of the fossil gas for domestic consumption provided electricity to run the CCS equipment. The burning of this gas emitted enough carbon dioxide to cause the net carbon dioxide emissions from the CCS facility to be positive. Thus, instead of reducing it, the Gorgon CCS plant likely increased carbon dioxide emissions and at a cost of $3 billion.

In sum, carbon capture is not close to a zero-carbon technology. For the same energy cost, wind turbines and solar panels replacing fossil fuels reduce much more carbon dioxide and also eliminate fossil-fuel air pollution and mining, which carbon capture increases. Using renewables to replace fossils also reduces oil-and-gas pipelines, refineries, gas stations, tanker trucks, oil tankers, coal trains, oil spills, oil fires, gas leaks, gas explosions, and international conflicts over energy. Carbon capture increases these by increasing energy use.

8.5 WHY NOT GRAY, BLUE, OR BROWN HYDROGEN?

In a WWS world, hydrogen should be used primarily for long-distance heavy transport, steel and ammonia manufacturing, grid-electricity storage, and electricity and heat production in microgrids. It should not be used for heating buildings, providing high-temperature heat for industry, or powering passenger vehicles. If it were to be used to heat buildings, it would need to be used either on its own or mixed with fossil gas in existing fossil-gas pipes or run through new hydrogen-specific pipes. However, hydrogen cannot be used in existing gas pipes without high leakage rates, and new pipes are expensive. What is more, burning hydrogen for heat in a building produces air pollutants (oxides of hydrogen and oxides of nitrogen, in particular) and, like burning fossil gas, is much less efficient than using an electric heat pump to provide heat. As such, hydrogen should not be used as a replacement for fossil gas to heat buildings, and proposals to do this have not materialized.[204] Similarly, burning hydrogen for high-temperature heat in industry creates air pollution. Using hydrogen in passenger vehicles can be clean if the hydrogen is produced from wind or solar electricity, but doing so requires up to three times the number of wind turbines or solar panels as using the same wind or solar electricity to power battery-electric vehicles.

For its useful applications, hydrogen should be only green hydrogen: produced by electrolysis or photoelectrochemical water splitting, where the electricity comes from WWS in both cases. This section discusses why hydrogen should not be so-called gray, blue, or brown hydrogen.

Three methods of producing hydrogen from fossil gas are steam methane reforming, autothermal reforming, and methane pyrolysis. When not coupled with carbon capture, the hydrogen resulting from steam methane reforming and autothermal reforming is referred to as **gray hydrogen**. When those two methods are coupled with carbon capture, the hydrogen is referred to as **blue hydrogen**. When hydrogen is produced from methane pyrolysis, the hydrogen is referred to as **turquoise hydrogen**. When hydrogen is produced from coal gasification, the hydrogen is referred to as **brown hydrogen**. Each of these techniques is discussed, in turn.

8.5.1. STEAM METHANE REFORMING

Steam methane reforming is a method of producing hydrogen by mixing methane with steam at a high temperature (700–1,000 degrees Celsius) and pressure (3–25 times atmospheric pressure). Fossil gas is needed for two purposes during steam methane reforming. The first is to provide the energy needed for high temperatures and pressures. The second is to provide the methane used in the chemical reactions that produce hydrogen. During the steam reforming process, methane in fossil gas reacts with two water vapor (H_2O) molecules to form four hydrogen molecules (H_2) and a molecule of carbon dioxide. About 96 percent of hydrogen production worldwide in 2019 was from steam methane reforming. This method of hydrogen production consumes about 6 percent of all fossil gas globally.[162]

During the mining, transporting, and processing of the fossil gas that arrives at a steam methane reforming plant, gaseous air pollutants, black carbon particles, carbon dioxide, and methane are released to the air. Gaseous air pollutants, black carbon, and carbon dioxide are emitted because mining requires construction equipment, which runs on diesel

8.5 WHY NOT GRAY, BLUE, OR BROWN HYDROGEN?

fuel that pollutes when burned. Mining also requires concrete production for wells, which results in pollution. Trees are also uprooted, and land is trenched for pipes. Those processes require diesel and gasoline machinery and vehicles, resulting in even more pollution. Tree-clearing releases stored carbon to the air as carbon dioxide.

Methane leaks from both active and abandoned fossil-gas wells. The United States has 1.3 million active wells and 3.2 million abandoned wells. Worldwide, 29 million oil-and-gas wells have been abandoned, and two-thirds are estimated to leak.[193]

Methane leak rates from two of the world's largest fossil-gas fields, in Turkmenistan, are estimated by satellite to be 4.1 percent.[187] Leak rates in the US are between 1.2 and 9.4 percent. Methane also leaks from pipes running between gas wells and the fossil-gas-processing center (where impurities are removed from fossil gas). More leaks occur at the processing center itself, from pipes running between the processing center and the hydrogen plant, and at the hydrogen plant itself. In addition, electricity from fossil gas is used to compress and transport fossil gas through pipes and to process the fossil gas at the processing station. Generating that electricity results in more emissions. An overall estimate of the worldwide average methane emission rate from fossil-gas mining and transport is 3.5 (1.5–4.3) percent.[162]

In sum, the production of hydrogen from steam methane reforming results in gaseous air pollution, black carbon particles, carbon dioxide, and methane emissions from mining, transporting, and processing fossil gas and from using the mined fossil gas for energy and as a feedstock.[162]

8.5.2 AUTOTHERMAL REFORMING

Another method of producing hydrogen from fossil gas is **autothermal reforming**. With this method, fossil gas is brought to an autothermal reforming facility. However, instead of methane reacting with water vapor, it reacts with pure oxygen to produce a carbon dioxide molecule and two hydrogen molecules. This reaction releases heat, which reduces the need for additional fossil-fuel energy to produce heat for this process.

One shortcoming of autothermal reforming is that it produces only two molecules of hydrogen per molecule of methane, whereas steam methane reforming produces four molecules of hydrogen per molecule of methane. As such, more methane is needed with autothermal reforming than with steam methane reforming to produce the same quantity of hydrogen. Autothermal reforming also produces carbon monoxide (CO), so additional equipment is needed to react carbon monoxide with steam (H_2O) at high temperature to produce carbon dioxide and one more hydrogen molecule. Even when this is done, though, a maximum of 2.89 molecules of hydrogen can be obtained per molecule of methane[205] (versus 4 molecules of hydrogen with steam methane reforming). Thus, autothermal reforming always requires at least 38 percent more methane mining (resulting in 38 percent more methane leakage) than steam methane reforming to produce the same quantity of hydrogen. A third disadvantage of autothermal reforming is that it requires pure oxygen, which must be separated from air. Separation requires equipment and energy. For these three reasons, steam methane reforming is generally preferred by industry over autothermal reforming.[206]

The inefficiency and emissions from autothermal reforming can be reduced slightly by using some of the hydrogen produced by the autothermal reforming process in a fuel cell to generate the electricity for oxygen separation from the air. The fuel cell also releases heat that can be used for the autothermal reforming process. However, this means a fuel cell must be purchased and more methane is needed to produce the additional hydrogen needed for the fuel cell. More methane means more fossil-gas mining, transporting, and processing, thus more upstream air pollution, carbon dioxide, and methane emissions. More methane consumption also means more carbon dioxide emitted during the autothermal reforming process.

Some have proposed to capture the carbon dioxide during steam methane reforming and autothermal reforming, resulting in the hydrogen produced becoming blue hydrogen. During steam methane reforming, carbon dioxide emissions occur during two processes: combustion to produce electricity and heat, and chemical reactions to produce hydrogen. Additional emissions arise from powering carbon-capture equipment.

During autothermal reforming (coupled with a hydrogen fuel cell), carbon dioxide emissions arise from the chemical reactions that produce hydrogen but not from combustion to produce energy. That is because heat is provided by both the chemical reaction producing hydrogen and by the fuel cell. So only carbon dioxide from chemical reaction needs to be captured. However, far more carbon dioxide is produced and needs capturing per molecule of hydrogen produced with autothermal reforming than with steam methane reforming. Also, much more equipment is needed with autothermal reforming than with steam methane reforming.

Whereas carbon dioxide capture rates from chemical reaction during steam methane reforming or autothermal reforming can be 90 percent or more when the capture equipment is fully operational, real annually

8.5 WHY NOT GRAY, BLUE, OR BROWN HYDROGEN?

averaged capture rates of pure carbon dioxide streams from steam reforming equipment have been reported as only an average of 78.8 percent.[198] The reasons are that capture equipment is often down for scheduled or unscheduled maintenance; the demand for carbon dioxide is temporarily low; or the capture equipment is less efficient than expected. Capture rates of carbon dioxide from energy production (relevant only to the steam reforming process) are even lower, ranging from 10 to 70 percent.[198]

The carbon dioxide that is captured with both steam methane reforming and autothermal reforming must be piped to either an underground storage facility or an industrial facility, where it is used. Piping requires additional energy and results in additional emissions. Of the carbon dioxide that is captured and piped for enhanced oil recovery, for example, 47–109 percent is released back to the air (Section 8.4). In addition, neither steam methane reforming nor autothermal reforming reduces upstream emissions or leaks of air pollutants, carbon dioxide, or methane.

Because of the need for additional equipment and energy required for the carbon-capture equipment, steam methane reforming with carbon capture (blue hydrogen) may reduce the warming impact of steam methane reforming with no carbon capture (gray hydrogen) by only 9–12 percent over a 20-year time frame.[162] Blue hydrogen, of course, always costs more than gray hydrogen because blue hydrogen requires carbon-capture equipment, pipes, and additional energy not required by gray hydrogen.

8.5.3 METHANE PYROLYSIS

Methane pyrolysis was discussed in Section 7.2 with respect to converting biological sources of methane to hydrogen. So long as the heat used for methane pyrolysis comes from a WWS source and the methane comes from a biological source, such as landfill gas, methane pyrolysis results in no carbon dioxide emissions, modest methane emissions (depending on the methane conversion efficiency during pyrolysis), and some air-pollution emissions. However, if the methane comes from fossil gas, the upstream emissions of methane, carbon dioxide, black carbon, and air pollutants are significant, just as with steam methane reforming and autothermal reforming.[162]

8.5.4 COAL GASIFICATION

Hydrogen can also be produced by **coal gasification**. With this process, coal reacts with oxygen gas and steam under high temperature and pressure to form a mixture of hydrogen gas, carbon monoxide, carbon dioxide, and other chemicals. After impurities are removed, the carbon monoxide reacts with steam to form more hydrogen. Coal gasification results in more carbon dioxide and other pollutant emissions than does steam reforming of methane. Continuous coal mining and transport result in additional fossil-fuel combustion emissions, leaked methane, and land degradation. Additional energy is needed to create the high temperatures and pressures for the gasification process. Hydrogen from coal gasification is referred to as **brown hydrogen**.

In sum, because the steam reforming, autothermal reforming, and coal gasification processes emit carbon dioxide, methane, and other pollutants and degrade land, they are not suitable candidates for producing hydrogen in a 100 percent WWS world. In a WWS world, all carbon and pollution emissions and continuous fuel mining are eliminated. The simplest and cleanest type of hydrogen in a WWS world is green hydrogen.

8.6 WHY NOT SYNTHETIC DIRECT AIR CARBON CAPTURE?

Synthetic direct air carbon capture and storage (SDACCS) is the removal of carbon dioxide from the air by its chemical reaction with other chemicals followed by sequestration of the carbon dioxide, either underground or in a material. **Synthetic direct air carbon capture and use (SDACCU)** is the same as SDACCS, except that in this case, the captured carbon dioxide is sold for use in industry.

SDACCS/U should not be confused with **natural direct air carbon capture and storage (NDACCS)**, which is the natural removal of carbon dioxide from air by planting trees or reducing permanent deforestation (reducing open biomass burning). Growing a tree removes carbon dioxide and water vapor from the air naturally by photosynthesis and sequesters the carbon within organic material in the tree for decades to centuries while releasing oxygen to the air. Reducing open biomass burning similarly sequesters carbon in trees and eliminates emissions of health-affecting air pollutants and climate-affecting global warming agents (methane, black carbon, and brown carbon, for example) aside from carbon dioxide at the same time. Trees also absorb air pollutants, helping to filter them from the air.

Whereas NDACCS is recommended in a 100 percent WWS world, SDACCS/U is not. SDACCS/U is a cost, or tax, added to the cost of energy generation. As such, it raises energy costs. Because SDACCS/U requires equipment and carbon dioxide pipelines, and because building and running equipment and pipes require electricity, SDACCS/U also increases

8.6 WHY NOT SYNTHETIC DIRECT AIR CARBON CAPTURE?

carbon dioxide emissions rather than decreasing them. The reason is that the electricity used to run the capture equipment, even if renewable, would otherwise reduce more carbon dioxide by replacing a fossil-fuel or bioenergy carbon dioxide source. Similarly, by not replacing a fossil-fuel or bioenergy source with that renewable electricity, SDACCS/U projects increase air pollution, fossil mining, and fossil or bioenergy infrastructure. They also enable the fossil-fuel industry to continue operating by demotivating the need to replace fossils with WWS.

In sum, SDACCS/U will always increase carbon dioxide until all carbon-based electricity sources are eliminated. It will also increase air pollution and allow fossil mining and infrastructure and bioenergy infrastructure to continue.

8.6.1 HOW DOES AIR CAPTURE REMOVE CARBON DIOXIDE FROM THE AIR?

In 1754, Joseph Black isolated carbon dioxide, which he named **fixed air**. He found that heating the odorless white powder magnesium carbonate ($MgCO_3$) released a gas (carbon dioxide) that could not sustain life or fire. The remaining solid, magnesium oxide (MgO), weighed less than the original magnesium carbonate. He similarly found that adding potassium carbonate (K_2CO_3) to magnesium oxide (MgO) in solution produced solid magnesium carbonate ($MgCO_3$). The mass of $MgCO_3$ exceeded that of MgO by the same amount that was lost when $MgCO_3$ was heated to form MgO. The difference in mass in both cases was the mass of carbon dioxide. As such, Black not only captured carbon dioxide, but he also quantified the mass of carbon dioxide for the first time.

Today, an important air-capture technique is to react carbon dioxide from the air with alkali metal oxides (Na_2O or K_2O, among others), alkali metal hydroxides (NaOH or KOH), alkaline earth metal oxides (MgO or CaO), or alkaline earth metal hydroxides [$Mg(OH)_2$ or $Ca(OH)_2$].[207] For example, a classic method of removing carbon dioxide from the air while recycling the material removing it is to expose the carbon dioxide to a large pool of calcium hydroxide [$Ca(OH)_2$,], also called slaked lime. The products of the reaction are calcium carbonate ($CaCO_3$) and water vapor. Heating the calcium carbonate to 427 degrees Celsius releases a concentrated stream of carbon dioxide that can be captured and used. The leftover calcium oxide (CaO) is then reacted with water to reform calcium hydroxide. The problem with this process is that it needs a continuous electricity input.

8.6.2 OPPORTUNITY COST OF DIRECT AIR CAPTURE

By removing carbon dioxide from the air, air capture does exactly what WWS generators, such as wind turbines and solar panels, do. This is because WWS generators replace fossil generators, preventing carbon dioxide from getting into the air in the first place. The impact on climate of removing one molecule of carbon dioxide from the air is the same as the impact of preventing one molecule from getting into the air in the first place.

However, WWS generators also (a) eliminate air pollutants aside from carbon dioxide from fossil-fuel combustion; (b) eliminate the upstream mining, transport, and refining of fossil fuels and the corresponding emissions; (c) reduce the oil-and-gas pipeline, refinery, gas station, tanker truck, oil tanker, and coal train infrastructure of fossil fuels; (d) reduce oil spills, oil fires, gas leaks, and gas explosions; (e) reduce international conflicts over energy; and (f) reduce the large-scale blackout risk associated with centralized power plants by decentralizing/distributing power. Air capture does none of that. Its sole job is to remove carbon dioxide from the air.

Transition highlight

To illustrate how synthetic direct air capture increases carbon dioxide, consider the world's second, and largest at the time, direct-air-capture plant. The plant, built in Iceland by Climeworks, is designed to capture up to 36,000 tonnes per year of carbon dioxide and store it underground. Large fans blow air over an adsorbent. The adsorbent reacts with carbon dioxide in the air, extracting it. Heat then releases pure carbon dioxide from the adsorbent. The released pure carbon dioxide is piped underground to a sequestration site. According to Climeworks, about 5,000–6,000 kilowatt-hours of electricity and electricity converted to heat are needed to remove one tonne of carbon dioxide from the air and store it.[208] This is more than twice the 2,000–3,000 kilowatt-hours per tonne claimed by industry[208] and translates to 167–200 grams of carbon dioxide removed from the air per kilowatt-hour of electricity consumed. In the best case, the electricity used for air capture is WWS electricity.

However, because WWS electricity is limited, if it is used for air capture, it cannot be used to replace a coal or fossil-gas electricity or heat source. Using 1 kilowatt-hour of WWS electricity to replace coal electricity eliminates about 1,380 grams of CO_2-equivalent emissions over a 20-year time frame (930 from coal combustion and 450 from coal mining).[201] Thus, using WWS electricity for air capture instead of for replacing coal electricity increases carbon dioxide in the air by a factor of 7–8. Using WWS for air capture instead of for replacing coal also prevents the elimination of air pollution, mining, and infrastructure associated with the coal plant.

8.6 WHY NOT SYNTHETIC DIRECT AIR CARBON CAPTURE?

In the present case, the electricity and heat generated for the Iceland air-capture plant are geothermal electricity and heat, and Iceland produces no coal electricity. However, Iceland is increasingly using coal for industrial heat. Coal and fossil gas for heat emit about 440 grams of CO_2-equivalent per kilowatt-hour of fuel energy.[162] Mining of coal and gas produces another 215 grams, for a total of 655 grams of CO_2-equivalent per kilowatt-hour of coal or gas over a 20-year time frame (ignoring the transport of coal or gas overseas to Iceland). Given that 1 kilowatt-hour of geothermal electricity produces about 1 kilowatt-hour of high-temperature heat to replace coal, using geothermal electricity for air capture instead of for replacing coal (or gas) for heat increases carbon dioxide emissions by a factor of 3.3–4.

On an absolute basis, using equipment and energy to remove carbon dioxide from the air results in little carbon removal. In one air-capture plant, for example, electricity for the equipment was provided by a fossil-gas combined cycle turbine. Mining, transporting, processing, and burning of the gas produced air pollutants, methane, and carbon dioxide. Such emissions were so large that, averaged over 20 and 100 years, they offset 89.5 and 69 percent, respectively, of all carbon dioxide captured by the plant.[201]

The social cost of air capture includes the costs of capture equipment, fossil-gas electricity used to run the equipment, and air-pollution health problems due to the electricity use, less the cost benefit of removing carbon dioxide from the air. The social cost of running the air-capture plant in this example is about eight times that of using the same electricity, but powered by wind, to replace part of a coal plant.[201] The reason is that wind replacing coal eliminates all air pollution and more carbon dioxide than does direct air capture, and incurs no carbon-capture equipment cost. Even when wind, instead of fossil gas, powers the air-capture plant, the social cost of using wind to power air capture is still five times that of using wind to replace part of a coal plant.[201] In fact, there is no case where wind powering air capture has a social cost below that of using wind to replace a fossil-fuel or biomass power plant.[201] This is because wind-powering-air capture always incurs an equipment cost, never reduces air pollution, and reduces only a modest amount of carbon dioxide. Wind replacing a fossil plant incurs no extra equipment cost, reduces air pollution and mining from the fossil plant, and reduces more carbon dioxide emissions than does air capture.

An argument for using air capture is that it will be needed to remove airborne carbon dioxide once all fossil fuels are replaced with 100 percent WWS. If all energy is provided by WWS at that point, air capture should reduce carbon dioxide without increasing air pollution. However, at that point, the question will be whether growing more trees, reducing open biomass burning; reducing agriculture and waste burning; or reducing halogen, nitrous oxide, or nonenergy methane emissions is more cost-effective for limiting further global warming than is air capture. Thus, air capture may or may not be useful even after fossil fuels are no longer used.

In sum, as with carbon capture, direct air capture is not close to a zero-carbon technology. For the same energy cost, wind turbines and solar panels replacing fossil or biomass combustion sources reduce far more carbon dioxide while also eliminating fossil air pollution, mining, and infrastructure, which air capture increases.

8.7 WHY NOT NONHYDROGEN ELECTRO-FUELS?

Electro-fuels are synthetic fuels that replace gasoline, diesel, methanol, ethanol, and jet fuel as transportation fuels. They are made from molecular hydrogen created by electrolysis (hence the prefix, "electro") and carbon dioxide from carbon capture or synthetic direct air capture. The simplest electro-fuel is hydrogen itself. When produced from clean, renewable WWS electricity and used in a fuel cell to power a long-distance aircraft or ship, green hydrogen is a useful electro-fuel. Gray, blue, and brown hydrogen, on the other hand, are not useful or recommended (Section 8.8).

Methanol (CH_3OH) is the next-simplest electro-fuel. It is produced from three hydrogen molecules (H_2) reacting with one carbon dioxide molecule (CO_2). A byproduct of this reaction is water (H_2O). Ethanol (C_2H_5OH) is similarly created from six hydrogen molecules and two carbon dioxide molecules. A byproduct is three water molecules. Synthetic methanol and ethanol are proposed to replace methanol derived from fossil gas and ethanol derived from corn, respectively, for ground transportation. Synthetic ethanol is also competing with other nonpetroleum hydrocarbon-based fuels to replace jet fuel for air travel. Such nonpetroleum fuels are referred to as sustainable aviation fuels (SAFs). Ethanol produced from corn is similarly vying to become an SAF.

Gasoline contains many long-chain hydrocarbons. Such hydrocarbons can be manufactured synthetically as an electro-fuel by first converting many carbon dioxide molecules to carbon monoxide molecules and then reacting the carbon monoxide with molecular hydrogen. The result is a distribution of hydrocarbon with different numbers of carbon atoms, similar to gasoline.

The argument for using these electro-fuels is that they eliminate the reliance on fossil-fuel mining and corn growing. However, aside from green hydrogen, electro-fuels should not be supported as a solution to air pollution, global warming, and energy insecurity. The reasons are (1) all such fuels are still combusted, creating air pollution, and in the case of SAFs, contrails; (2) since carbon capture and direct air capture are opportunity costs that increase air pollution, carbon dioxide, fossil mining, and fossil infrastructure (Sections 8.4 and 8.9), so are carbon-based electro-fuels; (3) carbon-based electro-fuels all require chemicals and energy, in addition to captured carbon dioxide, causing their emissions to be similar to or higher than the fossil fuels or biofuels they replace; and (4) biofuel-based electro-fuels (e.g., ethanol-based SAF) suffer from the same problems as ethanol with or without carbon capture (Section 8.9).

8.8 WHY NOT BIOMASS OR BIOGAS FOR ELECTRICITY OR HEAT?

Bioenergy fuels are solid (biomass), liquid (biofuels), or gaseous (biogas) fuels derived from organic matter. Most bioenergy is derived from dead plants, animal excrement, or microbial degradation of either. **Biomass**, such as wood, grass, agricultural waste, and dung, is burned directly for home heating and cooking in developing countries and for electric-power generation in most countries. **Biofuels** are generally used for transportation as a substitute for gasoline, diesel, jet fuel, or bunker fuel. **Biogas**, such as methane from a landfill, digester, or wastewater treatment plant, is used for either electricity, heat, or transportation.

Biofuels are discussed in Section 8.9. Biomass and biogas combustion are not recommended in a 100 percent WWS world for several reasons, discussed herein. Similarly, **biomass with carbon capture and storage** (**BECCS**) also represents an opportunity cost versus WWS, so is not recommended. One potentially acceptable application of biogas is methane pyrolysis, if the heat comes from a WWS source. These topics are discussed next.

8.8.1 BIOMASS AND BIOGAS COMBUSTION WITHOUT CARBON CAPTURE

The main sources of solid biomass burned to produce electricity or heat are as follows:[209]

  agriculture residues, which include dry crop residue (such as straw and sugar beet leaves), and livestock waste (such as solid or liquid manure);

8 WHAT DOESN'T HELP

forestry residues, which include bark, wood blocks, wood chips from treetops and branches, and logs from forest thinning;

energy crops, which include dry wood crops (such as willow, poplar, eucalyptus, and short-rotation coppice), dry herbaceous crops (such as miscanthus, switchgrass, reed, canary grass, cynara, cardu, and Indian shrub), oil energy crops (such as sugar beet, cane beet, sweet sorghum, Jerusalem artichoke, sugar millet), starch energy crops (such as wheat, potato, maize, barley, triticae, corn, and amaranth), and other energy crops (such as flax, hemp, tobacco stems, aquatic plants, cotton stalks, and kenaf);

industry residues, which include wood industry residues (such as bark, sawdust, wood chips, and cutoffs from saw mills), food industry residues (such as beet root tails, used cooking oils, tallow, yellow grease, and slaughterhouse waste), and industrial products (such as pellets from sawdust and wood shavings, bio-oil, ethanol, and biodiesel);

park and garden wastes, which include grass and pruned wood; and

contaminated wastes, which include demolition wood, municipal waste, sewage sludge, sewage gas, and landfill gas.

The primary reason biomass combustion is not recommended for use in a WWS world is that biomass combustion, like coal and fossil-gas combustion, produces air pollution. A 100 percent WWS energy infrastructure is designed to eliminate air pollution. The problem is greatest with the burning of municipal waste, which usually contains toxic chemicals. In sum, whereas biomass is partly renewable, it is not clean. A 100 percent WWS world requires both clean and renewable energy rather than just renewable energy.

The second reason for not including biomass for electricity production is that it causes more global warming per unit of electricity produced than WWS when all emissions are accounted for. Biomass grows by photosynthesis – it converts carbon dioxide and water vapor from the air into organic material and oxygen. Although growing biomass takes carbon dioxide out of the air to grow, that carbon dioxide is returned to the air when the biomass is burned, and any residue is decomposed by bacteria. Biomass causes additional emissions of carbon dioxide during fertilizing, watering, growing, collecting, transporting, separating, and/or incinerating the biomass. These processes all require fossil-fuel energy and emissions.

The overall emissions from biomass used for electricity production range from 86 to 1,788 grams of CO_2-equivalent per kilowatt-hour of

8.8 WHY NOT BIOMASS OR BIOGAS FOR ELECTRICITY OR HEAT?

electricity, or 10–373 times the emissions per unit electricity as onshore wind (Figure 8.1). These emissions are due largely to lifecycle emissions (43–1,730 grams of CO_2-equivalent per kilowatt-hour of electricity). A review suggests that the combustion of forestry and industry residues may result in the least emissions (43–46 grams) among biomass fuels.[209] Combustion of agricultural residues and energy crops may cause higher emissions (200–300 grams). Combustion of municipal solid waste may cause the most emissions (mean of 1,730 grams). The low emissions from forestry and industry residues are due to the fact that the feedstock in those cases does not need to be produced actively as it does with agricultural residues or energy crops. The high emissions from burning municipal solid waste are due to emissions from the energy required to collect, segregate, sort, transport, and incinerate the waste.

Biomass energy facilities have opportunity-cost emissions of 36–51 grams of CO_2-equivalent per kilowatt-hour of electricity generated because they take four to nine years between planning and operation versus two to five years for onshore wind or utility PV. During the additional time, the background grid is emitting.

Other sources of climate-affecting emissions due to using biomass for electricity are heat and water vapor emissions during biomass combustion. Because biomass combustion is less efficient than is coal combustion, biomass combustion releases more heat per unit of electricity than does coal combustion.

A third problem with some types of biomass, particularly energy crops, is that they require much more land than WWS does. Given that photosynthesis is only 1 percent efficient at converting sunlight to biomass energy, whereas solar PV panels are now 20–47 percent efficient at converting sunlight to electricity, a solar panel needs less than one-twentieth of the land to produce the same energy as a biomass crop.

Biogas from a landfill, digester, or wastewater treatment plant contains 45–75 percent methane by volume. The rest is mostly carbon dioxide. Burning biogas for electricity, heat, or transportation faces the same issues as burning biomass. An alternative to burning biogas is to use its methane to produce hydrogen by methane pyrolysis (Section 7.2). This process minimizes methane emissions and emits no carbon dioxide so long as the heat required for it comes from a WWS source. The hydrogen can be used to generate electricity in a fuel-cell vehicle (displacing the need for gasoline); in a fuel cell for electricity generation (displacing the need for fossil gas to do the same thing); or for steel or ammonia manufacturing (displacing the need for coal or fossil gas, respectively, for those processes).

In sum, burning forest and industry residue and other forms of biomass to provide electricity and heat, or burning biogas to provide electricity, heat, or transportation, results in higher CO_2-equivalent emissions and more air pollution than using WWS. Some forms of biomass also require much more land than WWS does. As such, using biomass or biogas for energy represents an opportunity cost. The exception is to use biogas to produce hydrogen by methane pyrolysis, where the hydrogen is subsequently used in a fuel cell or for steel or ammonia manufacturing.

8.8.2 BIOMASS AND BIOGAS COMBUSTION WITH CARBON CAPTURE

A proposed method of reducing biomass and biogas CO_2-equivalent emissions, and even creating negative carbon emissions, is to combine biomass and biogas combustion with carbon capture and storage to give **biomass (or biogas) with carbon capture and storage or use** (**BECCS/U**). Negative carbon emissions arise if a process removes more carbon from the air than it adds to the air.

BECCS would theoretically result in negative carbon emissions if, for example, forest wood residue (containing carbon dioxide from the air) were collected; little energy were used to collect, transport, and incinerate the wood; and the carbon dioxide were captured from the exhaust stream of the biomass electricity-generating facility and pumped underground. If successful, this method would be a one-way conduit for carbon dioxide to go from the air to underground, thereby resulting in negative carbon emissions.

The problems, however, are several-fold. As with fossil gas and coal with carbon capture, the carbon capture system with BECCS/U requires 13–44 percent more energy than without it.[200] If that energy comes from fossil gas, coal, or biomass, 13–44 percent more air pollution occurs with BECCS/U than without capture. Biomass and biogas combustion without carbon capture already produce substantial air pollution, whereas WWS produces none.

Similarly, as with carbon capture for coal and fossil gas, the carbon dioxide reductions with BECCS/U are 10 to 80 percent,[198] much lower than anticipated due to the high energy requirements of carbon capture equipment. Leakage of carbon dioxide from underground storage or from industrial use is also an issue.

Second, as with fossil gas and coal, few reliable underground storage facilities exist for BECCS. Because of the high cost of carbon capture, biomass with carbon-capture facilities are likely to couple with for-profit uses of the carbon dioxide, such as enhanced oil recovery or electro-fuel production, both of which require energy and produce combustion fuels, resulting in additional carbon dioxide and air-pollution emissions.

Third, the efficiency of biomass combustion for electricity (electricity output per unit energy in the fuel) is low (20–27 percent), even compared with the efficiency of coal combustion (33–40 percent). Thus, a large mass of biomass is needed to produce a small amount of electricity. As such, if BECCS/U were to provide negative emissions on a large scale, substantial land areas dedicated to biomass crops would be needed to maintain a continuous energy supply. Consequently, a share of agricultural land would be used for fuel instead of food, increasing the price of food. Higher food prices trigger deforestation by incentivizing people to turn high-carbon-storage forestland into low-carbon-storage agricultural land.

Fourth, removing agricultural residues usually means crops need to be fertilized more since the residues contain nutrients that are no longer available once they are removed. Fertilizers contain nitrous oxide, a greenhouse gas, and ammonia, a major air pollutant. Both are emitted to the air. Finally, the cost of BECCS/U is high, even compared with CCS for fossil fuels. In fact, as of 2023, no BECCS/U facility existed in the world for capturing carbon dioxide emissions from a biomass-to-electricity-and-heat power plant.[195]

In sum, paying for BECCS/U instead of WWS means less energy production, a longer time lag between planning and operation, more air pollution, greater land use (for some crops), and less carbon removal. As such, BECCS/U is not recommended.

8.9 WHY NOT LIQUID BIOFUELS FOR TRANSPORTATION?

Liquid biofuels are used primarily for transportation as a substitute for gasoline or diesel. The most common transportation biofuels are ethanol, used in passenger cars and other light-duty vehicles, and biodiesel, used in many heavy-duty vehicles. Ethanol is also being proposed for use in aviation fuel. Liquid biofuels should not be part of a 100 percent WWS transition since they result in substantial air-pollution deaths and illnesses, climate damage, land consumption, and water use relative to WWS. This conclusion applies whether or not equipment is added to ethanol refineries to capture carbon dioxide from them. This section discusses these issues.

Ethanol is produced in a refinery, generally from corn, sugarcane, wheat, sugar beet, or molasses. The most common among these sources are corn and sugarcane, resulting in the production of **corn ethanol** and

sugarcane ethanol, respectively. Microorganisms and enzymes ferment sugars or starches in these crops to produce ethanol in an **ethanol refinery**.

Fermentation of cellulose in switchgrass, wood waste, wheat, stalks, corn stalks, or *Miscanthus*, also produces ethanol, called **cellulosic ethanol**. However, the process of producing cellulosic ethanol is more energy intensive than is fermentation of sugar and starches because the breakdown of cellulose by natural enzymes (as it occurs in the digestive tracts of cattle) is slow. Faster breakdown of cellulose requires genetic engineering of enzymes.

Ethanol as a transportation fuel may be used on its own, as is done frequently in Brazil, or blended with gasoline. Common blends of ethanol are **E10, E15, and E85**. E10 contains 10–10.49 percent ethanol mixed with gasoline; E15 contains 10.5–15 percent ethanol mixed with gasoline, and E85 contains 51–83 percent ethanol mixed with gasoline.[210] Pure ethanol (E100) in Brazil contains no gasoline. However, in the United States, E100 must contain at least 2 percent gasoline as a **denaturant**, which is a poisonous or foul-tasting chemical added to a fuel to deter people from drinking it. Thus, if 15 percent gasoline is blended with 85 percent E100, the resulting mixture (E85) contains 83.3 percent ethanol and 16.7 percent gasoline, which is why the upper limit of the ethanol content in E85 is 83 percent.

Gasoline vehicles can use either pure gasoline (E0) or E10 fuel, but higher ethanol blends can damage the vehicles. Instead, higher blends (E15 and E85) must be used in a flex-fuel vehicle (FFV), which can also run on E0 or E10. By far, most ethanol today in the United States is blended as E10. However, due to the planned phase-out of gasoline based on climate concerns, the increased development of FFVs, and tax subsidies promoting ethanol, the use of E85 is increasing rapidly. In Brazil, the use of E100 is already widespread.

A proposed alternative to ethanol for transportation fuel is **butanol**. It can be produced by fermenting the same crops used to produce ethanol but with a different bacterium, *Clostridium acetobutylicum*. Butanol contains more energy per unit volume of fuel than does ethanol. However, unburned butanol also reacts more quickly in the atmosphere than does unburned ethanol, speeding up ozone production relative to ethanol. Ozone is harmful to those who breathe it. On average, ethanol for transportation produces more ozone than does gasoline for transportation in most regions of the United States.[211,212,213]

Biodiesel is a liquid diesel-like fuel derived from vegetable oil or animal fat. Major edible vegetable oil sources of biodiesel include soybean, rapeseed, mustard, false flax, sunflower, palm, peanut, coconut, castor, corn,

8.9 WHY NOT LIQUID BIOFUELS FOR TRANSPORTATION?

cottonseed, and hemp oils. Inedible vegetable oil sources include jatropha, algae, and jojoba oils. Animal-fat sources include lard, tallow, yellow grease, fish oil, and chicken fat. Soybean oil accounts for about 90 percent of biodiesel production in the United States. Biodiesel derived from soybean oil is referred to as **soy biodiesel**.

Biodiesel is produced by the chemical reaction of a vegetable oil or animal fat (both lipids) with an alcohol. It is a standardized fuel designed to replace diesel in standard diesel engines. It can be used as pure biodiesel or blended with regular diesel. Blends range from 2 percent biodiesel and 98 percent diesel (**B2**) to 100 percent biodiesel (**B100**). Generally, only blends B20 and lower can be used in a diesel engine without engine modification. The use of vegetable oil or animal fat directly (without conversion to biodiesel) in diesel engines is also possible; however, it results in more incomplete combustion, thus more air-pollution byproducts, as well as a greater buildup of carbon residue in, and damage to, the engine, than does biodiesel.

Significant efforts have been made to produce **algae biodiesel**, which is biodiesel from algae grown from waste material, such as sewage. However, these efforts have been hampered by the fact that algae can grow quickly only when exposed to the sun. As such, algae cannot grow quickly when one layer is piled on top of the other. Instead, algae must be spread, in a single layer, over a large unshaded area. Each volume of oil produced from algae also requires about 100 times that volume of water. Both factors have limited the growth of the algae biodiesel industry.

Liquid biofuels (corn ethanol, cellulosic ethanol, butanol, and biodiesel) are not recommended as part of a 100 percent WWS energy infrastructure. This conclusion applies whether or not carbon dioxide capture equipment is added to refineries producing the biofuel. The reasons liquid biofuels are not recommended are (1) nearly all biofuels are burned, resulting in air pollution similar to that from fossil fuels; (2) with or without carbon capture, liquid biofuels do not reduce CO_2-equivalent emissions nearly to the extent that WWS-powered battery-electric or hydrogen-fuel-cell-electric vehicles do; (3) without carbon capture some liquid biofuels increase CO_2-equivalent emissions relative to fossil fuels; (4) with or without carbon capture many biofuels require rapacious amounts of land; (5) many biofuels require excessive quantities of water; and (6) many biofuels are derived from food sources, increasing food shortages, food prices, and starvation.[178,214,215] Because liquid biofuels cause greater climate, pollution, land, water, and food problems than WWS technologies, biofuels, with or without carbon capture, represent opportunity costs, as discussed next.

8 WHAT DOESN'T HELP

8.9.1 CORN-ETHANOL REFINED WITHOUT CARBON CAPTURE

The main issues with liquid biofuels, even without considering carbon capture, are illustrated by comparing the impacts of using corn ethanol to power internal-combustion-engine vehicles with the impacts of using wind or solar to power battery-electric vehicles.

First, replacing gasoline vehicles with battery-electric vehicles powered by wind reduces energy requirements and carbon emissions substantially. It reduces energy requirements by about 75 percent due to the efficiency of electricity over combustion and reduces CO_2-equivalent emissions from gasoline production and combustion by 99.3–99.8 percent.[178] The remaining emissions are due to the fossil energy required to build and decommission the wind turbines. In comparison, using corn- or cellulosic-E85 vehicles increases or hardly changes energy needs or CO_2-equivalent emissions relative to using gasoline vehicles. This is quantified as follows.

First, proponents of corn ethanol argue corn-ethanol vehicles should reduce CO_2-equivalent emissions versus gasoline vehicles for three reasons: (1) carbon dioxide removed from the air by photosynthesis during corn growth offsets carbon dioxide emitted from fermentation and combustion during ethanol production and vehicle use, respectively; (2) carbon dioxide emitted during ethanol production is modest; and (3) land-use change emissions associated with corn production are small.

However, corn-ethanol production requires a lot of fossil-fuel energy, thus CO_2-equivalent emissions, to grow, fertilize, water, and cultivate the crop; transport the crop; refine the crop into fuel; and transport the fuel to market. Fertilizer use during corn production also emits nitrous oxide. Ethanol is too corrosive to be used in pipes, so ethanol must be transported by train, truck, or barge, all of which emit diesel exhaust, which contains black and brown carbon particles, the second-leading cause of global warming. Finally, using corn for fuel instead of food drives up corn prices, increasing deforestation, including of rain forests, to produce agricultural land, the least carbon-intensive use of land, further driving up emissions.

The US renewable fuels standard (RFS) is a policy that motivated the use of ethanol fuel in the US, from 2008 to 2016. A detailed study of the RFS concluded that it increased corn prices by 30 percent, expanded US corn cultivation by 8.7 percent, increased fertilizer usage by 3–8 percent, increased water quality degradants by 3–5 percent, and caused enough domestic land-use-change emissions such that "the carbon intensity of corn ethanol produced under the RFS is no less than gasoline and

8.9 WHY NOT LIQUID BIOFUELS FOR TRANSPORTATION?

likely at least 24 percent higher." In comparison, using wind to power battery-electric vehicles reduces gasoline-vehicle CO_2-equivalent emissions by 99.3–99.8 percent.[178]

The air-pollution mortality associated with both corn- and cellulosic-ethanol vehicles also significantly exceeds that associated with WWS-powered battery-electric vehicles.[178,211] The reason is that battery-electric vehicles have no tailpipe emissions. Thus, their only emissions are from the upstream production of wind turbines, solar panels, and the vehicles themselves; tire wear; and some brake-pad wear. Because electric vehicles use regenerative breaking, their brake pads are hardly engaged, so brake-pad emissions are very low.

On the other hand, E85 vehicles have high air-pollution emissions from their tailpipes, producing and transporting the ethanol, tire wear, brake-pad wear, and manufacturing the vehicle. E85's tailpipe emissions cause health impacts that often exceed those of gasoline vehicles,[217,218,219] especially at low temperatures.[212,220]

Another problem with using ethanol as a fuel is water consumption. The average US irrigation rate of corn is 13.2 percent.[178] Irrigating the corn crop needed to power an entire US on-road flex-fuel-vehicle fleet powered by E85 would require about 10 percent of the entire US water supply.[178]

Finally, because of the substantial land required for corn or cellulosic ethanol, neither can provide enough energy for more than a few percent of a US vehicle fleet. In fact, the land required to grow corn for an E85 flex-fuel vehicle fleet is 80–100 times that needed for solar PV to power a battery-electric-vehicle fleet. The reason is as follows. Photosynthesis is only 1 percent efficient. Most PV cells, on the other hand, are 20–26 percent efficient. Thus, a PV farm needs only one-twentieth to one twenty-sixth of the land to produce the same energy as a corn crop needs. Further, BEVs convert 80 percent of the electricity from a plug to battery electricity and then motion. The rest is waste heat. FFVs running on E85 convert only about 20 percent of energy in the E85 to motion. So, driving a BEV requires one-quarter of the energy of driving an FFV. Combining the PV-to-photosynthesis efficiency with the BEV-to-FFV efficiency suggests that driving a BEV powered by PV requires one-eightieth to one one-hudredth of the land compared with driving a FFV powered by E85 from corn ethanol.[178]

A wind turbine requires less than one five-thousandth of the footprint on the ground (accounting for only the pole in the ground plus a cement base) of a PV farm to provide the same electricity. As such, wind-powered BEVs may need less than one-four-hundred-thousandth of the land footprint as

FFVs powered by corn-E85.[178] Wind turbines do require space between them to prevent interference of the wakes of one turbine with another. However, even the spacing area for wind turbines powering BEVs may be one-tenth to one-twentieth of the land needed to grow corn for E85 powering FFVs.[178] Because most wind spacing area is open space between turbines, crops can still grow and PV panels can operate between wind turbines.

In sum, liquid biofuels are not recommended in a 100 percent WWS world because of their high climate, health, water-supply, and land issues. This conclusion applies even when carbon capture is considered, as discussed next.

8.9.2 CORN-ETHANOL REFINED WITH CARBON CAPTURE

Many argue capturing carbon dioxide from ethanol refineries, then piping it to an underground sequestration site, should reduce CO_2-equivalent emissions relative to corn-ethanol without carbon capture for running flex-fuel vehicles. To the contrary, corn-ethanol with carbon-capture equipment and carbon dioxide pipelines appear to be an opportunity cost that may damage climate and air quality, occupy land, and saddle consumers with high fuel costs for decades.[68]

In theory, capturing carbon dioxide during the fermentation process of ethanol production may reduce ethanol-FFV lifecycle CO_2-equivalent emissions to slightly below those of gasoline.[221] However, not only are the resulting emissions still high, but the comparison with gasoline alone ignores the fact that BEVs emit far less than both FFV and gasoline vehicles, and ignores the impacts of ethanol-fueled FFVs on air pollution, land use, and water supply.[178,211]

> **Transition highlight**
> To bolster the argument for using ethanol-fueled FFVs as a climate solution, several companies proposed capturing carbon dioxide during ethanol-refinery fermentation, then building over 3,000 kilometers of pipes under hundreds of landowners' properties across five US states (Iowa, South Dakota, North Dakota, Nebraska, and Minnesota) to transfer and store the carbon dioxide underground.[222] A relevant question is, What is the opportunity cost of such an "ethanol plan" (capturing fermentation carbon dioxide from ethanol refineries, building a pipeline, storing the carbon dioxide underground, blending the ethanol to produce E85, then using the E85 in FFVs) versus investing the same funds to build wind turbines to run battery-electric vehicles ("wind plan")? The wind plan eliminates the need to produce ethanol for vehicles, emit

vehicle exhaust, emit pollution from ethanol refineries, or sequester carbon dioxide from ethanol refineries.

Results suggest that if the same investment allocated for the ethanol plan ($5.6 billion) were spent instead on the wind plan, drivers in the five states would save $66.9 to $111 billion over 30 years in fuel costs alone due to the price difference between E85 and residential electricity, and the far better mileage per unit of energy of a battery-electric Ford *F-150* truck (the example used in the study) versus a Ford *F-150* FFV. Even with an upfront BEV cost $21,700 higher than the FFV cost, the savings from the wind plan are still $39.5 to $65.6 billion over 30 years.[68]

Further, the wind plan may avoid 2.4–4 times the CO_2-equivalent emissions as the ethanol plan. This accounts for emissions associated with building wind farms, batteries for BEVs, carbon dioxide pipelines, ethanol refineries, and carbon-capture systems; and for growing, cultivating, and transporting corn. Finally, the wind plan may significantly reduce air pollution and land needs compared with the ethanol plan.

In sum, investing in wind turbines to provide electricity for BEVs is far better in terms of cost, carbon emissions, land use, and air pollution than making the same investment in capturing carbon dioxide from ethanol refineries, piping the carbon dioxide to an underground storage facility, and using the ethanol to produce E85 for FFVs.[68]

8.10 WHY NOT NUCLEAR ELECTRICITY?

In evaluating solutions to global warming, air pollution, and energy security, two important questions that arise are, (1) Should new nuclear electricity-producing plants be built to help solve these problems? and (2) Should existing, aged nuclear plants be kept open as long as possible to help solve the problems? This section discusses these issues after nuclear power is explained.

All nuclear electricity today is generated by nuclear fission. **Nuclear fission** is the process by which tiny neutrons bombard and split certain fissile heavy elements, such as uranium-235 or plutonium-239 in a **nuclear reactor**. The 235 and 239 refer to the isotope, that is, the number of protons plus neutrons in the nucleus of a uranium or plutonium atom, respectively. A **fissile** element is one that can be split during fission upon neutron bombardment and whose neutrons released during splitting can split other fissile atoms in a chain reaction. Fissile elements do not spontaneously release neutrons to produce a chain reaction. Instead, they require outside neutrons to bombard them, thereby initiating the chain reaction. In most nuclear reactors, such outside neutrons are obtained from the decay of californium-252 and plutonium-238.

8 WHAT DOESN'T HELP

Uranium-235 is the only fissile element found in nature. Plutonium-239 is also a fissile element, but it is produced artificially in a nuclear reactor. It is the product of uranium-238 and a free neutron. This produces uranium-239, which decays to plutonium-239.

In a nuclear reactor, a moving neutron may either pass through or be absorbed by uranium-235. Slow-moving neutrons have a higher probability than fast-moving neutrons of being absorbed. If a neutron is absorbed, the resulting uranium atom's total energy is spread among all its 236 protons and neutrons now in its nucleus. The nucleus is now unstable, and some uranium atoms fragment into two smaller elements, whereas the remaining atoms form uranium-236. A variety of element pairs arise from fragmentation. Two of the most common are krypton-92 and barium-141. The fragmentation, with this product pair, also produces gamma rays and three free neutrons. The new neutrons may then collide with other uranium-235 atoms or with plutonium-239 atoms, splitting them in a chain reaction. When the fragments and the gamma rays collide with water, the collision converts their kinetic energy and electromagnetic energy, respectively, to massive amounts of heat.

In a **boiling-water reactor nuclear power plant**, the heat boils water directly. The high-pressure steam turns a turbine connected to a generator to produce electricity. The steam is then recondensed to liquid water in a condenser, and the liquid water is returned back to the reactor core. In the condenser, heat from the steam (but not the radioactive water vapor itself) is transferred to a separate (in an enclosed pipe) stream of cooling water that originates from a lake, a river, or the coastal ocean. The heated cooling water is then returned to where it originated from, warming the outdoor water body, creating thermal pollution. Other thermal power plants, such as those running on coal, oil, or gas, similarly heat water bodies.

In a **pressurized-water reactor** plant, the air pressure in the reactor is increased substantially, up to 155 bar. For comparison, air pressure at the Earth's surface is 1 bar. Because the boiling point of water increases with increasing pressure, water in the reactor doesn't boil, even when the reactor temperature reaches 282 degrees Celsius (at sea level and 1 bar of atmospheric pressure, water usually boils at 100 degrees Celsius). The hot water in the reactor, which is radioactive, passes through a pipe and exchanges its heat with a different batch of water maintained at normal air pressure, causing the latter water to boil. The boiling water creates steam that is pushed through a steam turbine to generate electricity. The water batches are kept separate to ensure radioactive material in the high-pressure reactor does not pass through to the water vapor running through the steam turbine.

8.10 WHY NOT NUCLEAR ELECTRICITY?

Boiling water reactors and pressurized water reactors are both called **light-water reactors**, which are reactors that use normal water.

Uranium in a nuclear reactor is stored in small ceramic pellets within a metal fuel rod, often 3.7 meters long. A conventional light-water reactor goes through one rod in about six years, and the rod and remaining material in it become radioactive waste. Reactors that use rods once are referred to as **once-through** reactors. The radioactive waste in the fuel rod must be stored for hundreds of thousands of years. In a typical once-through reactor, about 4 percent of uranium is uranium-235 and 96 percent is uranium-238 (3 percent of which gets converted to plutonium in the reactor). About one-third of the energy in a once-through reactor comes from the production and decay of plutonium. About two-thirds of the plutonium decays to fission products and one-third is left as waste. Overall, fuel-rod waste contains about 5 percent fission products, 1 percent plutonium, 1 percent uranium-235, and 93 percent uranium-238.

Thus, a fuel rod that has gone through a fission reactor once still has about 94 percent of its uranium left over, including a higher percentage of uranium-235 than in natural uranium. Plutonium-239 can be extracted from a fuel rod after two to three years to provide reactor-grade plutonium. Alternatively, all remaining uranium-238, uranium-235, and plutonium can be extracted and reprocessed for use in a **breeder reactor**, extending the life of a given mass of uranium and reducing waste significantly. However, the reprocessing increases the cost of energy. It also increases the production of plutonium-239 by the collision of uranium-238 with fast-moving neutrons. Breeder reactors can thus be optimized to produce plutonium-239 for use in nuclear weapons.[223] As such, they are a concern with respect to the proliferation of nuclear weapons. Only two reactors worldwide today are breeder reactors.

Nuclear fission became a source of electricity starting in the 1950s. The first nuclear reactor to produce electricity was an experimental reactor in Arco, Idaho. On December 20, 1951, it powered four light bulbs. On June 26, 1954, a 5-megawatt nuclear reactor was connected to the electric-power grid for industrial use in Obninsk, Russia. Subsequently, on August 27, 1956, a 50-megawatt reactor was connected to the grid for commercial use in Windscale, UK.

As of 2024, about 440 active nuclear reactors provided electricity in 33 countries for a combined nameplate capacity of 377 gigawatts. In 2023, their total energy output was 2,552 GWh of electricity. This was less than the world nuclear output of 2,616 gigawatt-hours per year in 2004.[224] Thus, world nuclear output has not grown in 20 years.

8 WHAT DOESN'T HELP

In 2022, mines worldwide produced about 49,400 tonnes of uranium. Most uranium was mined in Kazakhstan (43 percent), Canada (15 percent), and Namibia (11 percent).[225] Uranium reserves (aside from hard-to-extract uranium in seawater), as of 2019, were about 8.1 million tonnes.[226] This suggests that about 159 years of uranium are available for the number of once-through fuel cycle reactors operating in 2024. As such, even if the issues discussed below were not issues, uranium is a limited resource, and growing nuclear power will deplete uranium reserves faster.

An alternative fuel to uranium in nuclear reactors is thorium. **Thorium**, like uranium, can be used to produce nuclear fuel in a breeder reactor. The advantage of thorium is that it produces less long-lasting radioactive waste than does uranium. Its products are also more difficult to convert into nuclear-weapons material. However, thorium still produces uranium-233, which was used in one nuclear bomb core produced during the **Operation Teapot** bomb tests in 1955. Thus, thorium is not free of nuclear-weapons-proliferation risk. In addition, thorium reactors require the same lengthy time lag between planning and operation as do uranium reactors and likely longer because hardly any contractors or scientists have experience building or running thorium reactors. Thus, thorium reactors will produce greater emissions from the background electric grid than WWS technologies, which have a shorter time lag. Finally, lifecycle emissions of carbon from a thorium reactor are similar to those from a uranium reactor.

A proposed alternative to the large once-through reactor and the breeder reactor is the **small modular reactor (SMR)**. Small modular reactors are nuclear fission reactors that are on the order of one-third of the size of a traditional reactor. They have some parts that could be prefabricated in a factory, which could help to reduce construction time, costs, and mistakes during construction. However, as of early 2025, no small modular reactor has been commercialized worldwide, so it is difficult to determine whether any design will take advantage of prefabrication.

Many types of small modular reactors have been proposed, including miniature versions of current reactors. One type of new design is a **fast reactor**, in which the fuel is reformulated to allow fast-moving neutrons, rather than slow-moving neutrons, to split an atom. One way to do this is to increase the quantity of plutonium-239, which absorbs more fast-moving neutrons than does uranium-235. This is done by surrounding the core with uranium-238, which absorbs a fast-moving neutron to become uranium-239, which then decays to plutonium-239. By this mechanism, though, fast reactors become breeder reactors, increasing the weapons-proliferation risk.

8.10 WHY NOT NUCLEAR ELECTRICITY?

In sum, whereas slow reactors produce significant radioactive waste, fast reactors would produce less waste but would increase the risk of nuclear-weapons proliferation by producing more plutonium-239. Because all small modular reactors are indeed small and many are proposed to be transportable, many countries that don't currently have nuclear-energy facilities will want them, increasing weapons-proliferation risk. Most, but not all, small modular reactors also pose a meltdown risk. In addition, they require uranium, which must be mined. Small reactors have the same uranium-mining resource limitation, underground-mining lung-cancer risk, and land-despoilment risk as large reactors.

Many startup companies around the world are designing small modular reactors. However, no small modular reactor being designed in 2025 is expected to have a prototype available until past the year 2030 or a commercial product available until years after that. This is relevant, since the world needs to eliminate 80 percent of climate-affecting emissions by 2030 and the rest by 2035–2050 to avoid the harshest consequences of global warming. In addition, the world needs to eliminate 100 percent of its air-pollution emissions today to avoid the over seven million air-pollution deaths that occur yearly. As such, small modular reactors will not be able to help address global warning or air pollution in a rapid or meaningful way. Instead, money spent on them will prevent faster and less expensive solutions from being implemented, exacerbating the climate and air-pollution problems the world faces.

Historically, small nuclear reactors were planned and developed before large ones were conceived. However, nuclear plant developers abandoned small reactors in favor of large reactors because building one large reactor was much less expensive than building three small ones. Even today, the cost per unit of energy of a new small modular reactor is estimated to be higher per unit of energy than that of a large reactor. Further, the cost per unit of energy of a large reactor, in 2024, is five times that of new onshore wind or utility PV.[191] As such, it is expected that electricity from small modular reactors will be much more expensive than electricity from new WWS electricity generators.

Transition highlight
In late 2024, several companies proposed using small modular reactors in isolated microgrids to provide baseload electricity for computer data centers, which consume a lot of power at a relatively constant rate. However, such reactors are not expected to be commercial until the 2030s, and at an uncertain cost, planning-to-operation time, and security risk. Fortunately, WWS microgrids can power data centers much sooner and less expensively in at least two ways. The first is with a hybrid wind/PV farm

8 WHAT DOESN'T HELP

combined with a hybrid battery/green-hydrogen storage facility. Solar and wind are complementary in nature, so combining the two in one farm smoothens overall WWS supply. The storage fills in the gaps. When PV plus wind electricity output exceeds data center demand, the excess electricity is stored in the batteries and used to produce hydrogen by electrolysis. When PV plus wind output falls below demand, battery electricity is discharged, or the stored hydrogen is run through a fuel cell to reproduce electricity to supply the demand. Whereas batteries are more cost-effective to fill in WWS supply of minutes to days, the hydrogen system is more cost-effective to fill in gaps of days to weeks. The second way to power a constant-demand data center is with an enhanced geothermal system, which provides baseload power (Section 2.4).

The sun and all the stars in the universe are powered by a different type of nuclear process, nuclear fusion. **Nuclear fusion** involves the fusing together of light atomic nuclei (protium, deuterium, or tritium) into heavier elements. In theory, nuclear fusion could supply all power on Earth indefinitely. It could also do so without long-lived radioactive waste because its products are isotopes of helium, which are not harmful. Nuclear fusion has been explored for decades. However, its commercialization has always been about 30–100 years away. Recently, technical advances have been made in the development of nuclear fusion. But even the International Atomic Energy Agency acknowledges that a demonstration fusion reactor won't be available until at least 2040 and commercial electricity generation from fusion may or may not occur by the second half of the twenty-first century.[227] Given that we need to eliminate 80 percent of world emissions by 2030, that proposed date is too far away for fusion to be useful. As such, nuclear fusion will unlikely address global warning, air pollution, or energy insecurity at all.

8.10.1 RISKS AFFECTING NUCLEAR'S ABILITY TO ADDRESS GLOBAL WARMING AND AIR POLLUTION

The risks associated with nuclear electricity can be broken down into two categories: (1) risks affecting nuclear's ability to reduce global warming and air pollution and (2) risks affecting its ability to provide environmental security.

Risks under category one include the following: delays between planning and operation of a nuclear plant, emissions contributing to global warming and air pollution, and cost. Risks under category two include weapons-proliferation risk, reactor-meltdown risk, radioactive waste risk, underground-mining lung-cancer risk, and land-despoilment risk. These risks are discussed herein.

8.10 WHY NOT NUCLEAR ELECTRICITY?

8.10.1.1 DELAYS BETWEEN PLANNING AND OPERATION AND DUE TO REFURBISHING REACTORS

The longer the time lag between the planning and operation of an energy facility, the more the emissions of air pollutants and climate-damaging chemicals from the background electric grid. Similarly, the longer the time required to refurbish a nuclear plant for continued use at the end of its life, the greater the emissions from the background grid when the nuclear plant is down.

The time lag between planning and operation of a nuclear plant includes the times to secure a construction site, a construction permit, an operating permit, financing, and insurance; the time between construction permit approval and issue; and the time to build the plant, which includes the time between the end of construction and grid connection.

In March 2007, the United States Nuclear Regulatory Commission approved its first site permit in 30 years. This process took 3.5 years. The time to review and approve a construction permit is another 2 years and the time between the construction permit approval and issue is about 6 months. Thus, the minimum time for preconstruction approvals (and financing) in the United States is 6 years. An estimated maximum time is 10 years. The time to construct a nuclear reactor depends significantly on regulatory requirements and costs. Although nuclear-reactor construction times worldwide are often shorter than the 9-year median construction times in the United States since 1970,[228] they averaged 7.4 years worldwide in 2015.[229] As such, a reasonable estimated range for construction time prior to 2016 was 4–9 years, bringing the overall time between planning and operation of historical nuclear power plants worldwide to 10–19 years. However, since 2016, the range has expanded to 12–23 years worldwide and 17–23 years in North America and Europe. Below are examples.

The **Olkiluoto 3** reactor in Finland was proposed to the Finnish cabinet in December 2000 as an addition to an existing nuclear plant. Its construction started in 2005, and it first generated electricity in 2022, but it only began commercial operation on May 1, 2023, giving it a construction time of 18 years and a **planning-to-operation** time of 23 years. The total capital cost in 2024 was about $7.7 per watt-nameplate-capacity, about 3.7 times the original projected cost.

The **Hinkley Point C** nuclear power station in the United Kingdom was planned, starting in 2008. Construction began only on December 11, 2018. It has an estimated completion year of 2029 to 2031, giving it an estimate construction time of 11–13 years and planning-to-operation time of 21–23 years. The projected capital cost in 2024 was about $19 per watt-nameplate-capacity, or about 6 times the original projected cost.

The **Vogtle 3 and 4** nuclear reactors in the US state of Georgia were first proposed in August 2006 to be added to an existing nuclear power station site. Construction started for both in 2013. Unit 3 began operating commercially on July 21, 2023, and unit 4 began operation on April 29, 2024, giving them construction times of 10 and 11 years, respectively, and planning-to-operation times of 17 and 18 years, respectively. The final capital cost in 2024 was about $15.7 per watt-nameplate-capacity, about 2.5 times the original estimated cost.

The **Flamanville**, France, Unit 3 reactor was planned on an existing nuclear power station site starting in 2004. A contract was awarded in 2005. Construction started in 2007. The reactor began operating commercially in December 2024, for a construction time of 17 years and planning-to-operation time of 20 years. The final capital cost in 2024 was about $15.8 per watt-nameplate-capacity, about 5.8 times the original estimated cost.

The **Haiyang 1 and 2** reactors in China were planned in 2005. Construction started in 2009 and 2010, respectively. Haiyang 1 began commercial operation on October 22, 2018. Haiyang 2 began operation on January 9, 2019, giving them construction times of 9 years and planning-to-operation times of 13 and 14 years, respectively.

The **Taishan 1 and 2** reactors in China were planned, starting in 2006. Construction began in 2008. Taishan 1 began commercial operation on December 13, 2018. Taishan 2 began operation on September 9, 2019, giving them construction times of 10 and 11 years and planning-to-operation times of 12 and 13 years, respectively.

The **Shidao Bay** nuclear power plant in China includes the development of a 200-megawatt high-temperature gas-cooled (as opposed to water-cooled) reactor. Planning started in 2005. Construction began in December 2012. Grid connection occurred in early 2022. Thus, the construction time was 10 years and the planning-to-operation time was 17 years.

The **Barakah 1–4** nuclear plant in the United Arab Emirates contains four reactors, all planned, starting in 2009. Construction of the four reactors began during July 2012, April 2013, September 2014, and July 2015, respectively. Commercial operation began during April 2021, March 2022, February 2023, and in late 2024, respectively, giving construction times for all reactors of 9 years and planning-to-operation times of 12, 13, 14, and 15 years, respectively.

8.10 WHY NOT NUCLEAR ELECTRICITY?

Planning of and procurement for four reactors in **Ringhals**, Sweden, started in 1965. One took 10 years, the second took 11 years, the third took 16 years, and the fourth took 18 years to complete.

Some contend that France's 1974 Messmer Plan resulted in the building of its 58 reactors in 15 years. The **Messmer Plan** was a proposal, enacted without public or parliamentary debate, by the Prime Minister of France, Pierre Messmer, to build 80 nuclear reactors by 1985 and 170 by 2000. In fact, the plan had been in the works for years prior and was only proposed publicly following the international oil crisis of 1973.[230] For example, the Fessenheim nuclear reactor obtained its construction permit in 1967 and was planned before that. In addition, 10 of the reactors were completed only between 1991 and 2000. As such, the whole planning-to-operation time for the 58 reactors was at least 33 years, not 15. That of any individual reactor was 10 to 19 years.

In sum, planning-to-operation times for all reactors in history have been in the range of 10–23 years. For reactors operating prior to 2016, the range was 10–19 years. For reactors after 2016, the range is 12–23 years, with those in North America and Europe, between 17 and 23 years.

Planning-to-operation delays also occur during reactor refurbishment. Nuclear reactors have a lifetime on the order of 40 years. To run longer, they need to be refurbished. The time to refurbish can be three to four years. Refurbishment of the Darlington 2, Ontario nuclear reactor, for example, began in October 2016 and was completed in June 2020, taking just less than four years.

The background grid, which consists primarily of fossil fuels in most places worldwide, emits pollution when a nuclear plant is being constructed or down for refurbishing. The total opportunity-cost (background-grid) emissions due to nuclear not operating during either period average 64–102 grams of CO_2-equivalent per kilowatt-hour of electricity generated (Figure 8.1). These emissions are higher than the lifecycle emissions of nuclear.

Transition highlight
China's decision, prior to 2012, to invest in nuclear, which has a long planning-to-operation time, instead of wind or solar, may have resulted in China's carbon dioxide emissions rising 1.3 percent from 2016 to 2017 rather than declining by an estimated 3 percent during that period.[10] It may also have resulted in 82,000 more people dying from air pollution in China in 2016. The reason is that the nuclear plants planned prior to 2012 were not close to operating in 2016. Had China invested that same money in wind and solar, wind and solar would already have been operating by 2016, eliminating substantial CO_2 and air-pollution emissions from coal.

8.10.1.2 AIR POLLUTION AND GLOBAL WARMING RELEVANT EMISSIONS FROM NUCLEAR

Nuclear power contributes to global warming and air pollution in several ways: through (1) emissions of health- and climate-damaging pollutants from the background grid due to nuclear power's long planning-to-operation and refurbishment times; (2) emissions of health- and climate-damaging air pollutants during construction, operation, and decommissioning of a nuclear plant; (3) heat and water-vapor emissions during the operation of a nuclear plant; (4) carbon dioxide emissions due to the covering of soil or clearing of vegetation during the construction of a nuclear plant, uranium mine, and nuclear waste site; and (5) the risk of emissions arising from nuclear-weapons proliferation.

Each of these categories represents an actual emission or emission risk, yet all of these emissions, except for lifecycle emissions, are incorrectly ignored in virtually all studies of nuclear power impacts on climate. Studies that ignore these real emissions distort the impacts of nuclear power on climate and air-pollution health.

The estimated range of nuclear lifecycle emissions (9–70 grams of CO_2-equivalent per kilowatt-hour of electricity) in Figure 8.1 is well within the range (4–110 grams) from studies cited by the Intergovernmental Panel on Climate Change.[231] On top of those emissions are opportunity-cost emissions (64–102 grams); heat and water-vapor emissions (4.4 grams); emissions due to covering and clearing soil (0.17–0.28 grams); and emissions due to the risk of nuclear-weapons use arising from the spread of nuclear energy (0–1.4 grams). The total is 78–178 grams of CO_2-equivalent per kilowatt-hour of electricity. These emissions are 9–37 times the estimated emissions from onshore wind (Figure 8.1).

Although the emissions from nuclear are lower than those from coal or fossil gas with carbon capture, nuclear's high CO_2-equivalent emissions, coupled with its long planning-to-operation time, render nuclear an opportunity cost relative to the faster-to-operate and lower-emitting WWS technologies.

8.10.1.3 NUCLEAR COSTS

The third risk of nuclear electricity that affects its ability to reduce global warming and air pollution is its high cost. The cost of running existing nuclear reactors has also increased so much that many existing reactors are shutting down early. Owners of other reactors have requested large subsidies to stay open. This section discusses nuclear costs.

8.10 WHY NOT NUCLEAR ELECTRICITY?

The 2024 mean cost of electricity for a new nuclear plant in the United States is about 18.2 cents per kilowatt-hour.[191] This compares with about 5 and 6 cents per kilowatt-hour for onshore wind and utility-scale solar PV, respectively. Thus, new nuclear electricity is 3–4 times the cost per unit electricity of new wind and solar. A good portion of the high cost of nuclear is due to its long planning-to-operation time.

This cost of nuclear electricity does not account for the future cost of storing radioactive waste. For example, in the US alone, about $500 million is spent yearly to safeguard nuclear waste from about 100 civilian nuclear reactors.[232] Such waste must be stored for hundreds of thousands of years. The cost also does not account for the damage due to nuclear-reactor meltdowns. For example, the estimated cost to clean the damage from three Fukushima Dai-ichi nuclear-reactor core meltdowns in 2011 was $460–$640 billion.[233] This is equivalent to about 9–13 percent of the current-day capital cost of every nuclear reactor that exists worldwide.

The spiraling cost of new nuclear reactors in recent years has resulted in the canceling of several reactors under construction. For example, two reactors in South Carolina were canceled during July 2017. The high cost of nuclear has also resulted in threats by reactor owners to shut plants unless they receive a subsidy. The risk of shutting a functioning nuclear plant is that its electricity will be replaced by electricity from a fossil-fuel plant. However, the problem with subsidizing nuclear is that the funds could otherwise be used to replace the nuclear electricity with lower-cost and lower-emitting WWS electricity. Because the nuclear plant usually needs to be replaced within a decade of a subsidy request in any case, incurring the cost of new WWS immediately will almost always cost less than paying nuclear a subsidy each year for ten years plus incurring the cost of new WWS in ten years.

Transition highlight
In 2016, three upstate New York nuclear plants requested and received subsidies to stay open until 2028 using the argument that the plants were needed to keep emissions low. However, subsidizing such plants may have increased carbon dioxide emissions and costs versus replacing the nuclear quickly with wind or solar.[234]

In sum, before accounting for waste-storage or meltdown-damage costs, a new nuclear reactor costs 3–4 times that of a new onshore wind farm, takes 7–21 years longer between planning and operation than a wind farm, and produces 9–37 times as much carbon dioxide and air-pollution emissions per unit of electricity generated as a wind farm. Thus, funds spent on new nuclear means much less electricity, a much longer wait, and much more emissions than the same funds spent on WWS. The Intergovernmental Panel on Climate Change similarly concludes that the economic, social, and technical feasibility of nuclear power have not improved over time, "The political, economic, social and technical feasibility of solar energy, wind energy and electricity storage technologies has improved dramatically over the past few years, while that of nuclear energy and Carbon Dioxide Capture and Storage (CCS) in the electricity sector has not shown similar improvements."[235]

8.10.2 RISKS AFFECTING NUCLEAR'S ABILITY TO ADDRESS ENVIRONMENTAL SECURITY

The second category of risk related to nuclear electricity is the risk of a nuclear reactor not being able to provide environmental security. One reason is the risk of weapons proliferation. Others are the risks of meltdown, radioactive waste leakage, and uranium-mining lung cancer and land degradation. WWS technologies do not create such risks.

8.10.2.1 WEAPONS-PROLIFERATION RISK

The first risk of nuclear related to environmental security is that of weapons proliferation. The growth of nuclear energy has historically increased the ability of nations to harvest plutonium or enrich uranium to manufacture nuclear weapons:

> Peaceful nuclear cooperation and nuclear weapons are related in two key respects. First, all technology and materials related to a nuclear-weapons program have legitimate civilian applications. For example, uranium enrichment and plutonium reprocessing facilities are dual-use in nature because they can be used to produce fuel for power reactors or fissile material for nuclear weapons. Second, civilian nuclear cooperation builds-up a knowledge-base in nuclear matters.[236]

The Intergovernmental Panel on Climate Change recognizes this fact. They conclude, with "robust evidence and high agreement" that nuclear-weapons proliferation concern is a risk to the increasing development of nuclear energy: "Barriers to and risks associated with an increasing use of nuclear energy include operational risks and the associated safety concerns, uranium mining risks, financial and regulatory risks, unresolved waste management issues, nuclear-weapons proliferation concerns, and adverse public opinion."[231]

8.10 WHY NOT NUCLEAR ELECTRICITY?

The building of a nuclear reactor in a country with no reactor increases the risk of the country developing nuclear weapons. It allows the country to import uranium for use in the reactor. If the country so chooses, it can secretly enrich the uranium to create weapons-grade uranium and harvest plutonium from uranium fuel rods used in the reactor for nuclear weapons. This does not mean any or every country will do this, but historically some have.

Nuclear weapons can be produced from nuclear energy infrastructure as follows. Uranium ore is mined in an open pit or underground and contains 0.1–1 percent uranium by mass. The ore is milled to concentrate the uranium in the form of a yellow power called **yellowcake**, which contains about 80 percent uranium oxide. Uranium is then processed further into uranium dioxide or uranium hexafluoride for use in nuclear reactors. However, before the uranium can be used in a reactor, it must be enriched.

Of all uranium on Earth, 99.2745 percent is uranium-238, 0.72 percent is uranium-235, and 0.0055 percent is uranium-234. Uranium-238 has a half-life of 4.5 billion years. Most commercial light-water nuclear reactors use uranium consisting of 3–5 percent uranium-235. As such, the concentration of uranium-235 in a fuel rod must be increased to four to seven times its ore concentration. This is done by enrichment. **Uranium enrichment** is the process of separating the isotopes of uranium to increase the percentage of uranium-235 in a batch. Enriched uranium is useful for both nuclear energy and nuclear weapons.

Enrichment is done either by gas diffusion, centrifugal diffusion, or mass separation by magnetic field. Only gas diffusion and centrifugal diffusion are commercial processes, and most enrichment today is by **centrifugal diffusion** because it consumes only 2–2.5 percent of the energy of gas diffusion. Nevertheless, centrifugal diffusion still requires many centrifuges running for long periods, thus using lots of electricity. Centrifugal diffusion works by spinning a cylinder containing uranium. The heavier uranium-238 atoms collect toward the outside edge of the cylinder and the lighter uranium-235 atoms collect toward the inside.

Uranium with less than 20 percent uranium-235 is called **low-enriched uranium**. **Highly-enriched uranium** contains 20–90 percent uranium-235. A nuclear weapon can be made with highly enriched uranium. However, nuclear weapons increase their destructiveness with even more enrichment. Thus, 90 percent or more uranium-235 is considered **weapons-grade uranium** and is generally used together with enriched plutonium in a nuclear bomb. An estimated 9,000 centrifuges can produce enough weapons-grade uranium-235 for one nuclear weapon from natural uranium in about seven months. With 5,000 centrifuges, the process takes about one year.[237] Because uranium in a fuel rod used for nuclear energy has only 3–5 percent uranium-235 and even less once it goes through a nuclear reactor, spent fuel rods are not considered a useful source of weapons-grade uranium.

Plutonium is also used in nuclear weapons. Ten kilograms of plutonium-239 were used in the bomb dropped on Nagasaki. Plutonium can be obtained from a once-through nuclear reactor running on a uranium fuel rod. When uranium-235 decays and releases neutrons in a nuclear reactor, one of the neutrons can bind with a uranium-238 atom to produce uranium-239, which decays to plutonium-239. Plutonium that contains 93 percent or more plutonium-239 is considered weapons-grade plutonium. Plutonium with less than 80 percent plutonium-239 is reactor grade. Because plutonium can be used to make a bomb and is easier to obtain than is the process of enriching uranium (since plutonium can be harvested from a fuel rod running once through a nuclear reactor), plutonium is considered the element of even greater concern than uranium with respect to nuclear-weapons proliferation.

A large-scale worldwide increase in nuclear reactors for energy would exacerbate the risk of nuclear-weapons proliferation. In fact, producing material for a weapon requires merely operating a civilian nuclear reactor together with a sophisticated plutonium-separation facility. The historic link between nuclear reactors for energy and nuclear weapons is evidenced by the development or attempted development of weapons capabilities secretly under the guise of peaceful civilian nuclear energy or nuclear energy research programs in Pakistan, India, North Korea, Iraq (prior to 1981), Syria (prior to 2007), and Iran, among other countries.

If the world's all-purpose energy were converted to electricity and electrolytic hydrogen by 2050, the 9 trillion watts (TW) of resulting annually averaged end-use electricity demand would require about 12,500 850-megawatt nuclear reactors (28 times the number of active reactors today), or 1.4 installed every day for 25 years. Not only is this installation timeline impossible given the long planning-to-operation times of nuclear, but it would also result in all known reserves of uranium worldwide for once-through reactors running out in about three years. As such, there is no possibility the world will run solely on once-through nuclear energy by 2050.

Even if only 6.4 percent of the world's energy came from nuclear, the number of active nuclear reactors worldwide would need to nearly double, to around 800. Many more countries would possess reactors, increasing the

8.10 WHY NOT NUCLEAR ELECTRICITY?

risk that some countries would use the facilities to mask the development of nuclear weapons, as has occurred historically.

If a country were to develop a weapon as a result of its acquisition of one or more nuclear reactors for energy, the risk that it would use the weapons is not zero. Here, the emissions associated with a limited nuclear exchange are estimated.

The explosion of a hundred 15,000-tonne nuclear bombs (a total of 1.5 million tonnes, or 0.1 percent of the yield of a full-scale nuclear war) during a limited nuclear exchange in a megacity could kill 2.6–16.7 million people from the explosion and burn 63–313 million tonnes of city infrastructure, adding one to five million tonnes of warming and cooling aerosol particles to the atmosphere, including much of it to the stratosphere.[178] The particle emissions would cause significant short- and medium-term regional temperature changes. The carbon dioxide emissions, estimated at 92–690 million tonnes, would enhance long-term global warming. The warming impact of one such nuclear exchange over 100 years is equivalent to 1.4 grams of CO_2-equivalent emissions per kilowatt-hour of electricity produced by nuclear during that period. It arises from nuclear-weapons development facilitated by the spread of nuclear electricity. That is the high-end risk of carbon dioxide emissions from nuclear-weapons proliferation due to nuclear electricity in Figure 8.1. The low-end estimate is zero emissions because no exchange occurs.

8.10.2.2 MELTDOWN RISK

The second risk of nuclear power related to environmental security is reactor core-meltdown risk. The Intergovernmental Panel on Climate Change points to operational risks (meltdown) as a barrier and risk associated with nuclear power. In fact, about 1.5 percent of all nuclear reactors operating in history have had a partial or significant core meltdown. Meltdowns have been either catastrophic (Chernobyl, Russia, in 1986; three reactors at Fukushima Dai-ichi, Japan, in 2011) or damaging (Three-Mile Island, Pennsylvania, in 1979; Saint-Laurent, France, in 1980). The nuclear industry has claimed new reactor designs are safer. However, new designs are generally untested, and there is no guarantee that a reactor will be designed, built, and operated correctly or that a natural disaster or act of terrorism, such as an airplane flown into a reactor, will not cause the reactor to fail, resulting in a major disaster.

On March 11, 2011, an earthquake measuring 9.0 on the Richter scale and a subsequent tsunami knocked out backup power to a cooling system, causing six nuclear reactors at the **Fukushima 1 Dai-ichi plant** in northeastern Japan to shut down. Three reactors experienced a significant meltdown of nuclear fuel rods and multiple explosions of hydrogen gas that formed during efforts to cool the rods with seawater. Uranium fuel rods in a fourth reactor also lost their cooling. As a result, cesium-137, iodine-131, and other radioactive particles and gases were released into the air. Locally, tens of thousands of people were exposed to the radiation, and 170,000–200,000 people were evacuated from their homes; 1,600–3,700 people perished during the evacuation alone.[233,238] At least one nuclear plant worker died from lung cancer from direct radiation exposure.[239]

The radiation release created a dead zone around the reactors that may not be safe to inhabit for decades to centuries. The radiation also poisoned the water and food supplies in and around Tokyo. The radiation plume from the plant spread worldwide within a week.[240] Although radioactivity levels in Japan within 100 kilometers of the plant were extremely high, those in the rest of Japan and eastern China were lower, and those in North America and Europe were lower still. It is estimated that 130 (15–1,100) radiation-related deaths and 180 (24–1,800) radiation-related illnesses will occur worldwide, primarily in eastern Asia, during the decades after the meltdown.[240] The cost of the cleanup of the Fukushima reactors and the surrounding area is estimated at $460–$640 billion.[233]

The 1.5 percent risk of a nuclear-reactor meltdown is a high risk. Catastrophic risks with all WWS technologies, aside from the risk of a large hydropower dam collapsing, are zero. WWS roadmaps do not call for an increase in the number of large hydropower dams worldwide, only the more effective use of existing ones.

8.10.2.3 RADIOACTIVE WASTE RISKS

Another risk associated with nuclear electricity is the risk of human and animal exposure to radioactivity from fuel rods consumed by once-through reactors. Used fuel rods are considered **radioactive waste**. Currently, most used fuel rods are stored at the reactor site. This has given rise to hundreds of radioactive waste sites in many countries that must be maintained for hundreds of thousands of years, far beyond the lifetime of any nuclear power plant. The United States houses about one-quarter of all nuclear reactors worldwide. Plans to store the waste of all US reactors inside of Yucca Mountain, Nevada, never passed into law, so waste will continue to accumulate at reactor sites. The more that waste accumulates, the greater the risk that a leak of radioactive waste in water or the air will damage water supply, crops, animals, and humans.

8.10.2.4 URANIUM MINING HEALTH RISKS AND LAND IMPACTS

The use of nuclear electricity increases the risk that underground miners may contract lung cancer and that open-pit uranium mines degrade land. Such risks continue so long as nuclear reactors operate because the reactors need uranium to produce electricity. WWS technologies, on the other hand, do not require the continuous mining of any material, only one-time mining to produce the WWS equipment.

In 2022, 14 countries mined uranium. Of these, Kazakhstan, Canada, Namibia, Australia, Uzbekistan, Russia, and Niger produced the most uranium.[225] Mines can be open pit or underground. Open-pit mines cause the most land degradation. Underground mines cause the greatest lung-cancer risk.

Underground uranium mining causes lung cancer in large numbers because uranium mines contain natural radon gas, some of whose decay products are carcinogenic. Several studies have found a link between high radon levels and cancer.[241,242] One year-2000 study of 4,000 underground uranium miners between 1950 and 2000 by the US Centers for Disease Control and Prevention found that about 10 percent died of lung cancer, a rate six times that expected based on smoking rates alone.[10] Another 1.5 percent died of mining-related lung diseases, supporting the hypothesis that uranium mining is unhealthy. In fact, the combination of radon and cigarette smoking increases lung-cancer risks above the normal risk associated with smoking.[243]

WWS energy does not have this risk because (a) it does not require the continuous mining of any material, only one-time mining to produce energy generators and storage, and (b) the mining for materials related to WWS does not carry the same lung-cancer risk as does uranium mining.

8.11 WHY NOT GEOENGINEERING?

Geoengineering is the large-scale alteration of the natural properties of Earth or its atmosphere in an attempt to reduce global near-surface temperatures. The two main categories of geoengineering that have been proposed are techniques to remove carbon dioxide from the air (**carbon-capture techniques**) and techniques to increase the reflectivity of the Earth or its atmosphere in order to decrease sunlight hitting the Earth's surface (**solar radiation management** techniques).

8 WHAT DOESN'T HELP

Carbon-capture and synthetic direct air carbon-capture techniques have already been discussed. These are geoengineering techniques because they are intended to reduce the amount of carbon dioxide in the air to modulate the Earth's near-surface temperatures. They are not recommended. Natural direct air carbon capture is also a geoengineering technique because it involves humans actively increasing forestation. This technique is recommended.

The main solar radiation management techniques that have been proposed include (1) injecting reflective aerosol particles into the stratosphere to reflect sunlight directly, (2) injecting fine sea-spray particles into the air just above the ocean surface to increase the number, and decrease the average size of cloud drops, thereby increasing the overall cross-sectional area of cloud drops to increase their reflectivity, and (3) installing white roofs or roads.

The first problem with these techniques is that none reduces fossil-fuel or bioenergy emissions of gases or particles that cause global warming and over seven million air-pollution-related deaths per year. To the contrary, with geoengineering, the public and policymakers become complacent, no longer feeling an urgency to reduce global temperatures or emissions. As such, pollutant gases and particles continue to cause damage and, in fact, increase.

Second, geoengineering may temporarily mask some warming regionally. However, because long-lived greenhouse gases continue to accumulate with geoengineering, even more investment in geoengineering is needed to keep up with the increase in emissions. Any interruption or stoppage of the geoengineering results in an immediate worsening of the climate problem because of the accumulation of even more greenhouse gases during the period of geoengineering.

Third, since geoengineering does nothing to stop air pollution, death and illness from its use will continue to occur without abatement compared with no geoengineering. Such health impacts worsen since complacency allows more fossil fuels and bioenergy fuels to be burned. Health problems may also worsen due to the particles injected into the atmosphere to assist with the geoengineering. Such particles, when breathed in, are harmful for health.

Fourth, since geoengineering does not reduce fossil-fuel or nuclear use, it does nothing to help reduce energy insecurity associated with those energy sources.

Fifth, if the money spent on geoengineering were spent instead on WWS, not only would WWS eliminate greenhouse gas emissions (thus reduce temperatures, as geoengineering does), but WWS would also eliminate

8.11 WHY NOT GEOENGINEERING?

air-pollution emissions, death and illness resulting from the air-pollution emissions, mining of fossil fuels and uranium, and energy insecurity. As such, geoengineering is an opportunity cost compared with WWS.

A sixth problem with geoengineering is its unintended consequences. For example, reducing sunlight reaching the ground reduces photosynthesis, thus crop yields, which can cause starvation in some parts of the world. Reducing sunlight also reduces ultraviolet radiation, and many people in high latitudes rely on low levels of ultraviolet radiation to catalyze vitamin D production to avoid rickets and bow-leggedness. Injecting aerosol particles into the stratosphere also catalyzes stratospheric-ozone loss in the presence of halogens currently in the stratosphere. Injecting particles into the stratosphere or marine air changes weather patterns. Injecting particles into marine air increases the concentrations of particles entering populated coastal cities, increasing air-pollution death and illnesses. Particles injected into the stratosphere similarly sink, eventually to the surface, increasing air-pollution health and acid-rain problems as well.

An example of the unintended consequences of a geoengineering proposal is the potential impact of **white roofs** and roads on global climate. Although white roofs and roads reflect radiation, cooling buildings and the ground in cities locally, they may cause large-scale global warming.[244] The first reason is that, because white roofs cool the ground relative to the air locally, they reduce the ability of air to rise and thus clouds to form. Since clouds are reflective, reducing cloudiness increases sunlight to the surface, offsetting some of the reduced sunlight to the surface caused by the white roofs. However, since clouds travel and spread beyond a city, their reduction increases sunlight reaching the ground and temperatures outside of the city. Second, black and brown carbon pollution particles in the air absorb sunlight, then convert that sunlight to heat that is released to the air. In the presence of white roofs or roads, black and brown carbon absorb not only downward sunlight but also sunlight reflected upward by the white roofs, warming the air further. Finally, whereas white roofs cool buildings and thus reduce air-conditioning energy needs at low latitudes and during summers, their reflectivity increases heating needs during winters. In many places worldwide, heating demand exceeds cooling demand, so a white roof simply increases the fossil-fuel use required to heat the building.

A better solution than using a white roof is to install solar PV panels on a rooftop. The primary purpose of installing a PV panel is to generate electricity; however, panels also have several side benefits. Not only do rooftop PV panels remove 20 percent or more of incoming sunlight, converting it to electricity and cooling the underlying building, but the electricity they produce also displaces fossil-fuel use and its emissions. In addition, because solar panels do not reflect sunlight upward as white roofs do, solar panels do not allow absorption of upward reflected sunlight by black and brown carbon pollution particles. Similarly, because PV panels are warmer than a white roof is, PV panels don't increase air stability and thus don't reduce cloudiness as white roofs do.

In sum, with the exception of increasing reforestation and reducing deforestation, geoengineering is not recommended in a WWS world.

9

ELECTRICITY GRIDS

A 100 percent WWS energy infrastructure involves electrifying or providing direct heat for all energy sectors and then providing the electricity or heat with WWS. The solution also requires interconnecting geographically dispersed WWS generators on transmission grids and providing electricity and heat for isolated microgrids. Because electricity, grids, transformers, motors, and generators are such a large part of the solution, understanding how they work is important. This chapter discusses these phenomena and technologies along with the history of electromagnetism and the battle between George Westinghouse/Nikola Tesla and Thomas Edison to determine whether alternating current (AC) or direct current (DC) would predominate worldwide.

9.1 TYPES OF ELECTRICITY

Electricity is the free-flowing movement of charged particles, either negatively charged electrons, negatively charged ions, or positively charged ions. Electricity can travel in a wire or another medium, including the air. This chapter focuses on electricity through wires with application to the electric grid. Two other types of electricity are static electricity and lightning.

Static electricity arises by rubbing two materials, such as silk and glass, together so that one material strips electrons off the other. The result is that one material becomes negatively charged, and the other, positively charged. If the charge difference is small, and if neither material is a strong conductor of electricity, such as with glass and silk, the two materials merely stick to each other due to the attraction of the excess electrons in one material to the excess positive ions in the other. On the other hand, if one material is a strong conductor of electricity, a spark may occur. For example, a person walking on a wool rug scrapes electrons off the rug, giving his or her body a net negative charge. If the person's hand approaches a metal doorknob, which is a good conductor of electricity, the excess electrons in the person's hand induce positive charges in the metal doorknob to rise to the knob's surface. Just before contact, millions of electrons attracted to the positive charges on the doorknob fly from the person's hand, through the air, smashing into the doorknob, creating a spark.

Lightning occurs when a large charge difference arises between a cloud and the ground, between two parts of the same cloud, or between two clouds. If the charge difference builds up beyond a threshold, lightning strikes. Before cloud-to-ground lightning occurs, the bottom of a cloud builds up a net negative charge, and the top, a net positive charge. This happens because small ice crystals bounce off of heavy hail particles near the middle of the cloud, stripping the surface of hail particles of positive charges. The positively charged small crystals are then lifted by updrafts to the top of the cloud. The negatively charged hail falls to the base of the cloud.

The net negative charge at the base of the cloud induces positive ions at high points in the ground to migrate to the surface. Once the charge difference between the base of the cloud and the ground surface exceeds a threshold, a group of electrons stream from the base of the cloud toward the ground. As the electrons bash into air molecules (molecular oxygen and nitrogen), the electrons split the air molecules into positive ions and electrons. The additional electrons join the downward parade of electrons, but the positively charged ions slow the parade by attracting some of the streaming electrons upward. This temporary upward motion followed by the continued downward motion of electrons contributes to lightning's jagged shape.

As the electrons approach the ground, their attraction is so strong that a **streamer** of positive ions flows up from high points on the ground into the air to meet them. The connection creates a channel in which electrons in the air near the ground rush toward the streamer. These electrons bash into air molecules, forming a hot (30,000–60,000 degrees Celsius) **plasma** of ions and electrons. **Lightning** occurs because cooler molecules (3,700–7,300 degrees Celsius) on the outer edge of the flash emit photons of visible light. The thermal expansion of the air creates a sonic boom, or **thunder**.

Electrons backed up behind the first batch then rush to the ground, just as a line of cars released from a stoplight move one at a time, starting with the car in the front. In this manner, lightning starts near the ground and propagates up to the cloud as each successive batch of electrons behind the previous one accelerates toward the ground. This first major flash is the most luminous. Once the channel is open, additional electrons from the cloud come through it in a downward stroke of lighting.

Wired electricity is electricity that travels through a metal wire. In a metal wire, the metal's atomic nuclei are fixed in position and electrons

9.1 TYPES OF ELECTRICITY

carry charge freely through the wire to produce an electric current. An **electric current** is the number of electric charges transported per second past a given point. It is the flow rate of electrons or other charged particles through a wire and is the same at any point along a closed electrical circuit. An **electrical circuit** is a closed-loop electrical network with a return path through which an electric current runs. Flowing electrons in a wire in a circuit are sufficiently far from the nucleus of any specific atom in the wire that the attraction of the electrons to a nucleus is easily overcome. These conduction electrons wander from atom to atom, creating a current.

Negatively charged electrons balance positive charges in a wire. So, when an electron moves, it leaves behind a positive charge. Thus, the positive charges appear to move in the opposite direction from the negative charges. The direction of a current is defined, for historic reasons, as the direction of positive flow (or opposite the direction of negative flow). So, if electrons are moving to the right, then current (positive flow) is moving to the left. In a circuit, the direction of electrons is the physical reality of charge flow.

Direct current (DC) is an electric current flowing at a constant rate in one direction. An example of a DC electricity source is a battery. In a wire conducting DC electricity, electrons, when considered as a whole, move slowly, because each electron moves in a random direction. However, in the presence of an **electric field**, which is an electric force per unit charge, more electrons move in one direction than another, causing a net flow of electrons in one direction. Although electrons as a whole move slowly in a wire, each electron carries a lot of energy. As electrons collide with each other, they transfer energy to each other, sending a stream of electrical energy down a wire at nearly the speed of light.

Alternating current (AC) is a current that flows back and forth over time, in a sinusoidal manner. In one sinusoidal cycle, electrons first accelerate their flow to the right, reach a peak speed, decelerate to zero speed, accelerate to the left, reach a peak speed, then decelerate to zero speed. The number of complete cycles in a second is called the **AC frequency**. One cycle per second is defined as 1 **hertz**. So, 3 hertz means that three complete cycles of flow and counter flow occur in 1 second. In the United States, Canada, Central America, most of the Caribbean, most of South America, South Korea, and some other countries, the AC frequency for the electric-power grid is set at 60 cycles per second, or 60 hertz. **Nikola Tesla** (1856–1943) calculated that 60 hertz was optimal for the new US electric grid. In Europe and most of Africa, the Middle East, Asia and parts of South America, the AC frequency is set at 50 hertz. In a wire conducting alternating current at 60 hertz, electrons reverse themselves 60 times a second, so hardly move. Yet, they transfer energy to each other, sending a wave of electrical energy down a wire at nearly the speed of light.

9.2 VOLTAGE

Voltage is the energy per unit charge needed to move a single positive charge between two points in an electric field. Voltage can be understood through an analogy with potential energy, which equals mass multiplied by gravity and height. As a mass is lifted against gravity, it gains potential energy. Analogously, as the voltage of a charge is raised, the charge also gains potential energy, referred to as electrical energy.

A way to think of the difference between voltage and current is by using an analogy of water in a hose used to wash a sidewalk. The amount of water passing through the hose is analogous to current. The thrust of the water coming out of the hose is analogous to voltage.

A simple electrical circuit can be used to illustrate voltage and current. Such a circuit may consist of a battery, a light bulb, a wire from one end of the battery going to the light bulb, a wire going from the other side of the light bulb back to the other end of the battery, and a switch that connects and disconnects the circuit.

According to **Kirchhoff's voltage law**, "the sum of the voltages around any loop of a circuit at any instant is zero." Kirchhoff's voltage law is analogous to the changes in elevation along a circular path. A hiker walking up a hill on a trail eventually descends back to his or her original position. As such, the sum of elevation changes along the path always equal zero. Similarly, the sum of voltage changes around a circuit always equals zero.

Suppose a battery in a circuit is a 12-volt battery. Voltage is measured across components. This means that if the voltage at one end of the battery is zero volts, the voltage at the other is 12 volts, so the overall voltage across the battery is 12 volts. If the battery is connected to a light bulb in the circuit, the voltage drops 12 volts when measured from the wire going into the light bulb to the wire leaving the light bulb. Thus, the light bulb reduces the voltage by 12 volts, converting electrical energy into light energy through a resistance element. Because the voltage increases by 12 volts from one end of the battery to the other, and then decreases by 12 volts from the wire going into the light bulb to the wire going out of the light bulb and back to the battery, the net change in voltage through the circuit is zero volts, satisfying Kirchhoff's voltage law. Meanwhile, the current through the circuit is constant everywhere.

9.3 POWER

Power is the change in energy per unit of time. With respect to electricity, since energy per unit of time equals energy per unit of charge (voltage) multiplied by charge per unit of time (current), then power equals voltage multiplied by current. Power has units of watts, where 1 watt equals 1 joule of energy per second.

Electric current flows through wires to provide energy to an electric load. An **electric load** (or electric demand) is a component or part of an electrical circuit that consumes electrical energy, thereby reducing voltage. A light bulb, for example, is a load. Loads can also be heaters, cookers, dishwashers, car chargers, or arc furnaces. When loads consume energy, they reduce the voltage across the load. Loads convert electrical energy to heat, which is dissipated to the air.

An electric current is reduced along a circuit if the current passes through an electrical insulator. An **electrical insulator** is a material that impedes the flow of a current by preventing electric charges from flowing freely through the material. Good insulators are rubber, glass, air, plastic, paper, wood, wax, and wool. The opposite of an electrical insulator is an **electrical conductor**, through which charges flow freely. Good conductors are copper, aluminum, gold, and silver.

Electrical resistance arises when an electric load or insulator impedes the flow of an electrical current. Both loads and electrical insulators are forms of electrical resistance. The electrical resistance of a load or insulator is quantified as the voltage across the load or insulator divided by the current through it. Alternatively, the voltage across a load or insulator equals the current through it multiplied by resistance. Thus, a load or insulator drops voltage across it proportionally to the current through it. If voltage doesn't drop across a load or insulator, the resistance is zero. Both loads and insulators reduce voltage across them and convert the voltage to heat. However, because the current through an insulator is so low, the voltage drop across the insulator is also small, since voltage drop is proportional to current.

9.4 ELECTROMAGNETISM AND AC ELECTRICITY

Magnets, in addition to batteries, can produce an electric current through a wire. For example, a magnet moving toward or away from a coil of wire that is part of a circuit induces an electric current in the wire. Similarly, an electric current flowing along a straight wire creates a circular magnetic field around the wire. The relationship between electricity and magnetism is referred to as **electromagnetism**. Electromagnetism is the key physical process occurring in inductors, transformers, motors, and generators.

On April 21, 1820, the Danish physicist and chemist, **Hans Christian Orsted** (1777–1851) discovered electromagnetism when he noticed that the needle of a compass was deflected in the presence of an electric current originating from a battery that he switched on and off. He investigated and published this phenomenon the same year, concluding that an electric current flowing through a wire produces a circular magnetic field around the wire.

In 1825, **André-Marie Ampère** (1775–1836) found that a wire carrying a current exerted an attracting force on a second wire carrying a current in the same direction and a repelling force on a second wire carrying a current in the opposite direction.

Michael Faraday (1791–1867) made a third important advancement in electromagnetism. On August 29, 1831, he coiled two unconnected wires around opposite ends of a circular iron ring. He then connected one of the wires to a **galvanometer**, which measures current. Upon connecting the second wire to a battery, he noticed that the galvanometer briefly measured an electric current through the first coiled wire. Disconnecting the battery from the second wire also induced a brief current in the first wire.

What happened was that connecting the battery to the second coiled wire created a current through that wire. The sudden increase in current from zero created a pulse magnetic field around the second coil. The pulse-increase in magnetic field induced a temporary current through and voltage across the first coiled wire. Disconnecting the battery decreased the current through the second wire to zero, thus decreased the magnetic field around it. The sudden drop in magnetic field induced a current through and a voltage across the first wire.

Faraday similarly showed that moving a magnet toward or away from a coiled wire on a circuit induces a current through the circuit and a voltage across the coil.

In a third experiment, he showed that moving a coiled wire connected to a battery in and out of the center of a larger, second coiled wire induces a current in the larger coil. The current through the first coil creates a magnetic field, and the movement of that magnetic field induces a brief current through the larger coil. When the two coils are held still, no current flows through the second coil. A current can be induced only when a magnetic field fluctuates with time, not when it stays constant with time.

9.4 ELECTROMAGNETISM AND AC ELECTRICITY

The principle of electromagnetism was essential for the development of AC electricity. An alternating current arises when current flows in one direction for a short time, changes direction, then changes direction back again in a repeating cycle. With DC electricity, on the other hand, current flows constantly in one direction.

AC electricity can be produced with a magnet that spins clockwise inside of two stationary coiled wires placed on opposite sides from each other (Figure 9.1). The two stationary coiled wires are part of a closed circuit that has a load connected to it. When the magnet is at rest and its south and north ends are away from the coils, no current flows (Step 1 in Figure 9.1). As the magnet rotates clockwise, its south pole approaches the coiled wire on its right side (Step 2). The increase in magnetic flux experienced by the coiled wire induces a clockwise current in the circuit. As the south pole continues rotating clockwise away from the coiled wire on the right side, the current in the circuit decreases until it reaches zero (Step 3). As the magnet continues rotating clockwise, its south pole approaches the

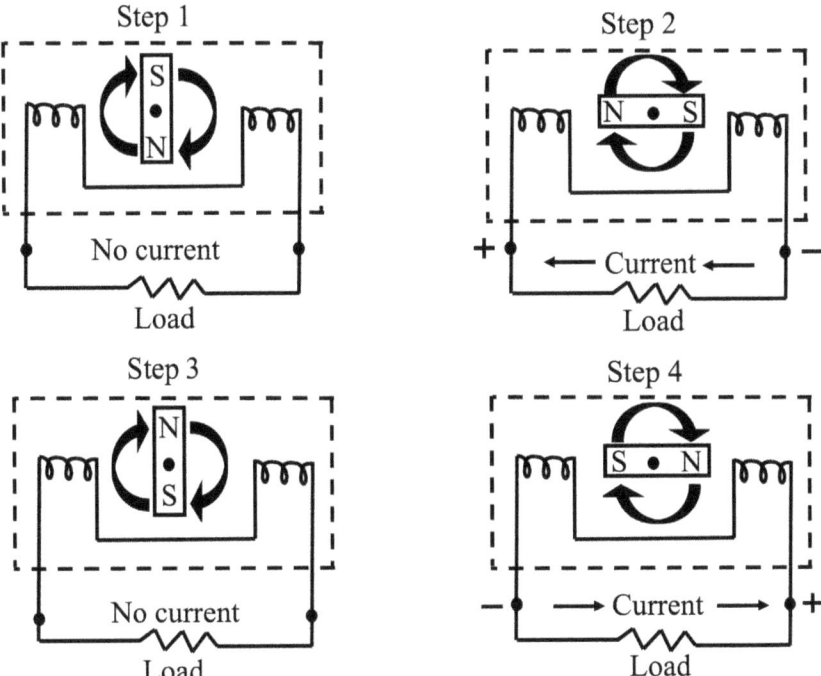

Figure 9.1 Production of single-phase alternating current (AC) by rotating a magnet in the presence of stationary wire coils. S identifies the magnet's south pole, and N identifies the magnet's north pole.

coiled wire on the left side, again inducing a current, but in the counter-clockwise direction (Step 4). As the magnet continues rotating clockwise, the current decreases again, ultimately to zero again (Step 1). Thus, one clockwise rotation by the magnet creates an alternating-in-direction current through the circuit. The faster the magnet turns, the greater the frequency at which the current alternates direction in the wire.

The current generated with one pair of coiled wires is called **single-phase AC electricity**. Such electricity is produced in an alternator, which is a type of generator. **Generators** use magnets and coiled wires to convert the rotational energy in a spinning shaft to electricity. Almost all rotational energy for generators worldwide is obtained from a fluid (steam, air, combustion gases, or liquid water) passing through turbine blades connected to a rotating shaft. **Motors** are generators that are run in reverse. They convert electricity into rotational energy.

The problem with single-phase AC electricity on a circuit is that the oscillating AC electricity current causes a continuous alternating increase and decrease in current. Thus, appliances, such as light bulbs, continuously increase and decrease in brightness. This flickering is ameliorated by the fact that the frequency of the increases and decreases is so rapid, 50 or 60 increases plus decreases per second, that most people cannot detect the flickering.

Three-phase AC electricity increases the apparent constancy of the current even further. It can be accomplished in a generator simply by rotating a rectangular magnet inside of three pairs of equally spaced coiled wires instead of one pair of equally spaced coiled wires. Another way to generate three-phase AC electricity is to rotate three equally spaced coiled wires inside of a circular magnet that has a north end and a south end. The advantage of three-phase AC electricity is that it results in the maximum positive current occurring three times per rotation rather than one time per rotation, decreasing flickering. Having more than three phases increases equipment cost with little additional benefit.

Three-phase generators smoothen current compared with single-phase generators and produce virtually all commercial AC electricity worldwide. Most three-phase AC electricity is sent along three-phase transmission lines. These have three wires to transmit each phase independently to the end load.

Since each phase in a three-phase AC system is transmitted along its own wire, different phases can be delivered to a household for different purposes. Because of the high frequency (60 or 50 hertz) of the waves, flickering is not noticeable by most people using even a single phase for home appliances.

9.5 CAPACITORS AND INDUCTORS

Capacitors store electric energy in an electric circuit. They are made of two flat conducting metal plates in parallel, separated by a thin nonconducting insulator in between. The simplest capacitor consists of two sheets of aluminum foil separated by air or paper. If a battery is connected to a circuit with nothing else connected to it but a capacitor, then one metal plate becomes positively charged, and the other, negatively charged. The difference in charge across the plates creates an electric field in which electrostatic energy is stored. Once fully charged, the capacitor holds the same voltage as the battery.

If a circuit contains a battery, a light bulb, and a large capacitor, the light bulb will first light up but become dimmer as the current flows to the capacitor, charging the capacitor. When the capacitor is fully charged, the light bulb will darken completely since the voltage drop across the capacitor is now the same as the voltage increase across the battery, so no more voltage is available for the light bulb. If the battery is then disconnected from the circuit, the capacitor will act like a battery and instantly provide current to the light bulb, lighting it, until the stored energy in the capacitor is depleted.

In sum, capacitors are like batteries in that they discharge stored electricity quickly. However, capacitors charge much faster than do batteries. They also weigh less, require fewer materials, cost less, and last many more cycles than do batteries. The disadvantage of a capacitor is that it stores much less total energy than does a battery. Thus, capacitors are used only when a small amount of energy is needed quickly, such as to maintain power in electronic devices when batteries are being charged or when voltage drops. Capacitors also smoothen voltage in DC power lines because they resist rapid changes in voltage. In other words, voltage cannot change instantly in a capacitor, so capacitors smoothen out sudden variations in voltage on a power line.

Because capacitors can be charged quickly and last long but store little energy, efforts were made starting in the late 1950s to improve them so that they could hold more energy. These efforts resulted in the first **supercapacitors**, which are capacitors with much larger plates, much smaller distance between plates, and made of different materials from traditional capacitors. Even the best supercapacitors today store only 10–20 percent of the energy that a battery stores. However, supercapacitors do have important applications. For example, a cordless drill powered by a bank of supercapacitors allows an astronaut to recharge the drill quickly.

Banks of supercapacitors are used in electrical and electronic equipment to smoothen out power supplies. They are also used to store temporarily the electricity produced from regenerative braking when battery-electric vehicles slow down, when cranes lower an object, and when elevators descend. The stored electricity is used immediately when vehicles move again, cranes lift an object, and elevators rise.

Whereas capacitors store energy in an electric field, **inductors** store energy in a magnetic field. An inductor consists of an insulated wire coiled around an iron core. When a current passes through the coil of an inductor, the inductor creates a magnetic field in which voltage is stored. Just as capacitors smoothen sudden changes in voltage across them, inductors smoothen sudden changes in current through them. When current that flows through an inductor suddenly increases, for example, the magnetic field becomes stronger. As the magnetic field strength increases, electrical energy is drawn from the current and converted to magnetic potential energy, increasing the voltage. Because electrical energy is taken from the current, the current decreases. In sum, inductors smoothen sudden changes in current. As such, inductors are useful in an electric-power grid for limiting the damage due to short circuits or other abnormal currents. Inductors are also integral components of transformers.

9.6 THE BATTLE OF DC VERSUS AC ELECTRICITY

When electricity travels down a transmission line, it loses power proportionally to the square of the current through the line. Thus, increasing current through a transmission line by a factor of 10 increases power loss by a factor of 100. As such, current passing through a transmission line should be kept low to minimize power loss. Since power equals voltage multiplied by current, decreasing current means increasing voltage to maintain the same overall power. Thus, transmitting electricity long distance can be accomplished without significant transmission line loss only if the current through the line is low and the voltage transmitted is high. This principle is the main reason the world's electricity grids today are AC grids, as described next.

9.6.1 THOMAS EDISON AND THE DC GRID

On September 4, 1882, **Thomas Edison** (1847–1931) and the Edison Illuminating Company began operating the first investor-owned electric utility in the world, the Pearl Street Power Station, in New York City. The electricity was produced by burning coal to boil water to produce steam

9.6 THE BATTLE OF DC VERSUS AC ELECTRICITY

for use in a steam engine. In the steam chamber of a steam engine, steam pushes a piston up. Liquid water is then squirted into the steam chamber to recondense the vapor, creating a vacuum that pulls the piston down. The up-and-down motion of the piston allows a lever attached to the piston to create a rotating motion. The rotating lever is connected to a **dynamo**, which is a generator that produces DC electricity when a coiled wire rotates continuously inside of a magnetic field.

Edison had invented the incandescent light bulb in 1879. His 1882 electric utility initially provided power for 400 light bulbs owned by 82 customers. By 1884, the number had increased to 10,164 bulbs, spread across 508 customers. The DC power provided to customers was at 110 volts. If the voltage delivered to customers had been substantially higher, it would have been dangerous. Because the voltage through the transmission lines was relatively low, the current needed to be high in order for the total power (energy per unit of time) delivered to the homes to be sufficient. The high current required thick copper wires, resulting in high transmission line losses and costs. The line losses were so high that the distance between the power generator and customers was limited to two kilometers. As such, the first coal-fired electricity-generating station in the world emitted pollution in the middle of one of the most densely populated cities in the world. However, if the coal plant had been sited outside of the city, transmission line losses would have been so high that customers would have received virtually no power.

The Pearl Street Station plant burned down in 1890. In the 1880s, however, Edison opened many other power stations and local grids based on DC electricity, including in Sunbury, Pennsylvania (1883); Shamokin, Pennsylvania (1883); Brockton, Massachusetts (1883); Mount Carmel, Pennsylvania (1883); Fall River, Massachusetts (1883); Cumberland, Maryland (1884); Tamaqua, Pennsylvania (1885); and Boston, Massachusetts (1886). By the end of the 1880s, in fact, Edison had built or licensed over 1,500 electricity-generating stations worldwide, many of which sent dedicated power to factories. The rest provided grid electricity for homes and commercial buildings.

9.6.2 NIKOLA TESLA

Several of Edison's utilities were in Europe. In 1882, the French branch of Edison Electric Light Co., Continental Edison Company, which was building power stations and grids in Paris and other European cities, hired **Nikola Tesla** (1856–1943; Figure 9.2). Tesla, born in Croatia, previously worked in

9 ELECTRICITY GRIDS

Figure 9.2 Nikola Tesla in 1890 at age 34. Photo by Napoleon Sarony (1821–1896), public domain, via Wikimedia Commons.

Budapest, Hungary, as an electrical engineer. Tesla's new job was to install incandescent lights along a Paris electric grid and to troubleshoot problems in that and other grids being built by Edison in France and Germany.

Tesla's skill as a technician and troubleshooter became apparent quickly. So much so, that Tesla was called on to investigate and repair what could have been the greatest disaster to date resulting from Edison's DC power grid. In early 1883, during the grand opening of a DC power station at the Strasbourg (in Germany at the time) rail station, the ceremonial switching on of power caused an explosion that nearly killed Emperor William I of Germany. Tesla was asked to find out what happened and to fix the problem. He determined that the explosion was due to a major short circuit resulting from poor dynamos and faulty wiring. He spent a year redesigning the system, particularly the dynamos. The new power system worked well. Tesla's talent as an electrical engineer was recognized. However, Tesla's boss and Edison's confidant in Paris, Charles Batchelor, who had orally promised Tesla a $25,000 bonus for fixing the problems in Strasbourg, backed out of paying him. Tesla was not happy but continued with his job.

While in Paris, Tesla recognized the problem with Edison's DC electricity grids – namely, that they were limited to low voltages and short distances and required thick copper-wire transmission lines to carry the high current. This meant that a power plant was needed every two-kilometer radius to provide electricity for a whole city. In addition, because thick copper wires were needed to carry the high current, even over short distances, copper requirements for transmission lines would be enormous for powering a large population. These issues created a practical problem that limited the spread of DC electricity. In 1882, just prior to moving to Paris, Tesla designed a prototype AC motor. Tesla felt that a solution to the DC grid problem may be AC electricity. However, his superiors at Edison's company in Paris did not want to hear about AC electricity, as they were heavily invested in DC.

Batchelor, though, recognized Tesla's talent. He also believed that Tesla should speak directly to Edison about Tesla's AC grid idea so that the two could work together on the idea rather than Tesla going off on his own. Batchelor thus convinced Tesla to move to the US, where Tesla would have a better opportunity to interact directly with Edison and design and improve dynamos and motors. Tesla agreed to go.

Having been robbed of money and luggage at the shipyard in Calais, France, Tesla traveled by ship to New York City with no money or clothes beyond what he was wearing. He did have a letter of introduction to Edison from Batchelor. Soon after arriving, on June 8, 1884, Tesla began working at the Edison Machine Works. His first job was to repair broken dynamos powering lights on the SS *Oregon* ship, which was docked in New York City. Edison was impressed with Tesla's successful repairs, which took less than a day. Thereafter, Edison had even more confidence in Tesla and decided to give him a more difficult task.

Edison asked Tesla to improve the design of the dynamo so that many dynamos could be linked to provide electricity for large buildings without causing lights to flicker or short circuits. Edison verbally offered Tesla a $50,000 bonus for this job. Not learning a lesson from his previous mistake, Tesla agreed orally, but not in writing. Over the next year, he developed 24 different improved dynamos. Edison applied for patents on the dynamos, and his company reaped substantial financial rewards from them. However, upon completion of Tesla's work in 1885, Edison refused to pay the bonus, telling Tesla, "You don't understand our American humor." This was the last straw. Tesla resigned.

9.6.3 TRANSFORMERS

Meanwhile, in 1882, **Sebastian Ziani de Ferranti** (1864–1930) working at the Siemens brothers firm in London with Lord Kelvin (William Thompson), designed a prototype AC power system. In the process, he developed a transformer. Working off the design of Ferranti, **Lucien Gaulard** (1850–1888) of France, with the financial backing of John Gibbs, invented a step-down transformer in 1883. Gaulard and Gibbs demonstrated the transformer, which they called a secondary generator, in London's Royal Aquarium. In 1884, Gaulard used his transformer in a demonstration 40-kilometer AC transmission line between Lanzo and Turin, Italy. The transformer reduced high-voltage alternating current to low-voltage alternating current to provide power for incandescent lights and arc lamps along the line. He also showcased the transformer at the Turin Italian National Exhibition that year, where **Otto Blathy** (1860–1939), a Hungarian engineer, saw it. In 1885, Blathy and two other Hungarian engineers, Miksa Deri and Karoly Zipernowsky, improved and patented a similar transformer, which Blathy named the ZBD transformer, ZBD being the initials of the three inventors.

A **transformer** consists of two or more inductors in proximity to each other. A transformer increases or decreases voltage between a powered inductor coil and an unpowered inductor coil. It works when an alternating current winding through one of the inductor coils creates a fluctuating magnetic field around the second inductor coil not connected to the first. The fluctuating magnetic field induces a current through and voltage across the second coil. The AC voltage induced across the unpowered coil equals that across the powered coil multiplied by the ratio of the number of coil-turns in the unpowered coil to those in the powered coil. So, if the unpowered coil has twice as many coils as the powered coil, the voltage in the unpowered coil is twice that in the powered coil. The current in the unpowered coil is the opposite: the current through the unpowered coil is half that in the powered coil. In other words, if the number of coils in the unpowered coil is higher than in the powered coil, a transformer increases voltage and decreases current (**step-up transformer**). If the number is lower, the transformer decreases voltage and increases current (**step-down transformer**). The power (voltage multiplied by current) stays the same between the two coils.

A transformer does not work with DC electricity. Whereas DC electricity passing through one coil creates a magnetic field around the second coil, the magnetic field stays constant over time because direct current in the first coil stays constant over time. A current occurs in the second coil only if the magnetic field fluctuates with time, and this requires the current in the first coil to fluctuate with time. Because AC electricity through the first coil continuously alternates direction, the magnetic field produced by it continuously fluctuates as well, inducing a current in the second coil.

9.6 THE BATTLE OF DC VERSUS AC ELECTRICITY

Transformers, which are made of inductors, are used not only to increase or decrease voltage along a transmission line but also to limit abnormal electrical currents along the line.

While still working for Edison, Tesla had offered to redesign Edison's grid from DC to AC. Tesla recognized that a transformer would allow AC electricity voltage to be increased (stepped up) for long-distance transmission, and a second transformer would allow the voltage to be decreased (stepped down) for end use. An increase in voltage of AC electricity would produce high voltage AC (**HVAC**) electricity and reduce current relative to DC electricity. Lower current would reduce line losses a lot over a long distance, increasing power output at the end of a transmission line.

The main implication of an AC grid was that AC electricity could be produced far from where the electricity was used. This would allow one electricity-generating station to serve far more customers than Edison's stations, which had to be near customers. A single, large generator station could operate at much lower cost than many small stations. In addition, although this wasn't a consideration at the time, a single large coal-fired power plant outside of a city would cause much less air-pollution exposure to a city's population than would many smaller coal plants within the city. Edison called Tesla's ideas "splendid" but "utterly impractical."[189] After Tesla resigned from Edison's company, Tesla set out to start his own AC utility, Tesla Light & Manufacturing.

9.6.4 GEORGE WESTINGHOUSE AND THE AC ELECTRICITY GRID

Coincidentally, in 1885, **George Westinghouse** (1846–1914), who started a DC lighting business in 1884, became interested in AC electricity. He was motivated and had the financial resources to invest in an AC grid. When he read an 1885 newspaper account about Gaulard's transformer, which had been used in the 1884 Turin Exhibition, Westinghouse purchased the American rights to Gaulard's design and ordered several of the transformers and a Siemens alternator (single-phase AC generator) to be shipped to Westinghouse's factory in Pittsburgh, Pennsylvania.

Westinghouse then asked an engineer working for him, **William Stanley, Jr.** (1858–1916) to develop an electric AC lighting system with the alternator and transformers. In 1885, Stanley began experimenting in Pittsburg. He also improved the transformer enough so that it became the world's first practical transformer.

In 1886, Stanley used his new expertise to build, with Westinghouse's funding, the world's first AC power system using both a step-up and a step-down transformer, in Great Barrington, Massachusetts. In 1886, Stanley, Westinghouse, and Oliver B. Shallenberger then built the first commercial AC power system, in Buffalo, New York. Westinghouse installed several other AC grid systems that year around the US, particularly in remote locations that Edison could not reach economically with his DC grid. By 1887, after only one year, Westinghouse had half the number of electricity generators as Edison. The same year, C. S. Bradley invented the first three-phase AC generator. Only single-phase generators were available until then. A problem at the time, though, was that an efficient AC motor had not yet been invented, so AC electricity could not be used to power any equipment that needed a motor.

In 1886, Tesla attempted to sell his AC power system idea to investors in New York City but found little interest, since New York City already had some of Edison's DC power. Nevertheless, Tesla kept working to develop AC technology. In an 1888 breakthrough, Tesla invented a three-phase AC induction motor, which was critical for powering equipment on an AC grid. After Westinghouse heard about Tesla's invention, Westinghouse visited Tesla's laboratory. The two began a collaboration to design a better AC grid system with three-phase generators and motors. Westinghouse paid Tesla well.

Transition highlight
George Westinghouse poured financial resources into developing an AC power system with step-up and step-down transformers, AC generators, and AC motors. In 1891, he built the first power plant (a hydroelectric plant) to supply AC power over a long distance (5.6 kilometers) for a gold mine near Ophir, Colorado.

9.6.5 THE FIGHT BEGINS

Edison considered going into the AC transmission business. In fact, he purchased options on Blathy's transformer. However, he ultimately decided against going forward with AC. Instead, the storm of development by Westinghouse gave Edison anxiety, which he soon reacted to. In November 1887, a dentist in Buffalo, New York, Alfred P. Southwick, wrote to Edison asking him to support the use of electricity as a humane way of executing criminals. When he became aware of the accidental August 1881 electrocution of a drunken dockworker by a dynamo, Southwick began experimenting with electrocuting dogs to euthanize them. He published work on his experiments in scientific journals and wanted to adopt electrocution as a way to execute criminals. In an early design, Southwick

9.6 THE BATTLE OF DC VERSUS AC ELECTRICITY

proposed using his dental chair to restrain a convict during an execution. This gave rise to the term, electric chair.

Edison did not believe in capital punishment but did believe that his competition should be punished. When pressed by Southwick about what kind of equipment should be used for executions, Edison eventually replied, "The most effective of these are known as alternating machines manufactured principally in this country by Mr. Geo. Westinghouse, Pittsburgh."[245] Edison then engaged Harold Brown, a senior engineer working for him, to publicly stoke fears about AC electricity.

Brown, who pretended not to be paid by Edison, stated correctly in front of audiences that the AC voltage needed to electrocute a dog was much less than the DC voltage. However, to incite fear, he went on to demonstrate a dog being electrocuted with AC electricity. In one stunt, he connected a sheet of tin to an AC dynamo and led a dog onto the tin to drink from a dish. The dog was electrocuted and died. This and other demonstrations often resulted in the smell of burned flesh and hair permeating a room. Despite protests, Brown ramped up electrocutions for show, but now of a horse and a calf, in addition to dogs. Edison and his helpers also leaked photographs of other animals they electrocuted to the press. They even attempted, unsuccessfully, to get a law banning AC electricity from being introduced.

When Westinghouse challenged Brown and Edison's advertising campaign, Brown came back proposing that Westinghouse "take through his body the alternating current while I take through mine a continuous current." Westinghouse refused. In 1889, Brown, funded by Edison, published a pamphlet stoking fears about the alleged dangers of AC electricity, calling it "the executioner's current." In fact, through Brown, Edison lobbied hard for the first electric chair in the United States to run on AC electricity. The lobbying was successful. Brown and Edison convinced the New York superintendent of prisons to adopt AC electricity for this purpose. The first use of the electric chair for capital punishment occurred on August 6, 1890, at Auburn Correctional Facility, New York. The criminal, William Kremler, was executed with AC electricity. The execution went awry, taking two and a half minutes and resulting in a charred and smoking body.

9.6.6 AC RISES VICTORIOUSLY

Nevertheless, within a year of the execution, AC had all but taken over. DC could operate only a few appliances, whereas AC could operate more. AC was also much cheaper and could run on larger, more distant power supplies. With the adoption of AC electricity at the Chicago World's Fair in 1893 and New York's Niagara Falls' power station in 1895, AC electricity crushed any hope of a large-scale DC grid. AC was ultimately adopted worldwide.

The major reason for AC's victory was that AC could be transmitted through long-distance HVAC lines without the same line losses as DC. AC line losses were minimized by using one transformer to step up the voltage (reducing the current) at the start of transmission and another transformer to step down the voltage (increasing the current) at the end of transmission. Transformers did not work with DC. Thus, transformers settled the battle of DC versus AC electricity.

Since then, a method has been developed to transmit direct current long distances. However, even this method requires an AC grid. The method is to boost AC voltage with a step-up transformer to HVAC, convert the HVAC to high-voltage DC (HVDC) current. The HVDC current then flows through a long-distance HVDC transmission line. At the end of the line, HVDC is converted back to HVAC. The HVAC voltage is then reduced with a step-down transformer. The world's first HVDC transmission line was installed in 1954 between mainland Sweden and the island of Gotland.

An advantage of using HVDC instead of HVAC transmission over long distances (greater than 600 kilometers) is that, although current is low and line losses are minimized in both cases, line losses are even less for HVDC than for HVAC over long distances. In addition, over long distances, HVDC is less expensive than HVAC. For distances less than 600 kilometers, HVDC transmission lines are more costly than HVAC lines because expensive AC-to-DC conversion equipment comprises a larger share of the overall HVDC cost than it does for long-distance transmission.

Today, generators produce electricity with voltages of between 12,000 and 25,000 volts. Step-up transformers boost the voltage up to between 100,000 and 1 million volts for HVAC transmission. If long-distance HVDC is used, voltage is stepped up further to between 100,000 and 1.5 million volts. Step-down transformers decrease the voltage down to between 4,160 and 34,500 volts for distribution to the local utility. Local utilities then distribute AC power to neighborhood power poles. Transformers on power poles step voltage down further to 120 or 240 volts.

In North, Central, and most of South America, home wall receptors receive 60-hertz single-phase AC power at 120 (110 to 127) volts. Clothes dryers, car chargers, electric water heaters, and some other appliances require three-phase power at 240 volts. In Europe and most countries, home power is 50-hertz single-phase AC power at 220 to 240 volts.

9.7 TRANSMISSION AND DISTRIBUTION LOSSES

In an AC **transmission and distribution system**, AC electricity flows along a transmission line between an electric-power-generating facility and a step-up transformer station. The station boosts the voltage to produce HVAC electricity to reduce long-distance AC transmission losses. Along the HVAC line, AC electricity may or may not be converted to DC electricity for extra-long-distance HVDC transmission. At the end of an HVDC line, the DC electricity is converted back to HVAC. The HVAC is then transmitted to a step-down transformer station in a neighborhood, where the voltage is decreased, and the electricity is sent to local distribution lines. This electricity then goes to a transformer near buildings, where the AC voltage is dropped further for use in the buildings.

Losses along transmission and distribution lines arise due to five factors. First, resistance along the lines converts some electricity to heat. Second, losses arise in step-up and step-down transformers. Losses similarly arise in local transformers. Third, losses occur in equipment converting HVAC electricity to HVDC electricity and back again. Fourth, losses arise from downed power lines. Fifth, in countries with transmission and distribution loss rates above 15 percent, electricity theft from power lines is a major source of loss.[246] Of all transmission and distribution losses in the current energy system in countries without power theft, about 16–33 percent are short- and long-distance transmission losses from the electricity generator station to the step-down transformer substation, 32–40 percent are distribution losses between the step-down substation and the end user, and 27–52 percent are transformer losses.[247]

When HVAC electricity is converted to HVDC electricity for extra-long-distance transmission, transmission losses are reduced compared with HVAC transmission alone beyond 500 kilometers.[248] This result applies although a portion of the long-distance HVDC transmission benefit is offset by losses from HVAC-to-HVDC-to-HVAC conversion.

Overall transmission and distribution losses worldwide in 2014 ranged from lows of 2 percent in Singapore, 2.3 percent in Trinidad/Tobago, 2.5 percent in the Slovak Republic, and 2.7 percent in Iceland, to highs of 60.1 percent in Haiti, 69.7 percent in Libya, and 72.5 percent in Togo.[249] In all, 54 percent of countries had transmission and distribution losses of 10 percent or higher. Losses in some large countries and regions were as follows: China (4%), the United States (5.9%), the European Union (6.4%), the Russian Federation (10%), Brazil (15.8%), and India (19.4%).

9 ELECTRICITY GRIDS

Much room exists for reducing transmission and distribution losses, particularly in countries with loss rates exceeding 10 percent (e.g., Brazil and India). Reducing line losses decreases energy needs. If a country's transmission and distribution loss rate is 10 percent, reducing it by one percentage point to 9 percent reduces electricity generation needs by 1.1 percent. If the loss rate is 20 percent, a one percentage point decrease reduces generation needs by 1.23 percent. If the loss rate is 72.5 percent, as in Togo, reducing it to 71.5 percent reduces generation needs by 3.5 percent.

Finally, in a 100 percent WWS world, a large portion of new electricity generation will be from offshore generators: offshore wind turbines, tidal turbines, ocean-current devices, wave devices, and floating solar panels. Offshore renewables often require short transmission distances to load centers because most people in the world live along the coasts, and most offshore generators will be within 200 kilometers of shore. As such, the growth of offshore renewables will increase the efficiency of the transmission and distribution system.

10

PHOTOVOLTAICS AND SOLAR RADIATION

Solar and wind will power the bulk of a 100 percent WWS all-sector energy system worldwide. The main types of solar generators are photovoltaics (PVs) for electricity, concentrated solar power for electricity and heat, and solar thermal systems for heat. The sun produces enough energy annually to power the world with PV alone for all energy purposes about 2,200 times over in 2050 if all energy sectors are electrified. Over land, PV can provide about 620 times the energy needed. Needless to say, the world needs to capture only a small portion of the sun's available energy. For example, only 0.08 percent of all sunlight hitting Earth annually is needed to provide half of the world's all-purpose energy from PV. Given the large potential of PV to power all energy needs, it is useful to understand PV panels and solar resources better. This chapter discusses both. The chapter starts with a history of solar photovoltaic cells, followed by a description of how modern cells work. It then discusses PV cell efficiencies, PV panels, and PV arrays. Lastly, it discusses solar resource availability and optimal tilt angles for solar panels.

10.1 SOLAR PHOTOVOLTAICS

A solar photovoltaic is a material or device that converts photons of sunlight to DC electricity. Each photon of sunlight breaks an electron in a PV material free of the atom that holds it. If an electric field is present, the field sweeps the free electron toward a metallic contact, where the electron becomes part of an electric current in a circuit.

PV cells are made of semiconductor materials that convert sunlight to electricity. A **semiconductor** material has an electrical conductivity between that of a metal and a nonmetal. A **metal** is material in which electrons can readily break free from atoms; they thus conduct electricity at any temperature, but less so with increasing temperature. A **nonmetal** has very low electrical conductivity at all temperatures, thus is an insulator. A semiconductor has very low conductivity at low temperature but much higher conductivity at moderate and high temperatures. At near zero degrees Kelvin (−273.15 degrees Celsius), for example, silicon is a perfect insulator (no electrons break free from its nucleus). As the temperature rises, some electrons break free of silicon's nucleus and flow in an electric current. The primary semiconductor materials used today are silicon (Si), germanium (Ge), gallium (Ga), and arsenic (As).

10.1.1 BRIEF HISTORY OF THE SOLAR PV CELL

Today's PV cells have evolved from early discoveries of the properties of light interacting with materials and the invention of early PV cells.

In 1839, the 19-year-old French physicist **Edmond Becquerel** (1820–1891), working in his father's laboratory, discovered that shining sunlight onto one of two parallel plates of platinum, gold, or silver placed in an electricity-conducting solution produced an electric current and voltage.[250] Visible or ultraviolet light hitting electrons on one metal plate caused the electrons to absorb the light and become energized. The added energy allowed the electrons to escape their attraction to the atomic nuclei they were bound to. The free electrons flowed into the conducting solution, where they joined other electrons moving to the second plate, creating electricity in a circuit attached to the second plate. The generation of electricity by exposing one material to visible or ultraviolet sunlight, thereby causing the release of electrons into a different material, is called the **photovoltaic effect**. The term photovoltaic was named after the Italian physicist and chemist **Alessandro Volta** (1745–1827), who contributed to understanding electricity.

Capitalizing on Becquerel's discovery, **Augustin Mouchet** (1825–1912) of France registered patents for solar-powered engines in the 1860s. He also unveiled a solar printing press at the 1878 University Exhibition in Paris.

In 1873, the English engineer, **Willoughby Smith** (1828–1891), found that selenium conducted electricity produced by sunlight. Shortly after this, in 1876, the English engineers, William Adams and Richard Day, discovered that exposing to sunlight a cylinder of solid selenium, with platinum wires connected to each end, generated electricity through the photovoltaic effect.[251] This was the first time a material had converted sunlight to electricity without heat or moving parts.

In 1883, New York inventor **Charles Fritts** (1850–1903) developed the first solar cell to generate electricity.[252] To do this, he pressed a thin film of selenium against a brass metal plate, then laid an even thinner layer of gold

10.1 SOLAR PHOTOVOLTAICS

on top. The gold layer was so thin that sunlight could penetrate through it. Exposing the gold leaf to sunlight resulted in electricity generation. The efficiency of the cell, however, was only 1–2 percent. Werner Siemens, who subsequently confirmed Fritts' experiment, commented,[253]

> In conclusion, I would say that however great the scientific importance of this discovery may be, its practical value will be no less obvious when we reflect that the supply of solar energy is both without limit and without cost, and that it will continue to pour down upon us for countless ages after all the coal deposits of the earth have been exhausted and forgotten.

Fritts' invention spurred others to develop solar photovoltaic cells and panels, and ways to improve exposure of a panel to sunlight. In 1894, for example, the American inventor **Melvin Severy** (1863–1951) received separate patents for a solar photovoltaic panel that generates electricity and for a two-axis tracker for the panel.

In 1904, the German physicist **Wilhelm Hallwachs** (1859–1922) produced a current when he exposed a cell consisting of a plate of pure copper coated with a thin layer of cuprous oxide (which can be obtained by heating the copper plate in the presence of oxygen) to sunlight. Sunlight passing through the cuprous oxide (a semiconductor material) excited and released electrons that were conducted away to a load by a wire.

Dozens of photovoltaic cells containing copper and cuprous oxide were experimented on through 1932 by several groups.[254] However, because of the low efficiency of these solar cells and the degradation of cuprous oxide upon its long-term exposure to sunlight, selenium became the material of choice in commercial cells during the 1930s. In 1932, Audobert and Stora discovered that cadmium sulfide, another semiconductor material, can help generate electricity through the photovoltaic effect.

The American electrochemist **Russell Ohl** (1898–1987), who was working at Bell Labs, in Holmdel, New Jersey, patented the first silicon-based PV cell in 1941. In 1939, Ohl discovered the p-n junction (see Section 10.1.4) and how it worked by adding different impurities to 99.85 percent pure monocrystalline silicon on each side of the junction. Ohl's PV cell was only 1–2 percent efficient. In 1954, **Daryl Chapin**, **Calvin Fuller**, and **Gerald Pearson**, working at Bell Labs, developed a more efficient silicon PV cell. This cell was 4 percent efficient and was the first to provide sufficient electricity to run basic electrical equipment. Hoffman Electronics improved the efficiency of silicon cells to 8 percent in 1957, and 14 percent in 1960. In 1985, the University of New South Wales, Australia, increased silicon PV cell efficiencies to 20 percent. In 2024, the record efficiency for a monocrystalline silicon solar cell was 26.1 percent. That of a multicrystalline silicon cell was 24.4 percent.[255]

In 1955, Western Electric in the US began selling licenses for products that ran on silicon PV cells. Such products included a dollar-bill changer and a device that decoded computer punch cards and tape. In 1958, several space satellites, including the Vanguard I and II, Explorer III, and Sputnik-3, used solar PV cells to power some equipment. Satellites after that had larger arrays. In 1980, ARCO Solar produced more than 1 megawatt nameplate capacity of PV panels, the first time a solar manufacturer had done that.

In 1981, the Solar Challenger, an aircraft with over 16,000 solar cells, was built by Paul MacCready. It was the first solar aircraft and flew from France to England. In 1982, Hans Tholstrup drove the first solar car 4,500 kilometers from Sydney to Perth, Australia, in 20 days. In 1982, the first megawatt-scale PV power station went online in Hisperia, California. It had 108 dual-axis trackers.

10.1.2 HOW SILICON PV CELLS WORK

Pure silicon crystals consist of silicon (Si) atoms bonded together. Silicon has 14 total electrons, including 4 in its outer shell, called the **valence shell**. **Valence electrons** are the electrons that occupy the outermost shell of an atom and determine the charge of the nucleus. With 4 electrons (minus-4 charge) in its valence shell, silicon's nucleus has a plus-4 charge, so the overall atom is neutrally charged.

Each electron in a valence shell can bond with another electron in the valence shell of another atom. Thus, one silicon atom can bond with four adjacent silicon atoms.

In 1888, Wilhelm Hallwachs, who later invented the copper–cuprous oxide solar cell, found that exposing a conducting metal to ultraviolet radiation caused the metal to obtain a net positive charge by gouging out an electron. This concept was the basis for what is called the **photoelectric effect**, which arises when photons of visible or ultraviolet radiation of sufficient short wavelength hit electrons in metals, nonmetallic solids, liquids, or gases and release the electrons to the air or a vacuum. The electrons emitted in this way are called **photoelectrons**.

The photoelectric effect is closely related to but differs from the photovoltaic effect. With the photoelectric effect, electrons are ejected out of the material into the air or a vacuum; with the photovoltaic effect, electrons are ejected into a new material to produce a current of electricity.

10.1 SOLAR PHOTOVOLTAICS

Modern photovoltaics are based on the photovoltaic effect rather than the photoelectric effect, although the main concept of electron release is the same in both cases.

In 1905, **Albert Einstein** (1879–1955) published a theoretical paper explaining the photoelectric effect. After Robert Millikan (1868–1953) proved the effect experimentally in 1916, Einstein won the Nobel Prize for his theory in 1921.

The photoelectric effect is explained with respect to semiconductors as follows. Semiconductor atoms have three main energy bands: a valence band, a forbidden band, and a conduction band. The **valence band** of an atom is the energy band occupied by the valence electrons. If excited with enough energy from sunlight, an electron in the valence band can jump into the conduction band, which is a band of energy that is otherwise vacant. Thus, the valence and conduction bands hold electrons of different energy levels, with the valence band holding electrons with less energy and the **conduction band** holding electrons with more energy.

The **forbidden band** is an energy gap between the conduction band and the valence band. When an electron obtains sufficient energy, it jumps from the valence band into the conduction band, bypassing the forbidden band. As such, no electrons reside in the forbidden band. In order to jump to the conduction band, an electron in the valence band must obtain enough energy, called **band-gap energy**. It is the energy required for a valence electron to free itself from the electrostatic force holding it to its own nucleus to jump into the conduction band. The band-gap energy is the energy needed for the photoelectric effect to occur.

Electrons in the conduction band contribute to current flow. For metals, the conduction band is partly filled naturally, even at low temperature, due to the fact that some electrons in metals are thermally excited enough to break free of their nucleus at low temperature. For semiconductors, the conduction band is empty at zero degrees Kelvin (−273.15 degrees Celsius), the coldest temperature possible in the universe. At room temperature, only one out of ten billion electrons in the valence band of a semiconductor is thermally excited enough to jump into the conduction band.

When an electron jumps out of the valence band of a silicon atom, it leaves behind a nucleus with four protons but only three electrons, so the silicon atom has a net plus-1 positive charge, or a **hole**. Soon, another electron recombines with the positively charged silicon atom to fill the hole. Since that electron comes from another silicon atom in the lattice, the hole (location of positive charge) shifts to the other atom. As such, holes in the lattice appear to move.

The band-gap energy is the energy needed for a semiconductor's electron to jump from its valence band to its conduction band. The unit of band-gap energy is the **electron volt**, which is the energy an electron acquires when voltage increases by one volt. Silicon has a band-gap energy of 1.12 electron volts. Thus, a wavelength of sunlight must supply at least 1.12 electron volts to propel an electron from silicon's valence band into its conduction band.

The shorter a wavelength of sunlight, the more energy the wavelength contains. Thus, at some long wavelength, the energy contained in the wavelength is less than the band-gap energy. The solar wavelength above which the energy in the wavelength is less than the band-gap energy, is the **band-gap wavelength**. For silicon, the band-gap wavelength is 1.11 micrometers (millionths of a meter). Thus, every wavelength of sunlight shorter than 1.11 micrometers can promote an electron from the valence band to the conduction band. Every wavelength longer than the band-gap wavelength contains too little energy to promote an electron.

At each wavelength shorter than the band-gap wavelength, only 1.12 electron volts is used to promote an electron into the conduction band. Since most wavelengths shorter than the band-gap wavelength contain more energy than the band-gap energy, the extra energy beyond the band-gap energy is wasted. This is one reason why solar panels can never be 100 percent efficient at converting sunlight to electricity: Only a fraction of the energy at each wavelength shorter than the band-gap wavelength is used to promote an electron into the conduction band. In addition, none of the wavelengths longer than the band-gap wavelength have enough energy to promote an electron.

The band-gap wavelength of 1.11 micrometers is a solar infrared wavelength. The solar spectrum includes ultraviolet wavelengths (below 0.38 micrometers), visible wavelengths (0.38–0.75 micrometers), and solar infrared wavelengths (0.75–10 micrometers). Thus, silicon PV cells generate electricity when exposed to ultraviolet, visible, and some solar infrared wavelengths. Ultraviolet wavelengths comprise only about 5 percent of total solar energy.

10.1.3 MAXIMUM POSSIBLE PV-CELL EFFICIENCY

Solar cells use the band-gap energy in wavelengths shorter than the band-gap wavelength to propel valence-band electrons into the conduction band. Only wavelengths shorter than the band-gap wavelength are used, and at each wavelength, only the band-gap energy is used. Thus, the solar energy for producing electricity in a PV cell is limited.

10.1 SOLAR PHOTOVOLTAICS

For silicon, a maximum of only 49.6 percent of total sunlight can be converted to useful band-gap energy. This is because 20.2 percent of radiation is at wavelengths longer than the band-gap wavelength, thus unavailable, and 30.2 percent of radiation that is at wavelengths shorter than the band-gap wavelength has energy above the band-gap energy. Of the remaining 49.6 percent, seven percentage points are lost due to high temperatures, which convert some energy in solar wavelengths to heat that cannot be used, and another nine percentage points are lost because some electrons recombine with positively charged atoms. The resulting maximum possible efficiency of a single junction PV cell is, therefore, about 33.7 percent, which is called the **Shockley–Queisser limit**.

The solar-cell efficiency is the actual power output of a solar cell obtained under standard test conditions divided by the maximum solar power available under those test conditions. **Standard test conditions** are conditions under which industry-wide solar PV panels are tested and compared to evaluate their efficiencies. As of 2024, the record efficiency for a single-junction solar cell, which is made of monocrystalline silicon, was 26.1 percent,[255] thus about 77.4 percent of the maximum possible efficiency.

10.1.4 CREATING ELECTRICITY IN A PV CELL

Electric fields are built into PV cells to carry away electrons boosted from the valence band to the conduction band, preventing the electrons from recombining with silicon atom holes. Electric fields push electrons in one direction and holes in the other. To create an electric field today in a silicon-based PV cell, one side of the cell is contaminated with one atom of phosphorus (P) per thousand atoms of silicon, and the other side is contaminated with one atom of boron (B) per thousand atoms of silicon. Phosphorus has five electrons in its valence shell. Boron has three electrons in its valence shell.

Each phosphorus and boron atom substitutes for one atom of silicon in the lattice of the PV cell (Figure 10.1). Since phosphorus has five electrons in its valence band, it can form five bonds. Silicon can form only four bonds. As such, when phosphorous replaces a silicon atom, phosphorous has a free fifth electron in its valence band. This free electron escapes the atom and roams, giving the immobile phosphorus atom a positive charge. Phosphorus, however, is called an **n-type** (negative-type) material since it donates a negatively charged roaming electron.

Since boron has three electrons in its valance shell, it can form only three bonds with silicon. When boron substitutes into the lattice of silicon,

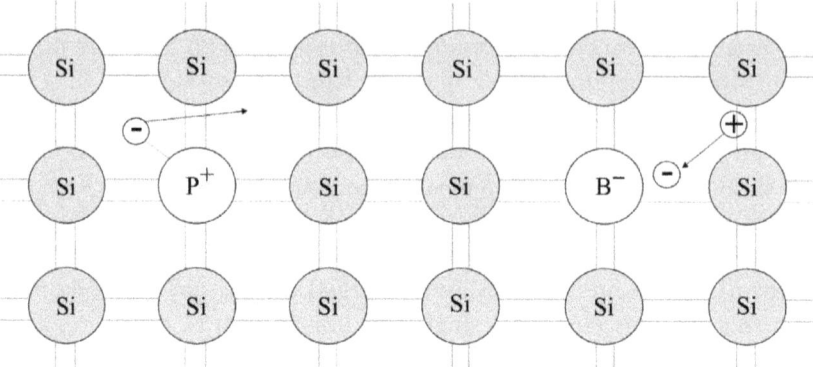

Figure 10.1. Diagram showing the substitution of a phosphorus (P) atom (left) and boron (B) atom (right) for a silicon (Si) atom in the lattice of a PV cell. The P atom, which attains an immobile positive charge, is an n-type material since it donates a negatively charged roaming electron. The B atom, which attains an immobile negative charge, is a p-type material since it results in a positively charged roaming hole. Adapted from Karolkalna at the English Wikipedia.

boron forms bonds with three of its four silicon-atom neighbors. The boron tries to bond with the fourth neighbor, first by moving its own three valence band electrons around, then by borrowing an electron from a nearby silicon atom. This results in the immobile boron becoming net negatively charged and a positive hole forming in a nearby silicon atom where the electron came from. Since a roaming electron from another silicon atom soon fills the hole, the hole roams among silicon atoms. Because the boron creates a roaming hole (positive charge), boron is called a **p-type** material.

Mobile electrons from the n-type (phosphorus) side of the PV cell drift by diffusion toward the p-type side. Similarly, mobile holes from the p-type side roam toward the n-type side. In the middle, mobile electrons fill mobile holes, creating uncharged, electrically neutral atoms. This no-charge **depletion region** is about one micrometer wide and is called a **p-n junction**.

Within the depletion region, the immobile positively charged phosphorus atoms on the n-type side of the p-n junction are now no longer balanced by roaming negative charges. As such, that region takes on a net positive charge. Similarly, the immobile negatively charged boron atoms on the p-type side of the p-n junction are no longer balanced by roaming holes, so that side takes on a net negative charge. In sum, the n-type side of the depletion region now has immobile positive charges and the p-type side now has immobile negative charges, giving rise to a charge gradient. When an external wire is connected from a cap at the end of the p-type region to a cap at the end of the n-type region, the charge gradient gives rise to an electric field that can sweep away conduction-band electrons to create a direct current. With metals, a p-n junction can't form, so free electrons move too randomly for a direct current to arise.

In a PV cell, the n-type layer can be on top (closest to the sun) or on the bottom. The most efficient solar cells generally have a p-type layer on top and an n-type layer on the bottom, with a p-n junction in between. In either case, photons of sunlight energize electrons in the valence band of silicon atoms in the layer on top, moving the electrons into the conduction band. If these drift into the depletion region, the electric field sends the electrons to the n-type layer. If the wire attached to the PV cell is connected to a load, electrons flow out of the n-type side into the wire and through the load and back to the p-type side. Electrons reaching the p-type side recombine with holes there, completing the circuit.

10.1.5 TYPES OF PV CELLS AND THEIR MATERIALS

Silicon is Earth's second most abundant element, comprising 20 percent of Earth's crust. The source of silicon for PV cells is silicon dioxide (SiO_2), also called silica and quartz, which comes from sand or a mine. Silicon processing begins with a high-temperature electric-arc furnace that currently uses carbon to reduce silicon dioxide to metallic grade (99 percent pure) silicon. Silicon is purified further until it is more than 99.9999 percent pure for use in PV cells.

First-generation PV cells, still widely used, are 160–240 micrometers thick. They are made of either **single-crystal silicon** (also called monocrystalline silicon) or **polycrystalline silicon** (also called multicrystalline silicon). Single-crystal silicon cells are uniform in structure and appear square but with clipped corners, so are really octagonal. They have a distinctive pattern of small white diamonds. A method of growing monocrystalline silicon was discovered in 1916 by the Polish chemist, **Jan Czochralski** (1885–1953) after accidentally dipping his pen in molten tin. This resulted in solid tin hanging from the nib of his pen. He verified that the crystal was **monocrystalline**, which is a solid material in which the crystal lattice is continuous and unbroken to the edges. Polycrystalline silicon cells resemble rock-like chunks of a multifaceted metal. Because single-crystal silicon cells are uniform in structure, they are more efficient but also more expensive to manufacture than polycrystalline cells.

Second-generation PV cells, called **thin-film PV** cells, are 1–10 micrometers thick. They are much thinner than first-generation cells, reducing the amount of expensive material needed. Thin-film PV cells usually consist of either **amorphous silicon** or a material with two to four elements, such as gallium arsenide, cadmium telluride, copper sulfide, cadmium sulfide, or copper-indium-gallium-diselenide. Thin-film materials are wedged between two panes of glass. Since glass is heavy and first-generation cells use only one pane of glass, second-generation cells are heavier than first-generation cells. However, the energy required to produce first-generation cells is greater.

Copper-indium-gallium-diselenide and gallium arsenide thin-film cells are efficient but expensive. They cost more per cell than first-generation silicon-based cells. Cadmium telluride cells, though, are less efficient than first-generation cells but cost less per cell and have an overall energy cost comparable with that of first-generation cells.

Thin-film solar cells were invented to power the first solar calculators in the late 1970s. They contained a small strip of amorphous silicon. In 1980, the University of Delaware developed the first thin-film solar cell for electricity generation. The efficiency exceeded 10 percent with a cell made of copper sulfide and cadmium sulfide. In 1992, the University of South Florida increased the efficiency of thin-film cells to 15.9 percent with a cell made of cadmium telluride. In 2024, the record efficiency of a thin-film cadmium-telluride cell was 22.6 percent. That of a thin-film copper-indium-gallium-diselenide cell was 23.6 percent and of a thin-film gallium arsenide cell was 27.8 percent.[255]

Amorphous silicon is made by chemical vapor deposition of silane gas and hydrogen gas. As such, it is noncrystalline. If tiny crystals nucleate and grow to about a micrometer in size within the amorphous silicon, the resulting material is called microcrystalline silicon. Amorphous silicon has a much lower band-gap wavelength and much greater band-gap energy than monocrystalline silicon. Thus, amorphous silicon absorbs across a much smaller portion of the solar spectrum than does monocrystalline silicon. As a result, amorphous silicon also has a much lower efficiency (14.0 percent) than monocrystalline silicon (26.1 percent).[255]

Third-generation solar cells are thin-film cells that are either emerging or still too expensive for widespread commercial use. Three of these are as follows:

Organic cells (cells that contain organic material) tend to be less expensive per cell than silicon-based cells but also have lower efficiencies. As of 2024, the record efficiency of a thin-film organic cell was 17.5 percent.[255] In addition, sunlight photochemically degrades organic-based cells over time.

Multijunction (tandem) cells are either individual thin-film PV cells containing multiple materials or a stack of thin-film PV cells, each

10.1 SOLAR PHOTOVOLTAICS

made of a different material, connected in series. In both cases, each material absorbs a different portion of the solar spectrum (thus has a different band-gap energy and band-gap wavelength). For example, with a two-band-gap stack of cells, the top cell may have a high band-gap energy, thus absorb more energy per photon, but only up to a limited wavelength. The bottom material may have a low band-gap energy, thus absorb less energy per photon, but over more wavelengths in the solar spectrum.

Whereas the theoretical peak efficiency of a single-junction PV cell is about 33.7 percent, that of a stack of two cells with different band gaps is around 47 percent. That of a stack of three cells is around 53 percent and of a stack of eight cells is about 62 percent.[256] To date, tandem cells have been used primarily in satellite PV arrays and in technologies that use lenses and mirrors to focus light onto a cell to increase the intensity of sunlight hitting the cell.

The record for a gallium arsenide double-junction cell in 2024 was 32.9 percent.[255] That for a double-junction perovskite silicon cell was 33.9 percent.[255] The record for a five-junction tandem cell was 39.46 percent.[255] A **perovskite** is a material with a crystal structure similar to the crystal perovskite, which contains calcium titanium oxide. Common perovskite materials used in solar cells are hybrid organic–inorganic lead or tin halide-based materials.

Concentrator cells are solar PV cells in which lenses or curved mirrors focus light onto multijunction cells. The efficiency of a concentrator cell is the largest among all PV cell types. The record efficiency for any cell measured in 2024, for example, was 47.6 percent, which was for a four-junction concentrator cell.[255] Concentrator cells, however, require direct sunlight, just as concentrated solar power does, so do not work well under cloudy or hazy conditions where direct light is scattered to become diffuse light. Thus, concentrator cells are limited to desert-like areas with lots of direct sunlight. Also, the concentrated light increases cell temperature more than in conventional cells, so a heat sink is needed. Concentrator cells may also need more frequent cleaning than conventional cells.

Because so many alternative PV cell types are available, it is unlikely that material limits will constrain the large-scale growth of PV. For example, in multijunction cells, the limiting material is germanium; however, substituting gallium, which is more abundant, would allow terawatt expansion. In addition, the production of silicon-based PV cells is limited not by crystalline silicon (because silicon is abundant) but by silver, which is used as an electrode. Reducing the use of silver as an electrode would allow the virtually unlimited production of silicon-based solar cells. In sum, the development of a large global PV system will not be limited by the scarcity or cost of raw materials.

10.1.6 PV PANELS AND ARRAYS

A **PV panel** consists of either 32, 36, 48, 60, 72, 96, or 128 prewired PV cells in series, fitted into a rectangular package. A **PV array** is a group of PV panels wired in series to increase the voltage, or in parallel to increase the current. For panels wired in series, the total voltage is the sum of the voltages across all individual panels, and the total current is the current running through any one panel. For panels wired in parallel, the total current is the sum of currents through each panel, and the total voltage is the voltage across one panel.

Alternatively, an array can be wired partly in series and partly in parallel to optimize power (voltage multiplied by current). In this case, panels are first wired in series to increase the voltage by as much as is safe, then each series string is wired in parallel to maximize power. Maximizing voltage also minimizes current, minimizing power losses along wires. When arrays are set up in this manner, the total voltage is the sum of voltages across one string of arrays. The current through the string is the current through one panel in the string. The total current through the array is the sum of currents through each parallel string.

10.1.7 UTILITY VERSUS DISTRIBUTED PV

Utility PV is large-scale ground-mounted PV connected to the transmission and distribution grid. Distributed PV is small-scale PV affixed to or incorporated in buildings (rooftops, walls, windows, carports), parking lots, parking structures, hillsides, yards, and vacant lots that service buildings. As discussed in Section 2.7, distributed PV may be behind-the-meter (BTM) or in-front-of-the-meter (FOM) PV. BTM distributed PV (hereafter, BTM PV) services buildings directly but is not owned or controlled by the utility managing the grid or subject to the same market and grid-connection rules as are utility-PV systems. FOM systems are connected to distribution lines, thus also serve buildings directly, but they are subject to the same market and grid-connection rules as utility-PV systems.

If BTM PV is colocated with battery storage, the system first provides electricity to a building it services. Any excess electricity is then stored in

10.1 SOLAR PHOTOVOLTAICS

the batteries. Remaining electricity is sent to the grid if the system is grid connected; otherwise, it is lost through curtailment. Thus, if all else is the same, BTM PV reduces immediate grid-electricity demand by supplying electricity directly to buildings, avoiding the need for grid electricity to supply those buildings.

Grid operators generally oppose BTM distributed PV because its first impact is to reduce demand for grid electricity. Utilities claim that the remaining customers must pay a higher cost for the remaining demand, mostly because the fixed cost of the transmission and distribution system is now spread over fewer customers. They further argue that only wealthy people can afford BTM PV, so the higher cost of grid electricity disproportionately affects low-income grid customers. Utilities have used this argument to stymie the expansion of BTM PV in many states (e.g., California, Hawaii, Nevada, Arizona, Utah, and Florida) and countries.

However, the opposite is true. BTM PV lowers electricity, health, and climate costs for everyone, for at least 11 reasons.

First, the argument that BTM PV reduces grid electricity demand and, therefore, increases costs to grid customers by spreading the fixed cost of transmission and distribution over fewer customers, ignores the realities of the current energy transition, where transportation, buildings, and industry are being electrified and the electricity is increasingly being provided by WWS. Upon a full transition, this results in an overall reduction in energy demand worldwide of about 54 percent.[43] However, all remaining energy will be electricity, so electricity needs will almost double.[43] With a doubling of electricity demand, even if 25 percent of the total electricity demand were met with BTM PV, overall grid electricity needs would still increase by 50 percent compared with today. Thus, the assumption by utilities that a large growth in BTM PV reduces grid demand holds true only for low levels of electrification, not for the large-scale electrification that is needed to address climate, pollution, and energy-security problems.

Second, BTM rooftop PV electricity requires no new land, whereas utility PV needs new land. Thus, most BTM PV reduces land requirements and habitat damage compared with utility PV, benefiting both BTM and grid customers.

Third, BTM PV reduces the need for transmission and distribution lines. BTM PV users connected to the grid still need transmission and distribution lines when their PV and colocated battery system do not produce sufficient electricity or when their PV system produces excess electricity, which is fed to the grid through the lines. In contrast, grid customers need transmission and distribution lines for 100 percent of their electricity consumption, and utility PV requires transmission and distribution lines for 100 percent of its generation.

Fourth, when a BTM-PV and colocated battery system produces more electricity than the building it serves consumes, the excess electricity is fed to the grid. This helps grid customers avoid blackouts during hot summer days in particular.

Fifth, transmission-line sparks have led to devastating wildfires, such as in California, Hawaii, and Australia. The costs of such fires and undergrounding transmission lines due to the fires have been passed down to customers in California, for example.[257] BTM PV reduces fire occurrence and these costs.

Sixth, by replacing grid electricity, BTM PV reduces the mining, processing, and burning of polluting fuels (fossil fuels and bioenergy) for electricity generation on the grid. Reducing polluting fuels reduces exposure of the general population to gases and particles that cause morbidity, mortality, and health costs. Thus, BTM PV reduces health costs for both distributed-PV and utility-PV customers. Since many electricity-generating power plants are located near low-income communities, the health-cost benefits of BTM PV accrue more to low-income residents than to high-income residents.

Seventh, by reducing greenhouse gas emissions from polluting fuels, BTM PV reduces climate damage to both BTM-PV and grid customers.

Eighth, by reducing the use of fossil fuels, BTM PV reduces energy-insecurity problems associated with fossil fuels, and this benefits both BTM-PV and grid customers.

Ninth, installing BTM PV creates more jobs than installing and running utility PV and other grid-scale electricity generation, and this benefits a state or country as a whole.

Tenth, because rooftop PV absorbs 20–26 percent of the sunlight that hits it, then converts the light to electricity, less light is absorbed by the building, cooling the building during the day, reducing daytime electricity demand for air-conditioning.[258] Such cooling is greatest during summer and during the day, when electricity prices are highest. By reducing demand in this way during the times of the day when grid electricity use is highest, BTM PV reduces strain on the grid and the risk of blackout to grid customers. At night, solar PV panels act as insulators, keeping buildings slightly warmer than they otherwise would be,[258] potentially reducing the demand for heating at night.

Eleventh, BTM PV facilitates the transition of a building to all-electric, thereby reducing occupant costs in the short and long run. Normally, two forms of energy, such as fossil gas and electricity, are used in buildings.

10.2 SOLAR RESOURCES AND HOW TO USE THEM EFFICIENTLY

However, there is nothing that fossil gas can do that electricity cannot do less expensively and more cleanly. An issue arises because the more that appliances in a building are switched from gas to electric, the greater the electricity demand in the building. Utilities often charge a higher rate for electricity use beyond a threshold. This disincentivizes customers from electrifying more appliances. BTM PV provides the additional electricity needed at a lower cost than does grid electricity in most places, so electrifying a building that has BTM PV reduces overall electricity cost to BTM-PV customers versus electrifying and using only grid electricity. Moreover, electrifying reduces outdoor and indoor air pollution from fossil-gas use in buildings and from mining the gas, benefiting both BTM-PV and grid customers.

BTM PV results in at least three additional benefits to PV owner and building occupants but not necessarily to users of grid electricity. First, BTM PV allows building occupants to keep their electricity on during a grid blackout. With one or two batteries colocated with the PV, the building can also continue to function using stored solar electricity during any night. Second, although the wholesale price per unit of electricity of utility PV is less than the cost per unit of electricity of BTM PV, utility customers do not pay the wholesale price of utility PV. Instead, they pay the retail price plus the price of transmission and distribution, which sums to about four times the cost of BTM PV in, for example, California. As such, a utility-PV customer who adds and uses BTM PV can reduce their daytime cost of electricity by up to a factor of four. Third, builders of new homes that have BTM PV but no fossil gas eliminate the costs of installing fossil-gas pipes and digging ditches for them ($3,000–$20,000) and of a fossil-gas hookup fee charged by the utility ($3,000–$15,000).

In sum, BTM PV should be installed as much as possible worldwide. It reduces not only pollution from current electricity generation but also the need for land and transmission and distribution lines, thereby reducing the costs associated with both, as well as wildfire risk. Due to the scale of the WWS transition needed, both distributed (BTM and FOM) and utility PV will be needed in large quantities. Thus, policies should encourage both and hinder neither.

10.2 SOLAR RESOURCES AND HOW TO USE THEM EFFICIENTLY

10.2.1 SOLAR RESOURCES

In the annual average, incident solar radiation reaching the Earth's surface peaks near the equator and generally decreases between the equator and the north and south poles. However, a slight increase in surface solar radiation occurs at the poles, relative to at 60 degrees north and 60 degrees south of the equator, due to lower cloud cover in polar regions than at those latitudes. Similarly, in the United States, more sunlight reaches the ground in the southwest than in the southeast due to the greater cloud cover in the southeast. The Sahara Desert also receives a substantial amount of sunlight at the ground due to low cloud cover there.

Worldwide, about 97,000 terawatts of solar power hit the Earth's surface in the annual average, after accounting for clouds, gases, and particles in the air, but before accounting for any reflection by the ground. Of this downward sunlight, about 28.7 percent hits land.

If the Earth's land and ocean were covered completely with horizontal solar PV panels with an efficiency of 20 percent (with no spacing between panels and ignoring any feedbacks of the panels to climate), the global electricity available from solar PV worldwide would then be about 19,400 terawatts. That over land would be about 5,600 terawatts.

If the world's all-purpose energy demand were electrified, the annually averaged end-use power demand in 2050 would be about nine terawatts.[43] As such, enough solar theoretically exists over land alone to provide the world's 2050 end-use power 620 times over. However, many locations over land, such as Antarctica and the Himalayan Mountains, are not easily accessible to transmission. In addition, panels require walking space between rows and space for inverters, which convert PV panel DC electricity to AC electricity that is sent to the grid. In addition, space is needed for agriculture and forests. These factors reduce the practical solar installation potential. The world's likely developable solar PV resource over land is thus about 1,300 terawatts. As such, about 144 times the world's all-purpose end-use demand is available from accessible solar PV over land.

Fortunately, the small amount of sunlight incident on flat (horizontal) panels in high northern and southern latitudes does not limit the installation of solar PV there. The reason is that incident sunlight on a panel is magnified by up to a factor of 2.6 if the panel tracks the sun or is tilted. As such, solar PV can be used to generate electric substantial power almost anywhere on Earth, including over water.

Floating solar PV panels have been used over inland waters, which are relatively calm, and ocean surfaces, which are rough, to generate electricity. The first floating solar panels were installed over a lake in Aichi, Japan, in 2007. In 2008, a winery in Oakville, California, installed 175 kilowatts of solar PV over an irrigation pond. Larger installations were installed subsequently over a reservoir in the Chiba Prefecture of Japan in 2016 (13.4 megawatts); over a quarry lake in Huainan, China, in 2017

10.2 SOLAR RESOURCES AND HOW TO USE THEM EFFICIENTLY

(40 megawatts); over a lake that is part of the Three Gorges Dam complex in China in 2018 (150 megawatts); over a reservoir in South Korea in 2021 (41 megawatts); over a reservoir in Singapore in 2021 (60 megawatts); and over a reservoir in Dezhou, China, in 2021 (320 megawatts), for example.

Ocean surfaces are rougher than are inland waters due to waves, and they corrode panels more, due to salt. Nonetheless, breakwaters and jetties can be constructed to create areas of calm water, just as in a harbor. In addition, solar arrays elevated a couple of meters above a rough ocean surface can reduce panel corrosion. Since most people worldwide live near the coast, many more offshore floating solar arrays are expected, particularly in countries such as Gibraltar, Malta, and Singapore, for example, that are densely populated and land constrained.

Transition highlight
During November 2024, Taiwan completed the world's largest offshore floating PV farm to date, a 440-megawatt PV array off its western coast. Also in November 2024, China connected the first part of what will be a 1-gigawatt offshore PV farm in the Yellow Sea. This expanded on China's first deep-water floating PV project, a 400-megawatt farm in Laizhou Bay, commissioned in August 2024. The PV panels were bifacial, thus generated electricity when sunlight hit either side of the panel.[259] To minimize damage due to corrosion, the panels included two layers of coated glass, ultraviolet-light-resistance film over the glass, and water-sealed rubber edges.

10.2.2 SUNLIGHT REACHING SOLAR PANELS

PV panel output depends on how much sunlight reaches the panel. The sunlight reaching a panel depends on the angle of the sun; cloud cover; the quantity of gases and particles in the air blocking the sun; shading by hills, mountains, trees, and buildings; and the orientation of the panel relative to the direct solar beam. In this section, these factors, aside from panel orientation, which is the subject of the next section, are discussed.

The amount of sunlight reaching the outside of the Earth's atmosphere varies according to the day of the year because the Earth's distance from the sun varies each day. The average Earth–sun distance is about 150 million kilometers. But, that distance is 147.1 million kilometers on December 22 (Northern Hemisphere winter solstice) and 152.1 million kilometers on June 22 (Northern Hemisphere summer solstice) because the Earth rotates around the sun in an elliptical orbit with the sun at one focus. This 3.4 percent difference in Earth–sun distance between December and June corresponds to a 6.9 percent difference in sunlight reaching the outside of Earth's atmosphere between these months. In other words, 6.9 percent more radiation falls on the Earth in December than in June. Although more sunlight reaches Earth during the Northern Hemisphere winter than summer, the north pole is tilted away from the sun during December, so most of that light falls on the Southern Hemisphere, making the Northern Hemisphere cold in December.

Sunlight is relatively unimpeded as it travels from the sun to the outside of the Earth's atmosphere. However, when it passes through our atmosphere, it interacts with gases, aerosol particles, and clouds. These obstacles may either scatter or absorb the light or allow the light to transmit through. **Scattering** is the process by which gases, aerosol particles, or cloud particles deflect radiation in a random direction, just as a tree deflects a rock thrown at it. All gases, aerosol particles, and cloud particles in the atmosphere scatter radiation to some degree. The amount of scattering depends on several factors and varies with wavelength of light. **Absorption** is the process by which sunlight enters a gas or particle in the air, and the light is converted to invisible heat, which is then reradiated to the surrounding air. Only a few gases, such as nitrogen dioxide and ozone, absorb sunlight. Very few particle components, such as black carbon, brown carbon, and iron-containing particles, absorb sunlight.

Sunlight that is unimpeded by scattering and absorption, and so makes its way directly to a solar panel, is **direct sunlight**. Sunlight that is scattered out of the direct solar beam by gases and particles but then encounters additional gases and particles that scatter some of the light to the solar panel is **diffuse sunlight**. For example, if the sun is close to the horizon, the direct solar beam may intersect the underside of a cloud, which scatters the light. Some of that scattered diffuse sunlight may then hit a solar panel.

Cloud particles, aerosol particles, and gases all decrease direct sunlight by scattering light out of a direct solar beam or absorbing light along the beam, but they all increase diffuse radiation as well. Hills, mountains, trees, and buildings also decrease direct sunlight to a panel. On average, the quantity of diffuse radiation to a PV panel is about 10 percent of the amount of direct sunlight reaching the panel. However, when the sun is blocked by heavy clouds, pollution, or shading by hills, mountains, trees, or buildings, nearly 100 percent of sunlight reaching the panel is diffuse.

10.2.3 SOLAR PANEL TILTING AND TRACKING

If a solar PV panel sits flat (lies horizontally) on the ground and the sun is directly overhead, the panel receives the maximum possible radiation from the sun. The sun can shine directly over a horizontal panel only if the panel

10.2 SOLAR RESOURCES AND HOW TO USE THEM EFFICIENTLY

lies between 23.5 degrees north of the equator (Tropic of Cancer) and 23.5 degrees south of the equator (Tropic of Capricorn). This is because the Earth's axis of rotation is tilted at 23.5 degrees relative to a line perpendicular to the sun, so the sun is directly overhead the Tropic of Cancer on June 22, the Tropic of Capricorn on December 22, and all other latitudes in between on every other day. If the Earth had no tilt, the sun's light would be directly over the equator every day. On the other hand, if the face of a solar panel swivels to follow the sun, the panel receives the maximum possible direct-beam sunlight, in addition to receiving diffuse sunlight, every minute of every day (but not at night).

At sunset and sunrise (and ignoring the slight bending of light that gives rise to twilight), a panel sitting flat on the ground receives no direct sunlight since direct radiation flies right over the panel, parallel to it, at those times. Instead, the panel receives only diffuse radiation. If the panel instead tracks the sun, direct sunlight, even at sunset, hits the panel straight on. As such, tracking the sun with a panel maximizes solar PV output.

Even tilting panels at an optimal tilt angle, without swiveling the panels east to west, improves solar output relative to flat panels. An **optimal tilt angle** of a solar panel is the estimated fixed tilt angle of the panel relative to the horizon that gives the greatest annually averaged incident sunlight on the panel compared with any other fixed tilt angle. The optimal tilt angle depends not only on latitude but also on cloud cover, elevation, and air pollution levels.

Rooftop PV panels are mostly placed at a fixed tilt angle relative to the sun although some lie flat. Utility solar PV panels are almost all placed at a fixed tilt or rotate to track the sun. Tilted panels generally, but not always, face south, southwest, or southeast in the Northern Hemisphere or north, northwest, or northeast in the Southern Hemisphere.

Panels that track the sun can track either on one axis vertically, one axis horizontally, or two axes. One-axis vertically tracking panels face south or north and swivel vertically (up and down) around a horizontal axis. Thus, they follow the sun vertically as it goes from the horizon to its highest point in the sky, but not horizontally, from east to west. One-axis horizontally tracking panels tilt at an optimal angle and swivel horizontally between east and west around a vertical axis. Thus, they follow the sun from east to west but not vertically from the horizon to the highest point in the sky. Two-axis-tracking panels combine one-axis vertical- and one-axis horizontal-tracking capabilities to follow the sun perfectly during the day.

Whether panel tracking or tilting is used depends on the annual amount of sunlight the panel receives, the cost of tracking equipment, and the cost of additional land needed to avoid shading. For example, two-axis tracking of solar panels in a utility-PV farm requires more shading behind each row, thus more land area than do one-axis-tracking panels or optimally tilted panels. Two-axis-tracking equipment is also more expensive than is one-axis tracking equipment or optimal-tilting equipment.

10.2.3.1 OPTIMAL TILT ANGLES

Optimal tilt angles increase with increasing latitude from equator to pole in both hemispheres. The higher the latitude, the less that sunlight is perpendicular to the Earth's surface. Thus, panels need to tilt more and more toward the low-lying sun with increasing latitude. The simplest estimate of the optimal tilt angle of a PV panel at a latitude is the latitude itself. Thus, at 45 degrees north latitude, a PV panel should face south with a tilt angle of 45 degrees relative to the horizon. At high latitudes, where clouds are prevalent, optimal tilt angles are not determined so simply.[260]

Transition highlight
At the same high latitude but at different longitudes, optimal tilt angles can differ a lot from each other. For example, Calgary (51.12 degrees north latitude) has a higher optimal tilt angle (45 versus 34 degrees) than does Beek, the Netherlands (50.92 degrees north latitude), which is at a similar latitude.[260] The reason is that Calgary is exposed to less cloud cover than Beek, so panels can take advantage of the overhead sun more efficiently in Calgary. In Beek, panels are exposed to less direct sunlight due to more cloud cover so must take advantage of diffuse light scattered by clouds above them. The lower the optimal tilt angle, the more PV can receive diffuse light under clouds.

10.2.3.2 IMPACTS OF TILTING AND TRACKING VERSUS HORIZONTAL PANELS ON SOLAR OUTPUT

In the global and annual average, two-axis-tracking panels receive about 35 percent more sunlight than do horizontal panels. Similarly, one-axis horizontal-tracking panels and one-axis vertical-tracking panels receive about 32 percent and 9 percent, respectively, more sunlight than do horizontal panels. Optimally tilted panels receive about 5 percent more sunlight than do horizontal panels.[260] These percentages vary with latitude.

At virtually all latitudes, one-axis horizontal-tracking panels receive almost the same (within 1–3 percent) sunlight as two-axis-tracking panels. Because one-axis-tracking panels require less land and cost less for nearly the same

10.2 SOLAR RESOURCES AND HOW TO USE THEM EFFICIENTLY

output as two-axis panels, one-axis horizontal tracking is preferred over two-axis tracking. Moreover, one-axis horizontal-tracking panels are more efficient than, thus preferred over, one-axis vertical-tracking panels. Optimally tilted panels receive less sunlight than do one-axis horizontal-tracking panels, but optimal tilting is still preferred on rooftops and in many solar arrays to minimize space needs and to eliminate the cost of tracking equipment. In sum, ignoring land needs and costs of tracking equipment, the recommendation for utility PV is for one-axis horizontal-tracking panels.[260] For rooftop PV, optimal tilting is recommended.

Transition highlight
Where is the sunniest place on Earth from a solar panel's perspective? Over the south pole, horizontal panels receive relatively little radiation. However, optimally tilted and tracking panels there receive more sunlight, averaged annually, than anywhere else on Earth.[260] This is primarily due to the fact that the south pole receives sunlight 24 hours per day during the Southern Hemisphere summer. Also, much of the Antarctic is at a high altitude, thus above more air and clouds than over the Artic or other latitudes. Tilted and tracking panels over the Arctic receive more radiation than between 40–80 degrees north latitude, but less than over the Antarctic, due to the lower altitude and greater cloudiness above the surface of the Arctic than Antarctic.

11

ONSHORE AND OFFSHORE WIND ENERGY

After sunlight, wind has the most practical potential to supply the world's all-purpose energy demand. Not only are wind resources abundant in almost every country, but the cost of onshore wind energy has declined so much that, in 2024, onshore wind (along with utility PV) is the least expensive form of new electricity in many countries. Its low cost has resulted in massive wind farms replacing fossil-fuel power plants and supplying new electricity demand. This chapter discusses the history of windmills and wind turbines, followed by a discussion of different types of wind turbines and of wind-turbine components. The chapter also describes how wind turbines work and how generators convert the mechanical energy from a rotating blade into electricity. Finally, the chapter analyzes the wind-farm footprint and spacing areas needed to power the world, wind resources available worldwide, and the impacts of wind turbines on global wind speeds, global temperatures, hurricanes, birds, and bats.

11.1 BRIEF HISTORY OF WINDMILLS AND WIND TURBINES

People often confuse windmills with wind turbines, but they are different. A **windmill** converts kinetic energy from the wind into mechanical energy. A **wind turbine** converts the wind's kinetic energy into electricity.

The first known use of a circular wheel driven by the wind to provide mechanical energy was during the first century CE. The engineer, **Heron of Alexandria** (c. 10–70 CE), living in Roman Egypt, invented a small windmill that rotated around a **horizontal axis** (parallel to the ground) and was connected to a piston that moved up and down with each full rotation of the windmill, forcing air through pipes in an organ. The resulting tweets and other sounds from the organ were similar to those from a flute.

In the fourth century CE, prayer wheels were invented in India, Tibet, and China that rotated horizontally around a **vertical axis** (an axis perpendicular to the ground) due to the power of the wind. Between the seventh and ninth centuries CE, the Persians, near the border of present-day Iran

11.1 BRIEF HISTORY OF WINDMILLS AND WIND TURBINES

and Afghanistan, developed a windmill with 6–12 rectangular sails rotating horizontally around a vertical axis to pump water and to grind corn into cornmeal and grain into flour. The use of windmills for these purposes spread across the Middle East, Central Asia, India, and China. By 1000 CE, the Chinese and Sicilians used vertical-axis windmills to pump seawater to make salt. Horizontal-axis windmills were developed to grind grain into flour in northwestern Europe beginning around 1180 CE.

Windmills were popularized by Miguel de Cervantes in his 1604 novel *Don Quixote*. In it, a man believes he is a knight named Don Quixote. He travels around Spain by horse with his squire, Sancho Panza. In one adventure, Don Quixote encounters 30–40 windmills on the plain in front of him but imagines they are giants. He tells Sancho he wants to slay the long-armed giants. Sancho tells him they are not giants but windmills that rotate in the wind, turning a millstone, which is a circular stone used to grind grain. Don Quixote doesn't believe Sancho. Instead, Quixote believes a magician changed the windmills to giants to hurt him. Quixote goes on to joust with one but is knocked off his horse by it. This interaction gave rise to the expression, "tilting at windmills," which means to attack imaginary enemies.

The first windmill in the Americas was built in 1621 by order of Virginia's governor. It was a horizontal-axis windmill built on the Flowerdew Hundred plantation of Sir George Yeardley and used to grind grain. By 1648, Virginia boasted at least four windmills.[263] In 1628, the Dutch built a wind-powered sawmill in New Amsterdam (now Manhattan). Later windmills were built by the French in Quebec City or Montreal (1629 or 1648), the English near Boston (c. 1630) and Maryland (c. 1651), the Swedes on the Delaware River (c. 1640), and the English in Barbados (c. 1654).[263] Windmills were used to grind grain, saw wood, pump water, extract oil from seeds, and crush sugarcane. During the American Revolution, the British set up a blockade to stop salt from entering the colonies. To evade the blockade, locals in Cape Cod, Massachusetts, built several windmills to pump seawater into evaporating vats to make salt.

On August 29, 1854, the inventor **Daniel Halladay** (1826–1916) patented the first commercially viable windmill worldwide, which he had built in his Connecticut machine shop. The differences between his and previous windmills were that he added a tail fin (**wind vane**) to allow his windmill automatically to change direction to face the wind, and his windmill maintained a constant speed by changing the **pitch** (steepness of slope) of its sails without human oversight. Halladay's windmills were sold by the thousands and were used to pump water and grind grain on farms and ranches. They helped to open the western United States to rail transport. Trains required water for their steam engines, and windmills were used to pump water along rail routes for the trains. Between 1854 and 1970, over six million windmills were installed in the United States.

In July 1887, **Professor James Blyth** (1839–1906), an electrical engineer at Anderson's College, Glasgow, built and operated the world's first wind turbine. The turbine used cloth sails to turn a rotor. It produced electricity that charged an energy storage device. The stored electricity lit up Blyth's cottage.

Around the same time, between 1887 and 1888, **Charles F. Brush** (1849–1929) of Cleveland, Ohio, built a much larger wind turbine with a unique design. It was 17 meters in diameter, had 144 rotor blades, and was made of cedar wood. The blades were mounted on an 18-meter tower and generated a peak of 12 kilowatts of electricity that he used to charge 12 batteries, which in turn powered 100 light bulbs, 3 arc lamps, and several motors in his mansion. The wind-turbine and battery bank operated for 20 years.

In 1891, **Paul la Cour** (1846–1908) of Denmark invented a wind turbine that he used to split water with an electrolyzer to produce hydrogen gas. The hydrogen powered several hydrogen gas lamps. Subsequently, between 1900 and 1940, hundreds of thousands of small wind turbines, with nameplate capacities of 5–25 kilowatts, were installed to produce electricity in rural areas of the United States that lacked access to grid electricity.

The first wind farm in the world was the **Crotched Mountain**, New Hampshire, farm, completed in December, 1980, by US Windpower. The farm consisted of twenty 30-kilowatt turbines, each 18.3 meters above ground at hub height. The farm was dismantled following bankruptcy of the subsequent owners. Nevertheless, Crotched Mountain led to the development of three enormous commercial wind farms in California that still exist. Between 1981 and 1986, California installed 15,000 wind turbines in the **Altamont Pass**, **Tehachapi Pass**, and **San Gorgonio Pass** wind farms. These three farms had a combined nameplate capacity exceeding 1 gigawatt. Today, Altamont Pass is still the largest wind farm in the world in terms of turbine number (almost 5,000), but no longer in terms of nameplate capacity. From 1990 to 2000, the center of wind-farm development moved to Europe. During that period, 10 gigawatts were installed in Europe, compared with 2.2 gigawatts in the United States.

In 1991, the first fixed-bottom offshore wind turbine, with a nameplate capacity of 225 kilowatts, was installed in 25-meter-deep water offshore of Sweden. In October 1991, Denmark installed the first offshore wind farm, off the coast of the town of Vindeby. It consisted of eleven 450-kilowatt

11.1 BRIEF HISTORY OF WINDMILLS AND WIND TURBINES

turbines in 2–4 meters of water, 3 kilometers from shore. It was decommissioned in 2017.

In 2007, the world's first prototype (80 kilowatt) floating offshore wind turbine was installed off the coast of Apulia, Italy, by Blue H Technologies of the Netherlands. That was followed by a 2.3-megawatt floating turbine installed off the coast of Norway in 2009; a 2-megawatt floating turbine installed off the coast of Portugal in 2011; and many individual turbines after that. The world's first floating offshore wind farm was the Hywind-Scotland farm (six 5-megawatt turbines) offshore of Peterhead, Scotland, commissioned in October 2017. The largest floating offshore farm in 2024 was the 88-megawatts Hywind Tampen farm off Norway's coast, commissioned in August 2023.

Since 2000, onshore and offshore wind-turbine development has taken off worldwide. In 2022, several countries produced 20–54 percent of the electricity they generated from wind. Denmark produced 54.2 percent of the electricity it generated from wind; Lithuania, 35.8 percent; Ireland, 33.4 percent; Uruguay, 32.3 percent; Portugal, 28.5 percent; Luxembourg, 26.5 percent; the United Kingdom, 24.9 percent; Spain and Curacao, 21.8 percent each; Germany, 21.7 percent; and Greece, 20.7 percent.[261] By the end of 2024, the world had installed 1,052 gigawatts (nameplate capacity) of onshore wind and 75.2 gigawatts of offshore wind. These totals represent 12.5 percent and 16.9 percent increases, respectively, above installations in 2022.[262]

> **Transition highlight**
> From October 1, 2023 to September 30, 2024, 12 US states produced the equivalent of 22–78 percent of the electricity they consumed from wind alone: South Dakota (77.5%), Iowa (76.6), Kansas (69.8%), Oklahoma (51.4%), Wyoming (49.7%), North Dakota (48.5%), New Mexico (46.3%), Montana (33.7%), Nebraska (33.5%), Colorado (28.9%), Texas (24.2%), and Minnesota (21.9%).[257]

As of 2024, almost all US wind farms installed were onshore. The only US offshore wind farms (all fixed-bottom turbines) were the 2016 Block Island farm off the coast of Rhode Island (30 megawatts), the 2020 Coastal Virginia pilot farm (12 megawatts), the 2024 South Fork Wind Farm off the coast of New York (132 megawatts), the 2024 Vineyard Wind 1 off the coast of Massachusetts (806 megawatts), and the 2025 Revolution Wind farm offshore of Rhode Island, which also serves Connecticut (704 megawatts). At the end of 2024, the US had installed or was constructing 1.7 gigawatts of offshore wind, versus 39.1 gigawatts installed in China.

11.2 TYPES OF WIND TURBINES

Wind turbines that spin vertically around a horizontal axis are **horizontal-axis wind turbines**. They are by far the most common type of turbine today. Wind turbines that provide grid electricity are usually mounted on tall towers, with a hub height of 40–250 meters. A turbine's **hub height** is the height above the ground or water surface where the horizontal axis, or **rotor**, of a turbine is located. Some horizontal-axis turbines have low hub heights (10–50 meters) for use in backyards and in locations where wind-turbine heights or nameplate capacities are restricted.

Almost all horizontal-axis wind turbines today are **three-bladed turbines**, and all face the wind (**upwind turbines**). In the past, some horizontal-axis turbines faced away from the wind (**downwind turbines**). The advantage of a downwind turbine is that the wind naturally rotates the turbine horizontally to the exact opposite direction from the wind. The horizontal rotation around its vertical axis of a horizontal-axis wind turbine is called **yaw control**. Yaw control with downwind turbines is automatic, so does not require mechanical equipment. With upwind turbines, however, mechanical equipment is needed to rotate the turbine blades continuously to face the changing wind direction. Wind tails, such as those used on the Halladay windmill, help with yaw control only for small turbines. The main problem with downwind turbines, though, is that they do not receive power when each blade passes by the tower, since the tower itself blocks the wind. Also, because wind speed is high, then zero, then high again as each blade passes the tower, the blades are subject to enormous **wind shear** (variation of wind speed with distance). Wind shear increases wear and tear on, and failure of, the blades. Because of these two problems (power loss and wear), downwind turbines are no longer built.

Another type of horizontal-axis turbine previously used but no longer manufactured is the **two-bladed turbine**. Advantages of the two-bladed over the three-bladed turbine are that the fewer blades a turbine has, the less expensive the turbine is to build and the less each blade's turbulence causes wear, tear, and efficiency loss, on the next blade. The reduced turbulence in a two-bladed system also causes the blades to spin faster, decreasing generator cost. For these reasons, a two-bladed turbine would seem to be less costly than a three-bladed turbine. However, three-bladed turbines absorb horizontal and vertical wind shear more evenly, thus are more efficient and require fewer repairs and less down time than two-bladed turbines. Finally, because of their lower spin rate and more evenly distributed wind shear, three-bladed turbines are quieter. For those reasons, only three-bladed turbines are used today in wind farms.

Wind turbines that spin horizontally around a vertical axis are called **vertical-axis wind turbines**. These turbines are usually short, so they can be used in urban regions or regions with height restrictions. An early vertical-axis turbine was the **Darrieus turbine**, which had two or more curved blades attached at each end to the top and bottom of a vertical rotating shaft. This turbine looked like an eggbeater and was named after the French aeronautical engineer, **Georges Jean Marie Darrieus** (1888–1979), who was granted a patent on it in 1927. Its main problem was that its blades fatigued easily. Today, many designs of vertical-axis turbines exist.

Advantages of vertical- over horizontal-axis turbines are that the former do not need to rotate to track the wind, they are easier to maintain because their gearbox is closer to the ground, they cost less to install because their towers are shorter, and they can be grouped more closely to each other in a wind farm. In addition, they can be deployed in places with legal height restrictions, such as near airports and in some urban areas, or where tall turbines are a disadvantage, such as near military bases within combat zones.

The main disadvantage of vertical-axis (and short horizontal-axis) turbines is, because their heights are low, they are exposed to slower winds than are tall, horizontal-axis turbines. Since instantaneous wind power is proportional to the cube of the instantaneous wind speed, and this means a lot less power for vertical-axis turbines, increasing their cost per unit of energy. A second issue is that vertical-axis turbine blades can fatigue easily due to the variation in forces acting on them during each rotation and because winds near the ground are more turbulent than are winds aloft. This can cause greater wear and tear and repair requirements for vertical-axis turbines than for tall horizontal-axis turbines.

11.3 WIND-TURBINE PARTS

A horizontal-axis upwind wind turbine consists of a tower, blades, rotor, nacelle, wind vane, and anemometer. The **nacelle** houses several additional components, including the shafts, gearbox, generator, controller, braking system, pitch system, yaw drive, and yaw motor. The **tower** holds up all components of the turbine and is generally hollow inside, except for wires and a ladder. The nacelle sits at the top of the tower. The rotor and blades are connected to the front of the nacelle. The wind vane and anemometer are attached to the back of the nacelle.

The kinetic energy in the wind rotates the **blades**, which are connected to the **rotor**. The spinning rotor turns a low-speed shaft, which sits inside the nacelle, 3–20 revolutions per minute for large turbines and up to 400 revolutions per minute for residential turbines. A gearbox connects the low-speed shaft to a **high-speed shaft**, which turns at 750–3,600 revolutions per minute. This is a sufficient spin rate for a **generator** to convert rotational mechanical energy into electricity. Many wind turbines today, called **direct-drive** or **gearless turbines**, operate without a gearbox. Instead, the low-speed shaft is connected directly to the generator, which is larger and heavier than is the generator in a geared turbine. With both geared and gearless turbines, the generators produce 50- or 60-hertz AC electricity (depending on country) although some generators produce DC electricity. A **heat exchanger** keeps the generator cool.

An **anemometer**, which sits outside the back (downwind side) of the nacelle, measures wind speed and communicates it to the **controller**, which lies inside the nacelle. When the wind speed first increases above a low threshold, the cut-in wind speed, the controller starts the rotor up from rest. The controller also shuts the rotor off when the wind speed increases beyond a high threshold, the cut-out wind speed. The **pitch system** changes the steepness of the slope (pitch) of the blades to control the rotor speed and stop the rotor from spinning when the wind speed exceeds the cut-out wind speed or drops below the cut-in wind speed. The **brake** stops the rotor in an emergency, as a backup to the pitch system, or as a "parking brake" during maintenance. The **wind vane**, which is near the anemometer outside the nacelle, measures wind direction and communicates the information to the yaw drive. The **yaw drive** rotates the turbine horizontally around a vertical axis to face the wind. The **yaw motor** provides power for the yaw drive.

The primary materials needed for a wind turbine include prestressed concrete (for the tower), steel (for the tower, nacelle, and rotor), copper (for generator coils and electricity conduction), aluminum (for the nacelle), wood epoxy (for the rotor blades), glass fiber-reinforced plastic (for the rotor blades), carbon-filament reinforced plastic (for the rotor blades), and neodymium (for the permanent magnet in the generator).

The large-scale growth of wind electricity will not be constrained by limits in the quantities of these materials. The major components of concrete – gravel, sand, and limestone – are abundant, and concrete can be recycled and reused. The world does have somewhat limited reserves of economically recoverable iron ore (on the order of 100–200 years at current use rate), but the steel used to make towers, nacelles, and rotors for wind turbines is 100 percent recyclable. The production of millions of wind turbines would consume less than 10 percent of the world's low-cost copper reserves. Other conductors of electricity could also be used instead of copper.

Most permanent magnets, used in wind-turbine generators, are made of a neodymium-iron-boron alloy. Neodymium magnets were developed in 1982 and are among the strongest permanent magnets available commercially. Although **neodymium** is abundant in the Earth's crust, it cannot be found isolated in nature. Instead, it appears, along with other heavy metals, called rare-earth metals, in the minerals **monazite** and **bastnasite**. These minerals contain small amounts of all rare-earth metals. **Rare-earth metals** are a group of 17 elements, including neodymium, in the periodic table, that tend to be found together. They all appear lustrous and silvery-white so are difficult to distinguish from each other. Although they are abundant in nature, rare earths are concentrated enough for mining in only a few geographically dispersed locations and are rarely found in economically viable ore deposits.

Transition highlight
Wind turbines with permanent-magnet generators require about 202 kilograms of neodymium oxide per megawatt of wind-turbine nameplate capacity.[264] With about 13.9 terawatts nameplate capacity of wind needed to provide 44 percent of the world's all-purpose end-use power demand in 2050, the neodymium needed among all wind turbines in 2050 is about 10.2 percent of the world's known reserves, the vast majority of which are in China.

An alternative to neodymium for permanent magnets is iron nitride. This compound contains both iron and nitrogen, thus no rare-earth metals. The magnetic energy from iron nitride is reported to be over two times the maximum of a magnet containing neodymium under optimal conditions.[91]

11.4 WIND-TURBINE MECHANICS

A wind-turbine blade rotates when the wind flows around it. A blade is like an airfoil. It is round (convex) on the top and flatter, or more concave, on the bottom. As such, the distance from the front tip to the back tip of the blade is greater on the top than on the bottom. Wind hitting the blade splits. Some of the air flows over the top, and the rest, under the bottom. Because the top distance is greater, air flowing over the top travels faster than air flowing under the bottom so that the two air parcels meet each other at the back tip. Because air flowing over the top must travel further and faster, it spreads out more, causing air pressure above the foil to drop relative to air pressure under the foil. The lower pressure on top and higher pressure on the bottom of the blade creates an upward **lift force** perpendicular to the airflow.

The lift force causes the blade to move in the direction of the lift force, just as the lift force acts upward over an airplane wing to keep the airplane afloat against the force of gravity. Because a wind turbine is constrained to rotate in a circle, the lift force accelerates the spin of the turbine in the clockwise direction so long as the more rounded top of the blade faces the clockwise direction. There is no advantage or disadvantage to turbines spinning clockwise versus counterclockwise, but all commercial turbines since 1978 have spun clockwise. Previous to that, wind turbines and windmills mostly spun counterclockwise. In order to impart a lift force on the blade, the wind must give up some kinetic energy, converting it to mechanical energy. The removal of kinetic energy slows the wind past the blade.

The other force the wind imparts on a wind turbine is the **drag force**, which consists of two components. First, viscous friction on the surface of the blade slows down the wind, removing more kinetic energy from it. That energy from the wind is converted into an equal and opposite drag force of the wind pushing the blade slightly in the direction of the wind. Second, because the wind speed downstream of the turbine is slower than the wind speed upstream, the air pressure downstream of the turbine is lower than that upstream. This results in a pressure force acting from the higher-pressure upstream to the lower-pressure downstream part of the blade. This force is also a drag force that is added to the first drag force in the direction of the wind.

Wind turbines operate best with high lift-to-drag ratios. Thus, they are built to maximize lift and minimize drag. This can be accomplished by increasing the angle of attack up to a point. The **angle of attack** is the angle between the airflow and the blade. If the front tip and back tip of the blade are parallel to the wind, the angle of attack is zero degrees. If the front tip is then lifted so that the wind slightly impinges on the underside of the blade, the angle of attack is positive. Increasing the angle of attack from 0 to 5 degrees increases the lift-to-drag ratio from about 50 to 123 for one particular turbine. However, increasing the angle of attack further to 10 degrees decreases the ratio to about 98.[265] Higher angles of attack decrease the ratio further.

The lift-to-drag ratio decreases at high angles of attack because turbulence builds up on the top of the blade with high angles of attack. This causes the air flowing over the top to no longer stick to the blade. As a consequence, air pressure above the top of the blade increases, reducing lift and slowing the spinning blade. This slowdown is referred to as **stall**. A benefit of stall is that, when wind speeds are above a threshold, the wind-turbine controller and pitch system can increase the angle of attack to increase stall until the blades slow down sufficiently or stop spinning. This method of reducing or stopping power output by a turbine is called **active stall control**.

Related to active stall control is pitch control. **Pitch control** is the computer-controlled change in the angle of attack of wind-turbine blades to maximize power output for all wind speeds. With pitch control, if wind power output exceeds the maximum nameplate capacity of the generator, but the wind speed is still below the maximum allowable wind speed set by the manufacturer of the turbine, the controller and pitch system can reduce the angle of attack. This reduces the lift to drag ratio, reducing output without stopping the turbine from spinning or producing power. Reducing the angle of attack in this manner is specifically called **feathering** the turbine's blades.

Finally, a method of limiting a wind turbine's power output to the maximum that the generator can handle without involving moving parts or electronics is passive stall control. With **passive stall control**, blades are designed so that they twist with increasing distance from the rotor so as to induce stall automatically when the wind speed exceeds the speed that generates the maximum allowable power. The turbine will continue to spin but not any faster than the speed that results in the maximum allowable power. The main problem with passive stall control is that it is difficult to design a turbine that ensures stall occurs under all conditions, so a safety margin is needed. As a result, passive stall control turbines do not operate under optimal conditions, and they are usually limited to smaller turbines.[266]

11.5 WIND-TURBINE GENERATORS

Wind-turbine generators convert the rotational energy of a wind turbine's high-speed shaft into electricity. The more the wind rotates the blade, the more the generator can produce electricity, up to the nameplate capacity of the generator. The three main categories of generators for electricity production are dynamos, AC synchronous generators, and AC asynchronous generators. Dynamos produce DC electricity and are often attached to small wind turbines. In an AC synchronous generator, the high-speed shaft operates at a fixed speed. Most generators worldwide are synchronous generators used to produce electricity in coal, fossil-gas, nuclear, and other power plants that generate electricity at a constant rate. In an AC asynchronous generator, the high-speed shaft operates at a variable speed. Wind turbines use AC asynchronous generators due to the variable nature of wind energy production.

Generators for wind turbines operate based on electromagnetic induction, thus are **induction generators**. Induction generators can also act as motors by operating in reverse, turning electricity into rotating motion. In fact, an induction generator often acts as a motor to start up a wind turbine. Once the wind itself begins rotating the turbine blades, the motor returns to producing electricity as a generator. All wind-turbine generators produce 50-hertz frequency (in Europe and most countries) or 60-hertz frequency (US and other countries) three-phase AC electricity.

Two important induction generators are wound-field generators and permanent-magnet generators.

A two-pole **wound-field generator** consists of two coiled wires mounted on opposite sides (180 degrees apart) of a rotating structure called a **generator's rotor**, not to be confused with a wind turbine's rotor. A wind turbine's rotor is connected to a low-speed shaft, which is connected to a gearbox, which is connected to a high-speed shaft, which is connected to the generator's rotor. When the wind is blowing, the high-speed, rotating shaft of a turbine turns the generator's rotor at 750–3,600 rotations per minute. The generator's rotor is surrounded by a stationary structure called a **stator** that has three sets of coiled wires spaced 120 degrees apart. The rotor first creates a magnetic field around its two sets of coiled wires when a direct current passes through the wires. This current must be created by an outside power source, such as from the electricity grid. When the rotor and its attached coils and their magnetic fields start spinning inside of the stationary stator, the stator coils see the magnetic fields of the spinning rotor coils approach them and then move away from them. The change in magnetic flux seen by the stator coils induces a three-phase alternating current in the coils. The faster the generator's rotor spins, the greater the voltage generated across the coils. The alternating current is then sent by wire to its destination.

A four-pole wound-field generator is similar, except that it has four coiled wires mounted 90 degrees apart on the generator's rotor. A four-pole generator needs half as many rotations per minute from the high-speed shaft as does a two-pole generator to obtain the same output frequency of 50 or 60 hertz.

A **permanent-magnet generator** is also similar. But in this case, permanent magnets, instead of coils, are connected to the generator's rotor. The magnets are spaced an equal distance apart. As the rotor spins at high speed, the stator coils see a changing magnetic field from the moving magnets on the rotor. The changing magnetic field induces an alternating current in the stator coils. The alternating current is then sent by wire to its destination. Unlike with a wound-field generator, a permanent-magnet generator does not need

11.6 POWER IN THE WIND AND WIND-TURBINE POWER OUTPUT

an external DC power source to initiate a magnetic field. This is useful particularly if the wind farm is not connected to an electric-power grid. Reducing the external DC power requirement also reduces the need for batteries and capacitors. In addition, magnets in a permanent-magnet generator weigh less than the copper wires in a wound-field generator that they replace.

On the flip side, controlling voltage is difficult with permanent magnets. Permanent magnets also don't operate well under high temperature and require more cooling than without magnets. Finally, most permanent magnets used today contain neodymium, which is subject to price fluctuations and shortages owing to the fact that it is mined in only a few places worldwide. Some concern also exists as to whether enough neodymium exists for permanent-magnet generators to be used on a large scale in wind turbines. Aside from the fact that other types of asynchronous generators, including wound-field generators, can be used instead of permanent-magnet generators, the answer is that enough neodymium does exist (Section 11.3). In addition, iron nitride magnets, which contain more common elements than neodymium magnets, are now being commercialized as an alternative.

11.6 POWER IN THE WIND AND WIND-TURBINE POWER OUTPUT

Like PV panels, wind turbines generate electricity variably over time. In other words, wind-turbine output fluctuates over timescales of seconds to months. This section discusses both the energy in the wind and wind-turbine output.

11.6.1 WIND-SPEED FREQUENCY DISTRIBUTIONS

Determining a wind turbine's average power output over a period requires knowing how often different wind speeds occur during the period. A **frequency (probability) distribution** of wind speed gives the percentage of all wind speeds measured over a month or year that are of a given wind speed. A simple frequency distribution, for example, is that 40 percent of all wind speeds during a year are less than 6 meters per second, and 60 percent are greater than 6 meters per second. Real frequency distributions, though, consider many more wind-speed intervals.

Most wind-speed frequency distributions can be characterized as a Rayleigh distribution. A **Rayleigh frequency distribution** is a probability distribution of wind speed that looks similar to a bell curve but stretched toward higher wind speed. It is characterized by an average wind speed and gradually decreasing probabilities of slower and faster wind speeds. Good wind-energy locations are those with mean wind speeds of 7 meters per second or faster at hub height. The Rayleigh frequency distribution is a specialized case of the **Weibull frequency distribution** of wind speed. Some wind speed frequency distributions are better described as Weibull distribution.[267]

11.6.2 BETZ LIMIT

The wind passing through the swept area of a wind turbine's blades at a given instant contains a certain amount of power, called the **power in the wind**. Wind turbines can extract only a portion of this power. The maximum percentage of power in the wind that can be extracted by a wind turbine, regardless of its design, is 59.3 percent. This is called **Betz's law**. The German physicist **Albert Betz** (1885–1968) derived this law in 1919.

The basic idea is that because a wind turbine extracts kinetic energy from the wind, the wind on the downwind side of the turbine has a lower speed and air pressure than that on the upwind side of the turbine. The pressure difference causes air passing through the turbine to spread out, expanding in volume. If a turbine extracted all the kinetic energy in the wind, air would stop behind the turbine, preventing any more wind from passing through the turbine's blades. If the wind speed downwind of the turbine equaled the wind speed upwind of the turbine, then no kinetic energy would be extracted. Betz's law states that the wind speed downwind of a turbine can be no less than one-third that upstream of the turbine. The resulting power extraction at this minimum ratio is 59.3 percent. This is the Betz limit. It means that no wind turbine can extract more than 59.3 percent of the power in the wind it is exposed to.

The best wind turbines today extract 45–47 percent of the power in the wind, thus they reach 76–79 percent of the Betz limit. A turbine's extraction efficiency depends largely on the number of blade rotations per minute. Blades that spin too slowly allow too much wind to pass. Blades that spin too quickly create turbulence that interferes with other blades, reducing lift and energy-extraction efficiency.

11.6.3 WIND-TURBINE POWER CURVE

A wind-turbine **power curve** is a graph of the instantaneous power output of a wind turbine as a function of wind speed. A power curve is characterized by a cut-in wind speed, a rated wind speed, a cut-out wind speed, and a destruction wind speed. Between zero wind speed and the **cut-in wind speed** (generally 2–3.5 meters per second), a turbine produces no power because

11.6 POWER IN THE WIND AND WIND-TURBINE POWER OUTPUT

the power generated is so low at those speeds that it is uneconomical to produce the power. Above the cut-in wind speed, the instantaneous power output increases roughly proportionally to the cube of the wind speed, up to the rated wind speed. At its **rated wind speed**, a wind turbine produces its maximum instantaneous power, called the rated power, which is also the turbine's nameplate capacity. A wind turbine's nameplate capacity is limited by the nameplate capacity of the generator in the nacelle of the turbine.

The turbine's power output stays at the rated power for all wind speeds above the rated wind speed, up to the cut-out wind speed, owing to pitch control or passive stall control. However, wind turbines operating under realistic conditions have difficulty maintaining exactly constant power with increasing wind speed because of the time delay associated with pitch control and because of the approximation of turbine-blade design in the case of passive stall control.

Above the **cut-out wind speed**, the turbine's power output decreases to zero. The shutoff is accomplished through pitch control or active stall control, and is needed to prevent damage to the turbine. Most turbines, even when shut off, are designed to survive wind speeds up to a **destruction wind speed**. The standard destruction wind speed of turbines historically has been 50 meters per second for up to 10 minutes. Some turbines today, however, have destruction wind speeds of upto 84 meters per second.

Transition highlight
In 2014, a class of turbines was designed to withstand sustained 10-minute wind speeds of up to 57.5 meters per second.[268] These **typhoon-class** wind turbines are placed primarily offshore in locations prone to hurricanes or typhoons. Sustained one-minute wind speeds in a Category 4 hurricane are, for example, 58.6–69.3 meters per second. In 2024, some new fixed-bottom offshore wind turbines, with nameplate capacities of 8–20 megawatts, and floating turbines, with nameplate capacities of 5.5–16 megawatts, were designed to withstand Category 5 hurricane wind speeds (greater than 70 meters per second).[269] The highest destruction wind speed among these was 84 meters per second, for a 7.5-megawatt floating offshore turbine. A 20-megawatt fixed-bottom turbine was designed to withstand wind speeds of up to 79.8 meters per second.[269]

11.6.4 WIND-TURBINE ELECTRICITY OUTPUT AND CAPACITY FACTOR

How much energy and power in the wind do wind turbines extract during the year? The **capacity factor** of a wind turbine is the annually averaged power produced by a turbine divided by the rated power (nameplate capacity) of the turbine. Alternatively, it is the energy produced per year divided by the maximum possible energy produced per year by the turbine. If a wind turbine ran for a full year at its rated power, its annually averaged power output would equal its rated power (nameplate capacity), and its capacity factor would equal one. However, the capacity factor of a turbine is always less than one because the annually averaged near-surface wind speed anywhere in the world is always less than the rated wind speed of a turbine. Real capacity factors range from 10 percent for old turbines at locations with low wind speeds to 56 percent for modern turbines at some high-wind-speed offshore locations. For example, the five-year (2017–2022) capacity factor of the world's first floating offshore wind farm, Hywind Scotland (five 6-megawatt turbines located 29 kilometers offshore of Peterhead, Scotland) was 54 percent.[270]

Transition highlight
Wind-turbine efficiency and size have all increased over the past three decades, as illustrated by US data. The annually averaged capacity factor in 2023 of wind turbines in the US built in 2022 was 38.2 percent, which compares with a capacity factor in 2023 of all wind turbines built between 1998 and 2022 of 33.5 percent. This is because older turbines had much lower capacity factors. For example, wind turbines built in 1998 and surviving to 2023 had a mean capacity factor of about 15 percent in 2023.[271] One reason for the higher capacity factors of newer turbines is that turbine hub heights increased from an average of 56.5 meters in 1998 to 103.4 meters in 2023. Higher hub heights increase output because wind speeds increase with increasing height in the lower atmosphere. Wind-turbine blade diameters also increased, from 48.1 meters in 1998 to 133.8 meters in 2023, on average. Average turbine nameplate capacities also increased, from 0.72 megawatts in 1998 to 3.4 megawatts in 2023.[271]

11.6.5 FACTORS REDUCING WIND-TURBINE GROSS ANNUAL ELECTRICITY OUTPUT

Some of a wind turbine's electricity output is lost due to four factors: transmission and distribution losses, downtime losses, curtailment losses, and array losses.

11.6.5.1 TRANSMISSION AND DISTRIBUTION LOSSES

As with electricity from all sources, wind electricity traveling through transmission and distribution lines suffers line losses. Given that many good land-based sites for wind energy development are far from population centers,

11.6 POWER IN THE WIND AND WIND-TURBINE POWER OUTPUT

wind electricity must often be transmitted long distances. If the distance is greater than 600 kilometers, high-voltage direct current (HVDC) transmission lines should be used. For shorter distance, high-voltage alternating current (HVAC) lines should be used. Most offshore wind will be located within 200 kilometers of a coastline, and most of the world's population lives near the coast. As such, most offshore wind transmission will be with HVAC lines.

11.6.5.2 DOWNTIME LOSSES

A wind turbine's electricity output is also reduced when the turbine is down for regularly scheduled maintenance, equipment failure, or refurbishment of the turbine.

An analysis of repair data from 1,500 onshore wind turbines over about 10 years found that the average downtime for repairs was 1.6 percent (6 days) of the year. Minor failures (such as failure of the electrical system, electrical controls, the hydraulic system for pitch control, and yaw system), which represented 75 percent of the problems, caused only about 5 percent of the downtime. Major failures (such as failure of the rotor blades, rotor hub, drive train, gearbox, and generator), which represented the rest of the problems, caused about 95 percent of the downtime.[272]

Offshore wind turbines generally require more repair-related downtime than onshore turbines because the harsh weather conditions offshore, including high waves and strong winds, prevent access to offshore turbines for several days per month during stormy months. In addition, merely scheduling a boat ride to an offshore wind turbine can take more time than driving to an onshore turbine. Third, the harsher weather conditions and faster wind speeds offshore increase wear and tear compared with onshore wind turbines. On the other hand, because of the lack of terrain and cooler daytime ocean surface than land surface, offshore turbines are exposed to less turbulence, thus experience less wear and tear than onshore turbines. Offshore turbines are down for scheduled maintenance for 4–6 days per year[273] and unscheduled repairs for 4–8 days per year, giving a total of 8–14 days per year (2.2–3.8 percent of the year).

In comparison, the average coal plant and combined-cycle fossil-gas plant in the eastern US were down 10.6 percent and 3 percent, respectively, of the days between January and June 2015 for unscheduled maintenance.[274] Coal plants were down an additional 6 percent of the year for scheduled maintenance.

A difference, though, between outages of **centralized power plants** (coal, nuclear, fossil gas) and outages of **distributed power plants** (wind, solar, wave) is that, when individual solar panels or wind turbines are down, only a small fraction of electricity production is affected. When a centralized plant is down, a large fraction of the grid is affected. When more than one large centralized plant is offline at the same time, an entire grid can be affected.

11.6.5.3 CURTAILMENT LOSSES

A third factor that reduces a wind turbine's output is curtailment. **Curtailment** is the deliberate reduction in output of an electricity generator below what the output could otherwise be. Grid operators may order curtailment under contract with an electricity producer when grid electricity supply exceeds demand. Curtailment is avoided if excess electricity is stored (e.g., in a battery) or used on site to produce heat, cold, or hydrogen, which are either stored or used immediately.

11.6.5.4 ARRAY LOSSES

Finally, competition for the same kinetic energy in the wind among wind turbines close to each other in a wind farm reduces aggregate wind-farm output compared with when the turbines in the farm are far away from each other. Losses due to competition among turbines for the same energy are commonly called **array losses**.

If wind powers 37.1 percent of the world's all-purpose end-use energy in 2050, competition among wind turbines for limited kinetic energy may reduce annually averaged wind-electricity output by about 6.7 percent.[181] Reductions in electricity output in densely packed farms or multiple farms close to each other are greater than in more sparsely packed farms or farms spread far apart from each other.[180]

11.6.5.5 OVERALL LOSS

The overall percentage loss in wind-electricity output due to the four processes discussed is estimated as follows. Transmission and distribution losses are expected to decline from a world average of 8.3 percent in 2014 to about 6.2 percent by 2050.[246] Downtime losses for onshore and offshore wind should decline slightly from today's losses to about 1.5 percent by 2050. Curtailment of wind, which is high in some locations today, should go to near zero in a 100 percent WWS world because excess wind electricity will either be stored in batteries or used to produce heat,

11.7 WIND-TURBINE FOOTPRINT AND SPACING AREAS

cold, and/or hydrogen that will be stored. Finally, array losses with 100 percent WWS in 2050 are expected to increase to about 6.7 percent. The overall loss in 2050 under these conditions thus averages to about 14 percent.

11.7 WIND-TURBINE FOOTPRINT AND SPACING AREAS

Two types of land or water areas associated with wind farms are their footprint area and spacing area.

11.7.1 FOOTPRINT AREA

The **footprint area** of a wind farm is the topsoil or water area touched by bases of all the wind-turbine towers in the farm.[178] It does not include the areas of turbines' bases under the ground or underwater because a wind farm's footprint area represents land above the ground that can be used for other purposes. The footprint area does include the areas of any permanent roads (those covered with pavement) installed due to the wind farm, but it does not include areas of unpaved roads, which can readily be reverted back to their natural conditions. Transmission lines between wind farm and grid that are underground also do not count toward footprint. However, land areas of transmission-line pads touching the ground due solely to the wind farm do count toward footprint. On the other hand, the areas of transmission-tower pads are trivially small, as indicated by the fact that vegetation can grow under transmission-tower lattices.

11.7.2 SPACING AREA

The **spacing area** of a wind farm is the land or water area between and around wind turbines required to (1) prevent one turbine's blades from touching another's, (2) prevent turbines from falling onto one another or nearby structures, (3) minimize wear and tear on a downstream turbine resulting from the turbulent wake of an upstream turbine, and (4) minimize array losses (competition among all turbines for limited available kinetic energy in the wind).[178] The spacing area between onshore turbines can be used for multiple purposes, including for agriculture, cattle grazing, ranching, solar arrays, forests, and open space. Water between offshore turbines is similarly open water. Because wind-farm footprint areas are small, wind farms leave 96–99.5 percent of land or water undisturbed or usable for other purposes.

11 ONSHORE AND OFFSHORE WIND ENERGY

Spacing area is an important parameter because it affects a wind farm's installed power density and output power density. **Installed power density** is the nameplate capacity (megawatts) per square kilometer of spacing area in a wind farm. **Output power density** is the average power output (megawatts) per square kilometer of spacing area. The larger the spacing area, the smaller the installed and output power densities. A wind farm's capacity factor is the ratio of the farm's output power density to its installed power density. Alternatively, the farm's capacity factor is its average power output divided by its nameplate capacity. As such, the capacity factor is independent of spacing area.

No unique way exists to define spacing area. At one extreme, it could be defined as the surface area of the Earth. This would give trivially small installed and output power densities. At the other extreme, it could be defined as the circular area on the ground around one wind turbine's tower, with the circle's radius equal to the turbine tip height, multiplied by the number of wind turbines in the wind farm. The **tip height** of a turbine equals its hub height plus the distance from the hub to the tip of one blade. The tip height is used to define the radius of the circle defining the minimum area needed because the tip height is the distance a turbine will extend to if it falls to the ground in any direction. Laws in many locations require structures to lie beyond the tip height of a turbine.

Although there is no correct method to determine spacing area, methods that assign large fixed rectangles, circles, or polygons to each turbine erroneously include areas that are not part of a wind farm. For example, they erroneously count, as part of a wind farm, space outside of a wind farm's boundary, space between clusters of turbines within a farm, and overlapping space that results when each turbine is assigned a large rectangle.

A way to define spacing that overcomes these three problems is as follows. The methodology involves defining and combining three areas. The areas assume that a wind farm consists of one or more clusters of individual wind turbines. The first area is simply a circular area around each turbine, where the radius of each circle is the tip height of the turbine, as described. Use of the tip height ensures that if a wind turbine is on the edge of a wind farm, no space beyond the tip height will be counted as part of the wind-farm spacing.

The second area is the spacing area of each cluster of turbines in the wind farm. The cluster area is obtained by tracing a line around the outer edges of the circles around each wind turbine that is on the outer edge of

11.8 WIND-TURBINE FOOTPRINT AND SPACING AREAS

the cluster. The area within this boundary line is the cluster area. A cluster of turbines within a wind farm is separated from another cluster when the distance from a turbine tower at the edge of the first cluster to the closest neighboring turbine tower exceeds three tip heights. The area in between clusters is not included as part of the spacing area of the wind farm.

The third area is the total spacing area of the farm. This is the sum of the spacing areas of all clusters within the farm.

The method just described avoids counting areas outside of cluster boundaries as part of a wind farm, avoids counting areas between clusters as part of a wind farm, and avoids counting overlapping areas, previously assigned to individual turbines, as part of a wind farm. It also accounts for actual distances between wind turbines. This method also ensures that the addition of a wind turbine to the farm increases the spacing area required by the farm, as it should in reality. In sum, the spacing area of a wind farm is important for determining the farm's installed and output power densities.

Transition highlight
A study of 16 onshore and 7 offshore wind farms among 13 countries across five continents using the spacing area method just described found that installed power densities ranged from 7.2 megawatts per square kilometer for offshore European farms to 19.8 megawatts per square kilometer for onshore European farms to 20.5 megawatts per square kilometer for onshore farms in four other continents.[275] The output power densities in the three respective cases were 2.94, 6.64, and 6.84 megawatts per square kilometer. Taking the ratio of the output to installed power densities gives the capacity factor in each case of 40.8%, 33.5%, and 33.4%, respectively.

11.8 GLOBAL AND LAND WIND RESOURCES

Large regions of fast onshore winds worldwide include the Great Plains of the United States and Canada, northern Europe, parts of Russia, the Gobi and Sahara Deserts, much of the Australian desert areas, New Zealand, parts of South Africa, parts of Peru, and southern South America. Windy offshore regions include the North Sea, the east and west coasts of North America, offshore of Australia, and offshore of the east coast of Asia, among other locations.

In order to determine the maximum energy available for wind turbines worldwide over land plus ocean or over land alone, it is necessary to account for competition among turbines for limited kinetic energy in the wind. When one wind turbine converts energy in the wind to mechanical energy to spin a turbine's blades, less kinetic energy is available for other wind turbines in the wind farm and in the world. As more wind turbines become operational, each turbine is able to extract less and less energy from the wind. At some point, the addition of one more wind turbine worldwide results in no additional energy extraction, thus electricity generation. At that point, the annually averaged power produced by all existing wind turbines is called the **saturation wind power potential**.[180] The saturation wind power potential is important because it gives the upper limit to how much average power is available from wind turbines installed worldwide (over land plus ocean) or over land alone at a given hub height.

As the cumulative nameplate capacity of wind turbines installed over land plus ocean worldwide increases, the total extractable power among all wind turbines increases, but with diminishing returns. Above 3,000 terawatts nameplate capacity of turbines installed evenly worldwide, for example, no additional power can be extracted from the wind at 100-meters hub height. The resulting worldwide limit to extractable power (not nameplate capacity) is about 253 terawatts. Over land outside of Antarctica, the limit is about 72 terawatts.[180]

> **Transition highlight**
> The world's near-surface winds contain far more power than is needed to power all of humanity. The world needs only about 4 terawatts of annually averaged wind output in 2050 for wind to provide 45 percent of world end-use power demand after all energy sectors have been electrified.[4] As such, only one-eighteenth of the power available over land and one sixty-third of the power available worldwide are needed to meet this demand. Thus, there is no wind resource barrier to obtaining even 100 percent of the world's 2050 all-purpose electric power from wind.

Finally, when wind farms are far from each other, their output can increase by up to 4.6 times for the same nameplate capacity, summed among all farms, versus when wind farms are close to each other. This is because, when wind farms are far apart, each farm faces less competition for available energy in the wind than when farms are close together.

11.9 WIND-TURBINE IMPACTS ON CLIMATE, HURRICANES, AND BIRDS

This section examines the impacts of wind turbines on global and local climate, hurricanes, and birds. Briefly, wind turbines cool the globe on average by reducing water vapor, a greenhouse gas. Second, lots of offshore wind turbines can help to dissipate a hurricane from the outside in, thereby reducing hurricane wind speeds and storm surge. Third, while

11.9 WIND-TURBINE IMPACTS ON CLIMATE, HURRICANES, AND BIRDS

wind turbines do kill birds, they kill far fewer birds than fossil-fuel generators, buildings, and cats.

11.9.1 WIND-TURBINE IMPACTS ON CLIMATE

Whereas wind turbines can cause a local warming of the ground downstream of a wind farm,[276] they cause a net cooling of the Earth in the global average.[180,181,277] The reasons are as follows.

Wind turbines extract kinetic energy from the wind, reducing wind speeds downwind of a wind farm. A reduction in wind speed reduces evaporation of liquid water from soil or a water body downwind of the farm. Since evaporation is a cooling process, reducing evaporation warms the ground, ocean, or a lake. Some studies have thus found an increase in ground temperature downwind of a wind farm.[276]

However, because wind turbines reduce evaporation and thus reduce water vapor in the air, they also reduce condensation of vapor to form clouds, reducing cloudiness. Because condensation releases heat to the air, less condensation cools the air. As such, these two processes caused by wind turbines – less evaporation at the surface and less condensation in the air – cancel each other out.

On the other hand, water vapor is a greenhouse gas that traps heat emitted by the surface of the Earth in the lower atmosphere. Less water vapor means that more heat can now escape to the upper atmosphere or outer space. This loss of energy from the lower atmosphere caused by wind turbines through this mechanism results in wind turbines causing a net global near-surface cooling.[180,181,277] Such cooling is measurable only with a large number of turbines.

> **Transition highlight**
> Suppose 2.5 million 5-megawatt turbines are distributed over 139 countries to provide a projected 37.1 percent of the world's all-purpose end-use energy in 2050 after all energy sectors have been electrified. In this case, wind turbines may reduce globally averaged water vapor in the atmosphere by 0.29 kilograms per square meter of ground and near-surface air temperature by about 0.03 degrees Celsius.[81] In other words, global cooling due to this number of wind turbines in a WWS economy could offset about 2 percent of 1.5 degrees Celsius global warming.

In sum, wind turbines serve an additional benefit beyond replacing electricity from polluting fossil fuels. In the global average, they cool the Earth's surface; thus, they reduce global warming beyond the reduction they cause by replacing fossil-fuel energy generators and eliminating their pollution.

11.9.2 WIND-TURBINE IMPACTS ON HURRICANES

Because wind turbines extract kinetic energy from normal winds, wind turbines also extract kinetic energy from hurricane-force winds. However, if hurricane winds are too strong, they may topple an individual wind turbine. On the other hand, some new offshore wind turbines are built to withstand the fastest (Category 5) hurricane wind speeds,[269] and regardless, thousands of turbines near each other may slow a hurricane's winds sufficiently to prevent wind speeds from reaching the turbines' destruction wind speed.[278] A hurricane contains so much kinetic energy that thousands of turbines are needed to measurably reduce hurricane damage to coastal or inland cities.

About half of a hurricane's damage is due to high winds, and the other half is due to storm surge. **Storm surge** is the rise in sea level generated by a storm above and beyond the sea level arising from normal tides and waves under normal conditions. Storm surge in a hurricane is enhanced by three factors. First, warm ocean temperatures due to global warming thermally expand water and melt ice. Both factors increase the background sea-level height. Second, the low surface air pressure in a hurricane raises sea level compared with normal air pressure. Third, the strong winds around a hurricane's core create tall waves that lift water to great heights.

Wind turbines in the past were designed to withstand only 50 meters per second wind speeds. Some turbines now can withstand 70–84 meters per second wind speeds.[269] Wind speeds of 50 meters per second correspond to Category 3 hurricane wind speeds on the Saffir–Simpson scale. Wind speeds greater than 70 meters per second correspond to Category 5 wind speeds. An important question is whether many wind turbines can extract sufficient energy from a hurricane to dissipate it while keeping hurricane wind speeds below the turbine's destruction wind speed.

Because hurricane paths are not predictable, offshore wind farms will always be built primarily to generate electric power year-round in order to pay for themselves. However, if enough offshore farms are built, some farms can be placed strategically in front of a city to maximize protection of the city as well, thus serving a dual benefit.

It is impossible to know how a real wind farm might diminish a hurricane, because once a hurricane has passed the farm, it is not possible to replicate the hurricane in the absence of the farm. However, such an experiment can be carried out with a computer model that predicts hurricane formation and accounts for extraction of kinetic energy by wind turbines. In such a case, two simulations of a hurricane are needed – one that

11.9 WIND-TURBINE IMPACTS ON CLIMATE, HURRICANES, AND BIRDS

includes wind turbines and another that does not.[278] Here, analyses of the impacts of turbines on two hurricanes, Katrina and Sandy, are discussed.

Hurricane Katrina occurred from August 23–31, 2005. It damaged New Orleans, Louisiana, and caused 1,836 deaths. Peak wind speeds offshore were 78.2 meters per second, making it a Category 5 hurricane offshore. During Hurricane Katrina, storm surge caused flooding in New Orleans, which is below sea level. The flooding was exacerbated due to the breach of many levies built to prevent flooding.

Hurricane Sandy formed on October 22, 2012, and dissipated on November 2, 2012. It hit the US east coast between Washington, DC, and New York. Its peak wind speed was 35.8 meters per second. It was one of the widest hurricanes to hit the US and one of the few to hit the northeast coast. Its damage was due primarily to flooding from storm surge.

Based on computer-model simulations, large arrays of offshore wind turbines within 100 kilometers of the coast could have reduced peak wind speeds and storm surge significantly in both hurricanes. In the case of Katrina, large wind farms (totaling 78,000 7.58-megawatt wind turbines) located to the southeast of New Orleans could have reduced peak wind speeds by about 36 percent and storm surge by 6–71 percent, depending on location along the coast. In the case of Sandy, large offshore wind farms (totaling 112,000 7.58-megawatt turbines) between Washington, DC, and New York City could have reduced peak wind speeds by 36 percent and storm surge by 12–21 percent.[278]

Wind turbines reduce hurricane wind speeds for the following reason. Wind turbines first reduce the outer rotational winds of the hurricane. This increases the hurricane's central pressure, which in turn, decreases peak wind speeds in the hurricane's eye wall (which rotates around the hurricane's center, or eye). As such, wind turbines dissipate a hurricane from the outside in. The reason can be elucidated further as follows.

Wind turbines are exposed first to the hurricane's outer rotational winds, which are slower than eye-wall winds. The reduction in outer rotational wind speeds by wind turbines decreases wave heights there because wave heights are somewhat proportional to wind speed. Since waves are a source of friction, decreasing wave height reduces surface friction. The angle at which winds that are moving around a hurricane's eye wall tilts toward the eye wall depends on friction. The less the friction, the more circular the flow. The more the friction, the more the winds converge inward toward the eye wall. Thus, the decrease in friction due to wind turbines decreases wind convergence toward the eye wall, causing the winds to flow more in a circle around the hurricane's eye wall.

The resulting decrease in the mass of air moving toward the eye wall decreases both the upward spiraling of the air around the eye wall and the outward flow of air aloft away from the eye wall. Because fast outflow aloft relative to inflow at the surface drops a hurricane's surface pressure, strengthening the hurricane, a decrease in air mass spiraling upward and flowing away aloft increases surface air pressure, weakening the hurricane.

The increase in surface air pressure due to wind turbines then reduces the horizontal pressure difference between the center (eye) of the hurricane and the outside of the hurricane. The reduction in pressure difference slows hurricane winds. Slower winds, in turn, reduce wave heights further, reducing convergence to the center, reducing the spiral lifting of air and outflow aloft, increasing central pressure further, in a positive-feedback loop. In this way, slowing the outer rotational winds of a hurricane with wind turbines dissipates the hurricane from the outside in.

Because wind-turbine arrays dissipate a hurricane from the outside in, hurricane winds reaching the arrays never exceed a turbine's destruction wind speed. By reducing hurricane wind speeds, wind turbines also reduce storm surge.

Transition highlight
Any wind farms offshore can help reduce hurricane damage. For example, in one scenario, 53,000 offshore 5-megawatt turbines may be needed to provide 7.1 percent of US 2050 all-purpose end-use energy from offshore wind.[4] Dividing this number into wind farms of 100–300 turbines, placing those farms in separate clusters (where each farm is separated by distance to minimize competition for available kinetic energy), and placing each cluster strategically in front of major cities at risk of a hurricane may reduce future hurricane damage.

Because offshore wind farms will be built to generate year-round electricity but may also reduce damage due to hurricanes, they are cost-effective in comparison with other techniques of reducing hurricane damage, such as building sea walls. Whereas turbines pay for themselves from the sale of the electricity they produce, sea walls have no other function than to reduce storm surge. Sea walls also do not reduce hurricane winds, whereas wind turbines do.

11.9.3 WIND-TURBINE IMPACTS ON BIRDS AND BATS

Birds and bats can fatally collide with the spinning blades of a wind turbine. Some bats also die from the drop in pressure behind the wind turbine,

11.9 WIND-TURBINE IMPACTS ON CLIMATE, HURRICANES, AND BIRDS

which causes their organs and blood vessels to expand to equalize the pressure. Of particular concern is the collision of **raptors** (birds of prey) with wind turbines. Raptors include eagles, ospreys, kites, hawks, buzzards, harriers, vultures, falcons, caracaras, and owls.

Estimates of avian mortalities, based on 2017 wind electricity production in the US range from 76,000 to 990,000 per year. These appear to be large numbers until other sources of avian death are compared. For example, cats are the number one threat to birds in the US, killing an estimated 2.4 billion per year.[279] Windows of buildings kill another one billion birds per year. Transmission and distribution lines kill 8–57 million birds per year, and communication towers, another seven million per year. These sources all cause more bird deaths than do wind turbines. Bird deaths from wind turbines in the US in 2017 ranged from 0.002 to 0.028 percent of all other bird deaths.

Coal and fossil-gas electricity-generating plants also kill more birds per unit of energy and in total than wind turbines.[280] As such, transitioning from coal or gas to wind reduces bird deaths. Coal and fossil gas used for electricity generation kill birds in three major ways. One is through the destruction of bird habitat due to invasive mining, such as mountaintop removal in the case of coal, and fracking in the case of fossil gas. The second is through air and water pollution, acid rain, and climate change caused by fossil-fuel power plants. The same air pollution from fossil fuels that kills birds contributes to the 7.4 million human air-pollution deaths each year worldwide. The third way coal and fossil gas kill birds is through fossil-fuel infrastructure, which birds collide with or are electrocuted by.

Even nuclear energy kills a similar number of birds per kilowatt-hour of electricity produced as wind turbines.[280] Nuclear kills bird in two major ways. One is from contaminated ponds associated with uranium mining and milling. Abandoned open-pit uranium mines also form hazardous lakes. Second, birds collide with nuclear power infrastructure.

Raptor deaths from wind turbines comprise between 5 and 15 percent of all bird deaths. Bat deaths per unit of energy are about 50 percent higher than bird deaths.[281,282] Many bat deaths occur near wind turbines in the northeastern and upper midwestern US. A non-wind-turbine major source of bat death since 2006 has been white nose syndrome, which has decimated populations of bats in North America.

Wind turbines today are safer for birds and bats than were those built in the 1980s and 1990s. Early wind-turbine towers often had lattices that birds could perch on, increasing their proximity to wind-turbine blades. Also, spin rates were not previously controlled, resulting in higher tip speeds.

The removal of lattices and the control of spin rates at high wind speed have contributed to safer turbines. Because many birds fly low, tall turbines also permit birds to fly under turbine blades. In addition, radar systems are now used to alert wind-farm operators of approaching flocks of birds, including raptors. Some endangered birds, such as condors, are fitted with a global positioning satellite transmitter. Wind-farm operators can slow down or stop wind turbines if they are alerted that a condor is approaching. Devices that emit high-frequency ultrasonic sounds are now being tested to deter bats from entering a wind farm. Another way of reducing bird mortalities is to site wind turbines out of migratory paths of birds.

Transition highlight
An encouraging study found that painting just one of a wind turbine's three blades black (and keeping the rest white or gray) reduced bird mortalities by 72 percent compared with bird deaths from neighboring unpainted turbines.[283] Raptor fatalities were reduced the most. The reduced deaths were due to the increased contrast in shades of color, thus visibility, caused by painting a single blade black.

12

STEPS IN DEVELOPING 100 PERCENT WWS ROADMAPS

So far, this book has examined the main technologies needed for a 100 percent clean, renewable energy and storage system. Virtually all of these technologies exist today, and none is a miracle technology. This chapter focuses on combining the technologies together in countries, provinces, states, cities, and towns to provide end-point roadmaps for a transition. Such roadmaps provide targets for meeting energy demand among all energy sectors with 100 percent WWS in the annual average by some year, often 2050, but ideally sooner, such as 2035. Chapter 13 discusses methods of matching power demand continuously (rather than in the annual average) with WWS supply, storage, and demand response. Roadmaps and grid-stability analyses are helpful for giving policymakers, utilities, and the public confidence that a transition will not cause grid failures, particularly during extreme weather events.

12.1 PROJECTING END-USE ENERGY DEMAND

In a 100 percent WWS world, all energy sectors will use only electricity or direct heat, where the electricity and heat are provided by WWS. Some electricity is used to produce low-temperature heat and cold in buildings with electric heat pumps. Some electricity is used to produce high-temperature heat for industry that is used immediately or stored in firebricks. Some electricity is used to produce green hydrogen, which is stored and then used either in fuel cells for transportation, fuel cells for grid and microgrid electricity backup, or steel and ammonia manufacturing. Some electricity is stored in one of many electricity storage options. Some solar and geothermal heat and electric-heat-pump heat and cold are stored in district heat and cold storage. Energy-efficiency and energy-reduction measures are adopted to reduce demand. Grid operators also shift the times of peak demand with demand response to help stabilize the grid.

The first step in developing a roadmap to transition the energy infrastructure of a country, province, state, city, or town to 100 percent WWS is to project current annually averaged end-use energy demand across all energy sectors to a future year in a business-as-usual case.

A **business-as-usual case** is a future scenario describing how energy demand may change over time in a country if technologies and policies do not change much. It may assume moderate economic growth, some population growth, the use of renewable energy but growing only at historic rates, the use of modest energy-efficiency measures, and modest reductions in energy use between the present and the future.

End-use energy demand (also called energy consumption) is energy directly used by a consumer. It is the energy embodied in electricity, fossil gas, gasoline, diesel, kerosene, and jet fuel that people use directly. It equals primary energy minus the energy lost in converting primary energy to end-use energy, including the energy lost during transmission and distribution and due to waste heat. **Primary energy** is the energy naturally embodied in chemical bonds in raw fuels, such as coal, oil, fossil gas, biomass, uranium, and renewable (hydroelectric, solar, wind, and geothermal) electricity, before the fuel has been subjected to any conversion process.

The conversion from primary energy to end-use energy differs for different energy sectors and types of fuels. In the electricity sector, for example, end-use energy equals primary energy minus the energy lost during the generation, transmission, and distribution of electricity. When coal is burned to produce electricity, only about one-third of the energy embodied in the coal is converted to electricity. The rest is waste heat. Further, some of the electricity produced is lost during transmission and distribution. The end-use electricity in this case is the electricity that consumers use in the end. It does not include the heat lost during the conversion of coal to electricity or the energy lost during transmission and distribution of the electricity from the power plant to where it is used.

In another example, solar electricity produced by a PV panel is primary energy. Some of that electricity is lost during transmission and distribution. The solar electricity that actually reaches a consumer after transmission and distribution losses is end-use energy. Similarly, electricity produced at a hydropower plant is primary energy. The electricity remaining after transmission and distribution losses is end-use energy.

In the transportation sector, the energy embodied in crude oil is primary energy. Converting crude oil to end-use products, including gasoline, diesel, kerosene, refinery gas, and jet fuel, involves little loss of the primary energy in crude oil, so the end-use energy in these cases is close to, but not the same as, the primary energy in crude oil.

Fossil gas used for heating and cooking is similar to fossil gas recovered from a well, so primary energy and end-use energy are similar. On the other hand, when fossil gas is burned for electricity, only a portion of the fossil gas is converted to electricity and some of that electricity is lost during transmission and distribution. As such, the end-use energy in the delivered electricity is much less than the primary energy in the fossil gas used to create the electricity.

> **Transition highlight**
> In 2020, the annually averaged, end-use power demand (which equals energy consumption, in terawatt-hours per year divided by the number of hours per year) among all energy sectors in 149 countries representing 99.75 percent of world emissions, was about 12.6 terawatts (trillion watts). Of this, 22.7 percent was electricity demand. By 2050, this overall demand may grow, in a business-as-usual case, to 18.9 terawatts (by 50.6 percent) if no large-scale transition to WWS occurs.[5] The growth is due to a population increase, partly mitigated by lower energy use per person resulting from some modest shifts from coal to fossil gas, biofuels, bioenergy, some WWS, and some energy-efficiency improvements.

12.2 POWERING FUTURE ENERGY WITH WWS

The second step in roadmap development is to transition 2050 business-as-usual end-use energy for each fuel type in each energy sector to electricity, some hydrogen created from electricity, and some direct solar and geothermal heat. The electricity and heat are produced by WWS. Such a transition also involves employing energy-efficiency measures.

The energy sectors to transition are the residential, commercial, transportation, industrial, agriculture/forestry/fishing, and military sectors.

Energy in the **residential sector** includes electricity and heat consumed by households, but not for transportation.

Energy in the **commercial sector** includes electricity and heat consumed by commercial and public buildings, but not for transportation.

Energy in the **industrial sector** includes energy consumed by industry. Industry includes businesses that manufacture iron, steel, cement, chemicals, petrochemicals, iron-free metals, nonmetallic minerals, transport equipment, machinery, food, tobacco, paper, pulp, wood, wood products, textiles, and leather. Industry also includes printing, construction, and mining businesses.

Energy in the **transportation sector** includes energy consumed for any type of transport by road, rail, pipeline, air, and water, and for the use of highways by agricultural and construction machines. For pipelines, the energy required is for the support and operation of the pipelines, including for pushing of fossil gas and oil through pipes. The transportation category excludes fuel needed for local and international fishing vessels and non-highway use of agricultural machines, since such fuel is included under the agriculture/forestry/fishing category.

Energy in the **agriculture/forestry/fishing sector** includes energy consumed by agriculture, hunting, forestry, or fishing. For agriculture and forestry, it includes consumption of electricity and heat for buildings and equipment, and energy for vehicles aside from agricultural machines using the highway. For fishing, it includes energy for inland, coastal, and deep-sea fishing, including fuels delivered to ships of all flags, for local and international shipping, that have refueled in the country, and all other energy used by the fishing industry.

Finally, energy in the **military sector** includes fuel that originates in the country of interest used by the military for all transport (ships, aircraft, tanks, on-road, and nonroad transport) and bases (forward operating bases, home bases), regardless of whether the fuel is used by the country or another country.

About 95–97 percent of the technologies needed to transition all sectors are available today. The main technologies still missing are those for medium- and long-distance air and marine transport. Figure 2.1 summarizes the components of a WWS system. The main WWS electricity-generating technologies available include onshore and offshore wind turbines, concentrated solar power (CSP) plants, geothermal electricity plants, solar photovoltaics on rooftops and in power plants, tidal and ocean-current power devices, wave-power devices, and hydropower plants.

Vehicles used for transportation in a WWS world include battery-electric vehicles and hydrogen-fuel-cell-electric vehicles, where the hydrogen is green hydrogen produced from WWS electricity. Whereas battery-electric and hydrogen-fuel-cell-electric vehicles are available now for most types of land transport and short-distance air and marine transport, they are still being developed for medium- and long-distance air and marine transport. Battery-electric vehicles will dominate two- and three-wheel ground transport; short- and long-distance light-duty ground transport; and most truck, construction-machine, and agricultural-equipment transport. Battery-electric vehicles will also dominate short- and medium-distance trains, short-distance boats and ships (ferries, speedboats, tugboats, dredgers), short-distance military equipment, and aircraft traveling less than 1,500 kilometers. Some short-distance trains will run on overhead-wire electricity.

Of all commercial aircraft flights worldwide, about 85.2 percent by number and 53.9 percent by distance are short-haul flights (flights less than three hours in duration, with a mean distance of 783 kilometers).[79] As such, all but about 15 percent of aircraft flights worldwide by number will be electrified with batteries. Hydrogen fuel cells coupled with electric motors will power long-haul aircraft as well as long-distance ships, truck, trains, and military equipment.[81]

Electric heat pumps will provide building air heating and cooling. Heat pumps will also be used in domestic water heaters, clothes-washing machines, clothes dryers, and dishwashers. Cooking stoves will be electric induction. Lawn mowers, leaf blowers, chain saws, and other machines will run on electricity.

Firebricks, heated by resistance heating, will store and provide most medium-to-high-temperature heat for industry. Remaining medium-to-high-temperature heat will be provided by electric-arc furnaces; electric-induction furnaces; electric-resistance furnaces, kilns, and boilers; electric crackers; electron-beam heaters; dielectric heaters; and solar heaters. Electric heat pumps and solar heaters will provide low-temperature heat for industry. In sum, all fossil-fuel and bioenergy combustion for energy will be replaced with WWS electricity and heat.

12.3 CHANGES IN ENERGY NEEDED UPON A TRANSITION TO WWS

The third step in developing a 100 percent WWS roadmap is to calculate the decrease in annually averaged, end-use energy demand due to moving to WWS electricity and heat for everything. The decrease occurs for five main reasons:

1. the higher efficiency of electricity and electrolytic hydrogen over combustion for transportation;
2. the higher efficiency of electric heat pumps over combustion for air and water heating in buildings;
3. the higher efficiency of electricity over combustion for high-temperature industrial heat;
4. eliminating the energy needed to mine, transport, and process fossil fuels, bioenergy, and uranium; and
5. improving end-use energy efficiency and reducing energy use beyond what will occur with business-as-usual.

These reductions are discussed in turn.

12.3.1 EFFICIENCY OF ELECTRICITY AND ELECTROLYTIC HYDROGEN OVER COMBUSTION FOR TRANSPORTATION

First, replacing end-use energy supplied by fossil fuels (fossil gas, gasoline, diesel, kerosene, jet fuel, bunker fuel) and biofuels for transportation with that supplied by WWS electricity and green hydrogen eliminates waste heat of combustion. This is because electricity and electrolytic hydrogen have higher energy-output-to-work-input ratios than do fossil fuels and biofuels. This factor is embodied in the **electricity-to-fuel ratio**, which is the energy required for a battery-electric or hydrogen-fuel-cell-electric vehicle to perform the same work (move the same distance) as a fossil-fuel or biofuel vehicle. This ratio is calculated here for battery-electric vehicles and hydrogen-fuel-cell vehicles versus **internal-combustion-engine** vehicles.

12.3.1.1 EFFICIENCY OF BATTERY-ELECTRIC VEHICLES OVER FOSSIL-FUEL VEHICLES

An example of the greater efficiency of electricity over combustion arises with battery-electric cars in comparison with internal-combustion-engine cars. Only 17–20 percent of the end-use energy embodied in gasoline, for example, is used to move a gasoline passenger car. This is the **tank-to-wheel efficiency** of the car. The rest of the energy (80–83 percent) is lost as waste heat.

A battery-electric passenger car, on the other hand, converts 64–89 percent (plug-to-wheel efficiency) of electricity at the plug (before charging the car) into motion, and the rest is waste heat (Section 4.1.1).

The tank-to-wheel efficiency of a fossil-fuel car divided by the plug-to-wheel efficiency of a battery-electric car is the fraction of a fossil-fuel car's end-use energy consumption that is needed to move an equivalent battery-electric car the same distance with electricity. This electricity-to-fuel ratio ranges from 0.19–0.31. In other worlds, a fossil-fuel car requires 3.2–5.3 times the energy in gasoline that a battery-electric car needs in electricity at the plug (before charging) to drive the same distance. By 2050, the average electricity-to-fuel ratio is expected to be 0.21, closer to the low end, due to improvements in vehicle-charging efficiencies and battery efficiencies and the greater use of permanent-magnet motors.

12.3.1.2 EFFICIENCY OF HYDROGEN-FUEL-CELL-ELECTRIC VEHICLES OVER FOSSIL-FUEL VEHICLES

A portion of future transportation will use hydrogen fuel cells. Hydrogen-fuel-cell-electric passenger cars are much less efficient than

12.3 CHANGES IN ENERGY NEEDED UPON A TRANSITION TO WWS

battery-electric passenger cars but are more efficient than internal-combustion-engine cars.

Hydrogen fuel cells convert hydrogen to DC electricity. The DC electricity is then converted with an inverter to AC electricity, which is used in an AC motor to produce rotation to turn wheels or a propeller.

The overall plug-to-wheel efficiency of a hydrogen-fuel-cell-electric passenger car accounts for multiple energy losses. These include losses in the electrolyzer used to produce hydrogen, in the compressor used to compress hydrogen, in the fuel cell within the car, in the DC-to-AC inverter in the car, and in the AC motor in the car. The efficiency also accounts for hydrogen leaks. The overall efficiency ranges from 34.6 to 46.6 percent (Section 4.2.3). Thus, 34.6–46.6 percent of the electricity used to produce hydrogen for a passenger car actually moves the vehicle.[65]

Dividing the low and high tank-to-wheel efficiencies of a fossil-fuel passenger car by the high and low plug-to-wheel efficiency of a hydrogen-fuel-cell passenger car, respectively, gives the electricity-to-fuel ratio of a hydrogen-fuel-cell car replacing a gasoline car as 0.37–0.58. Given that fuel cells are improving more rapidly than gasoline cars, the ratio 0.4 is more appropriate for 2050.

Since the plug-to-wheel efficiency of a hydrogen-fuel-cell passenger car is only 42–90 percent that of a battery-electric car, a hydrogen-fuel-cell passenger car needs proportionately more wind turbines or solar panels than does a battery-electric passenger car to travel the same distance. Thus, most short- and moderate-distance transport in a 100 percent WWS world will be battery-electric transport. On the other hand, for heavier and longer-distance vehicles, such as long-distance aircraft and ships and some trains, trucks, and military vehicles, hydrogen-fuel-cell-electric vehicles are more efficient than battery-electric vehicles.[81]

12.3.2 EFFICIENCY OF ELECTRIC HEAT PUMPS FOR BUILDINGS

Air-source heat pumps move heat from one place to another rather than create new heat. As a result, they are much more efficient than fuel-burning heaters or electric-resistance heaters for low-temperature heat in buildings. Air-source heat pumps generally have a coefficient of performance of 3.2–4.5, whereas ground-source heat pumps have a coefficient of performance of 4.2–5.2.[120] Thus, only 1 joule (unit of energy) of electricity is needed to move 3.2 to 5.2 joules of hot or cold air with an electric heat pump. This compares with a coefficient of performance of 0.97 for electric-resistance heaters and 0.8 for a typical fossil-gas boiler. In those cases, 1 joule of electricity creates only 0.97 or 0.8 joules of heat, respectively. As such, a heat pump reduces energy demand substantially compared with an electric-resistance heater or a fossil-gas boiler.

For example, the electricity-to-fuel ratio of a WWS-powered electric-heat-pump air heater replacing a fossil-gas boiler in the residential sector is 0.2. This is calculated assuming that the average coefficient of performance of all (ground-, air-, and water-source) heat pumps is 4. It also assumes that the average coefficient of performance of a fossil-gas boiler is 0.803, which itself assumes that 98 percent of combustion heaters have a coefficient of 0.8 and 2 percent have a coefficient of 0.95. Dividing 0.803 by four gives 0.2.

12.3.3 EFFICIENCY OF ELECTRICITY OVER COMBUSTION FOR HIGH-TEMPERATURE HEAT

Replacing fossil fuels and bioenergy for high-, medium-, and low-temperature industrial heat with WWS electricity or heat also reduces energy requirements. In a WWS world, oil, gas, coal, biomass, and waste combustion for medium- and high-temperature heat (above 150 degrees Celsius) will be replaced primarily by electric-arc furnaces; electric-induction furnaces; electric-resistance furnaces, kilns, and boilers; electric crackers; electron-beam heaters; dielectric heaters; and solar heaters. Electric heat pumps and solar heaters will replace fuels used for low-temperature industrial heating processes (below 150 degrees Celsius). Firebricks heated by resistance heating will store heat of all temperatures and thus can avoid the need for most all the technologies just listed.

Most fossil-gas furnaces used for producing high- and medium-temperature heat are about 80 percent efficient (coefficient of performance of 0.8). In other words, about 80 percent of the energy in the chemical bonds of fossil gas is converted to useful heat; the rest is lost either as waste heat or incompletely combusted fossil gas that escapes as exhaust. Some new fossil-gas boilers that recapture waste heat and latent heat of condensation have an efficiency of up to 107 percent. An efficiency above 100 percent is due to the capture of the latent heat. Assuming 80 percent of conventional boilers have an efficiency of 80 percent and the rest have an efficiency of 107 percent gives a mean efficiency of business-as-usual fuels for industry of 85.4 percent.

On the other hand, an electric-resistance furnace for producing high-temperature heat is about 97 percent efficient. Thus, about 97 percent of the electricity going into an electric-resistance furnace gets converted to useful heat for industry. The rest is waste heat that is lost due to conduction of heat out of the furnace without the heat being used. The efficiency of an

12.3 CHANGES IN ENERGY NEEDED UPON A TRANSITION TO WWS

average electric heat pump is about 400 percent (since the coefficient of performance is 4). Thus, heat pumps move four units of heat for every one unit of electricity they consume.

In sum, the electricity-to-fuel ratio for WWS replacing business-as-usual fuels for high- and medium-temperature industrial heating is 0.88. That for low-temperature industrial heating is 0.21. Assuming 85 percent of heating is for high-temperature heat and 15 percent is for low-temperature heat (an estimate for the electricity sector) gives a mean electricity-to-fuel ratio for WWS electricity replacing business-as-usual fuel for industry as 0.78.

12.3.4 ELIMINATING ENERGY TO MINE, TRANSPORT, AND PROCESS CONVENTIONAL FUELS

Fourth, producing all energy with WWS eliminates the need to mine, transport, and process fossil fuels, bioenergy, and uranium. Worldwide, about 11 percent of all energy is used for these purposes.[5] A WWS energy economy eliminates the need for such energy. Instead, wind comes right to the wind turbine and sunlight comes right to the solar panel, so no energy is needed to mine wind or sunlight.

Energy for the mining and transport of fuels is consumed in the industrial and transportation sectors. In the industrial sector, the energy consumed is for fossil-fuel and uranium mining operations and for petroleum refining. In the transportation sector, energy is needed to push fossil gas and oil through pipes and to move oil tankers, coal trains, gasoline and diesel trucks, trucks carrying biomass crops, and trucks and barges carrying liquid biofuels. About 2 percent of oil, 50 percent of coal, and 80 percent of fossil gas in the transportation sector are used just to transport fossil fuels, bioenergy, and uranium.[284] This need for energy is eliminated upon a transition to 100 percent WWS.

12.3.5 INCREASING ENERGY EFFICIENCY AND REDUCING ENERGY USE

Finally, increasing energy efficiency and reducing energy use beyond what will occur in a business-as-usual scenario reduce end-use demand further. Most business-as-usual projections of energy demand between today and 2050 are based on moderate economic growth and account for some end-use energy-efficiency improvements and reductions in energy use. However, such improvements are modest. Additional policy-driven energy-efficiency measures and incentives can reduce end-use energy demand further.

Most additional efficiency improvements will be due to increasing energy efficiency in the residential and commercial sectors. For example, more efficiency can be obtained by insulating pipes and weatherizing homes better. It can also be obtained by replacing more incandescent lights with LED lights and implementing energy-efficient appliances faster than with business-as-usual. Further improvements in vehicle design and in lightweight materials will also help.

12.3.6 OVERALL REDUCTION IN END-USE DEMAND

When business-as-usual energy is converted to WWS, annually averaged end-use energy demand decreases substantially due to the five reasons just discussed. For example, among 149 countries, representing 99.75 percent of all world fossil-fuel carbon dioxide emissions, a transition in 2050 reduces end-use demand by 54.4 percent, from 18.9 to 8.6 terawatts.[5] Of this reduction, 19.7 percentage points are due to the efficiency of WWS over business-as-usual transportation; 13.1 percentage points are due to the efficiency of heat pumps for heating and cooling buildings; 4.1 percentage points are due to the efficiency of WWS electricity for industrial heat; 10.9 percentage points are due to eliminating energy in the mining, transporting, and refining of fossil fuels, bioenergy, and uranium; and 6.7 percentage points are due to end-use energy-efficiency improvements and reduced energy use beyond those in the business-as-usual case. In sum, transitioning from fossil fuels, bioenergy, and uranium to 100 percent WWS has the potential to reduce energy demand worldwide significantly by reducing waste heat, eliminating unnecessary energy in finding and processing fuels, and improving energy efficiency.

12.4 PERFORMING A RESOURCE ANALYSIS

The next step in developing a 100 percent WWS roadmap is to perform a resource analysis for each WWS energy-generating technology. The purpose is to determine the maximum renewable energy resource available to meet demand in each country, state, city, or town. Resource analyses can be performed through data analysis, computer modeling, or both.

Types of data include local measurements or remotely sensed data. Local measurements include, for example, measurements of wind speed and sunlight at a weather station. Remotely sensed data are mostly satellite products. Satellite products are derived from measurements combined with model predictions. For example, satellite-derived wind speeds near the

12.4 PERFORMING A RESOURCE ANALYSIS

surface of the Earth are obtained by combining data from a satellite instrument, such as a microwave radiometer, a microwave scatterometer, or an ultraviolet laser, with results from a computer model.

Whereas local measurements are considered the most accurate, they are available only at a few locations. Satellite products are less accurate, particularly in the presence of clouds, but satellite products are available over most of the world and at high resolution. Two major problems with both local data and satellite products are (1) they cannot be used to predict WWS energy resources in the future, and (2) they cannot be used to predict how, for example, wind resources change due to extracting energy from the wind by a new wind farm.

Three-dimensional atmospheric computer models, on the other hand, can cover high spatial and time resolutions, predict the future, account for competition among wind turbines for limited energy in the wind, and estimate the wind and solar resource worldwide. However, model predictions, while covering greater areas, are less accurate than either local measurements or remotely sensed satellite products.

Analyses of local data, satellite products, and computer-model results suggest that enough solar and wind energy exist, independently, to power the world many times over for all purposes. The world solar resource is much larger than is the world wind resource. CSP resources are lower than are solar PV resources because CSP is limited to locations with significant direct sunlight, such as in deserts. In theory, sufficient heat can be extracted from deep in the Earth with enhanced geothermal systems to provide enough heat for the world's electricity and heat production, but the cost of obtaining such electricity is still higher than is the cost of solar and wind. No other WWS technology aside from solar, wind, and geothermal can practically power the whole world on its own, but all can play a role.

Another important component of a resource analysis is to determine the land or rooftop area that can feasibly be used to harvest renewable energy. A quantification of the rooftop area available for solar PV is needed in each country, province, state, city, and town. Rooftop areas include areas over buildings, parking canopies, parking structures, and road canopies. Rooftop areas suitable for PV exclude areas facing north in the northern hemisphere or south in the southern hemisphere. They also exclude areas continuously shaded and areas between rows of panels needed for walking and for avoiding shading of one row by another.

Land available for onshore wind and utility PV must also be determined. In considering available land, it is necessary to determine not only which land is exposed to fast winds or lots of sunlight but also which land is not excluded by law or practical constraint. Legal exclusions arise due to zoning ordinances that prevent development in sensitive areas or areas close to homes or other buildings.[285] Practical exclusions arise due to topography. For example, building a wind farm near the top of the Himalayan Mountains is not practical.

12.5 SELECTING A MIX OF WWS ENERGY GENERATORS TO MEET DEMAND

The next step in developing a 100 percent WWS roadmap is to quantify a mix, for each region of interest, of WWS electricity and heat generators that can supply the end-use all-purpose energy demand in the annual average. The penetration of each WWS electricity generator in each region is limited by the following constraints:

1. Each type of renewable generator cannot draw more energy than is available based on the renewable resource in the region.
2. The land area required by all WWS generators should be no more than a few percent of the total land area of the region.
3. No new conventional hydropower dams should be installed, if possible, but existing ones can be improved.
4. Wind and solar, which are complementary in nature, should both be used in similar proportions to the extent possible.
5. Wind farms should be installed, where possible, in cold regions.
6. The average cost among generators, storage, and transmission/distribution should be kept low.
7. A technology's use should be minimized in a region if its capacity factor is low there.

With these basic guidelines, it is possible to provide a first estimate of the WWS resources needed to power a town, city, state, province, or country for all energy purposes in the annual average. First estimates must then be refined, as described in Chapter 13, in order to ensure that energy demand is matched with supply continuously.

12.6 ESTIMATING AVOIDED ENERGY, AIR-POLLUTION, AND CLIMATE COSTS

Transitioning to 100 percent WWS will reduce the private and social costs of energy. The **private (business) cost of energy** is the marketplace cost of energy. It does not account for health or climate costs. The **social**

12.6 ESTIMATING AVOIDED ENERGY, AIR-POLLUTION, AND CLIMATE COSTS

(economic) cost of energy is the total cost of energy to society. It is the private cost of energy plus all externality costs associated with energy. An **externality cost** is a cost not captured in the market. The most relevant energy-related externality costs are health, climate, and nonhealth and nonclimate environmental costs. In this section, these costs are discussed.

12.6.1 PRIVATE ENERGY COST

The private (nonexternality) cost of energy includes the costs of electricity generation and storage; heat generation and storage; cold generation and storage; hydrogen generation and storage; transmission and distribution; and the appliances and machines that use electricity, heat, cold, and hydrogen.

Two metrics commonly used to calculate electricity-generation costs are the cost per unit of energy generated (in this case, the cost per kilowatt-hour-electricity produced) and the annual cost of energy (cost of energy per year). The annual cost of energy is simply the cost per unit of energy multiplied by the energy consumed per year. A 100 percent WWS system may reduce annually averaged energy demand worldwide, compared with a business-as-usual system, by about 54.4 percent in 2050.[5] As such, even if the cost per unit of energy between a business-as-usual and a WWS system is the same, WWS will reduce the annual energy cost to consumers by 54.4 percent.

Here, the method of calculating the cost per unit of energy, also called the levelized cost of electricity for electricity generators, is described. The **levelized cost of electricity** is the value, in today's dollars, of the total cost of electricity over the life of an electricity-generating plant, divided by today's value of the total energy generated by the plant over its life. The levelized cost of electricity is the average price an electricity generator must be paid before taxes for it to break even over its lifetime. It accounts for the generator's upfront capital cost, its annual operation and maintenance cost, its annual fuel cost (if any), its lifetime, and its decommissioning cost at the end of its life. It also accounts for the discount rate. For renewable energy generators (wind, solar, geothermal, hydro, tidal, and wave), the fuel cost is zero. For hydrogen, the fuel cost is the cost of water used to produce the hydrogen by electrolysis. There are two types of discount rate – a private and a social discount rate.

A **private discount rate** is the interest rate that banks charge builders and consumers for taking out loans. Such loans may be used to pay for the construction of a power plant or to build a house. The private discount rate is also the opportunity cost of capital. In other words, it is the rate of return that can be obtained by investing capital in a market. Private discount rates are appropriate only for short-term public projects that, dollar-for-dollar, crowd out private investment.[286] The private discount rate from 2022 to 2024 in the United States ranged from 3 to 8 percent.

A **social discount rate** is the discount rate used in a social cost analysis, which is a cost analysis that includes externality costs, such as health or climate costs. The **social cost** of an investment is the investment's private cost plus its externality costs. A social discount rate is used primarily when the costs and benefits of a project occur at different times and over more than one generation. Such projects are called intergenerational projects.

Social discount rates are smaller than private discount rates, because society, as a whole, cares more about the welfare of distant future generations than does the average consumer or investor, who is mostly concerned with near-term impacts during his or her lifetime. As a result, social discount rates appropriately weigh the present value of future impacts higher than do private discount rates. The (incorrect) use of a high private discount rate in the evaluation of long-term climate-change mitigation undervalues future social benefits and thus biases present-day investments away from efforts that provide long-term benefits to society. In order to properly evaluate long-term costs and benefits from the perspective of society, a social discount rate must be used. The consensus among experts of the mean social discount rate that should be used is 2 percent.[286,287,288]

A study of 149 countries found that a transition from business-as-usual energy to WWS energy may reduce the 2050 levelized cost of energy by about 11.4 percent. Because WWS also reduces all-purpose end-use energy requirements by 54.4 percent, WWS may reduce the world annual energy cost by 59.6 percent, from $16.5 to $6.7 trillion per year.[5]

Transition highlight
The lower private cost of energy due to WWS is being borne out in the real world. For example, in the United States, the cost of new renewable electricity generation in 2024 was less than the cost of new fossil-fuel or nuclear generation, even when storage cost was included.[191] Utility PV and onshore wind had the lowest levelized costs of electricity among all electricity-generating technologies, including combined-cycle fossil gas. What is more, new onshore wind and utility PV with storage had a much lower average cost than new nuclear or fossil-gas-peaking plants.[191] Similarly, in 2023, the world-average fossil-fuel-electricity-generation cost was $100 per megawatt-hour (MWh), but the costs of all new major WWS electricity generators were lower. New utility PV ($44/MWh), onshore wind ($33/MWh), offshore wind ($75/MWh), conventional

12.6 ESTIMATING AVOIDED ENERGY, AIR-POLLUTION, AND CLIMATE COSTS

geothermal ($71/MWh), and hydro ($57/MWh).[289] Enhanced geothermal is expected to cost $60–$70 per megawatt-hour by 2030 and $45 per megawatt-hour by 2035.[23] All of these except conventional geothermal and hydro were costlier than fossil-fuel electricity in 2010.[289] The low solar cost today is largely because the 2024 PV-panel price dropped to $0.05 per watt in Europe and most of the world. Thus, a standard 500-watt panel cost only $25. US panel prices were higher because of tariffs. In all countries, the overall cost of buying plus installing a PV panel varies due to differences in labor cost.

Transition highlight
Although California has the second-highest electricity prices in the US after Hawaii, and a lot of renewable electricity, the two facts are only a correlation, not a cause and effect. In fact, data indicate that, on average, higher renewable penetrations today reduce electricity costs and prices. First, with respect to the correlation between renewable electricity generation and electricity prices, California is an anomaly. Of the 12 US states with the highest WWS supply as a percentage of demand from October 1, 2023 to September 30, 2024, 6 were among the 10 states with the lowest residential electricity prices in March 2024; 10 were among the 20 states with the lowest electricity prices; and only 2 (California and Maine) had high prices.[257] As such, WWS is correlated more with low, rather than high, electricity prices. For example, South Dakota, Montana, and Iowa provided 109.8, 86.5, and 79.4 percent, respectively, of the electricity they consumed from October 1, 2023 to September 30, 2024 with WWS. Yet, those states, respectively, had the ninth-, eighth-, and twelfth-lowest March 2024 residential electricity prices among states.[257]

Second, the average spot price of electricity on California's main grid from March 7 to June 30, 2023 decreased by 53 percent compared with the same period in 2023 despite more WWS generators and batteries on the grid in 2024.[257] The **spot price** is the estimated market price of electricity during a given five-minute period and is calculated from marginal prices in numerous locations throughout a grid. These "locational marginal prices" represent marginal costs of serving the next increment of demand at pricing nodes on the grid and account for marginal energy costs, marginal costs of losses, and marginal costs of congestion. In 2023, California's fossil-gas price for producing electricity was more than twice the average US price. Because fossil-gas electricity generation in California has a higher cost than WWS generation, California's main grid-average spot prices should decrease when fossil-gas generation decreases and WWS generation increases. The data bore this out: California main-grid fossil-gas use decreased by 40 percent, and system-average spot prices decreased by 53 percent between the two periods.[257]

So why were California's electricity prices high in 2024? Aside from the high cost of fossil gas in California, a big factor was wildfires. For years, utilities had passed on to customers the cost of wildfires due to transmission-line sparks, the cost of undergrounding transmission lines to reduce such fires, and other wildfire-related mitigation costs. Utilities had also passed on to customers the costs of the San Bruno and Aliso Canyon fossil-gas disasters, the cost of retrofitting gas pipes following San Bruno, the cost of upgrading the aging transmission and distribution system, and the cost of keeping an aging nuclear plant, Diablo Canyon, open.[257] In sum, available data on price and cost indicate that increasing the share of WWS tends to reduce electricity price. When high prices occur, they are not due to WWS.

12.6.2 AVOIDED HEALTH COSTS FROM AIR POLLUTION

Transitioning homes, cities, states, provinces, and countries to WWS reduces air-pollution health problems immediately. The combustion of fossil fuels and bioenergy releases air-pollutant gases and particles. In the presence of ultraviolet rays of sunlight, some of the gases cook to form ozone and aerosol particles. Particles smaller than 2.5 micrometers in diameter penetrate deeper into people's lungs than larger particles. Particles, mostly small ones, are the most dangerous component of air pollution, causing up to 90 percent of air-pollution deaths and illnesses.[5] Ozone gas causes much of the rest.

Today, over 7.4 million people die each year from air pollution arising from fossil-fuel burning, bioenergy burning, open biomass and waste burning, and all other human and natural pollution sources. Air pollution also causes injury to billions more and harms animals, crops, vegetation, materials, works of art, and visibility. About 90 percent of air-pollution deaths and illnesses are due to human needs for energy.[5] Because emission-control technologies are expected to improve between today and 2050, human-caused and natural air-pollution deaths and illnesses, in a business-as-usual case, may decline to about 5.3 million per year in 2050.[5] Reductions may occur in almost all world regions despite higher populations in all regions. The main exception is Africa, where population growth may be so high that it outpaces technological improvement, resulting in more air-pollution deaths and illnesses in 2050 than today in a business-as-usual case.

The cost and emotional damage of over five million deaths and billions more illnesses per year worldwide in 2050 are enormous. Air pollution costs are due to hospitalizations, emergency room visits, lost workdays, lost school days, higher workman's compensation premiums, higher insurance premiums,

12.6 ESTIMATING AVOIDED ENERGY, AIR-POLLUTION, AND CLIMATE COSTS

higher taxes, and loss of companionship. Costs are also due to agricultural losses, visibility losses, and structure damage arising from air pollution.

Conversely, eliminating such air-pollution damage with WWS results in a better quality of life and lower costs. The cost reduction is usually estimated by first determining the value of statistical life. The **value of statistical life** is a widely used metric determined by economists to assign a cost to a reduction in the risk of dying. It is the value of reducing one statistical death in a population. For example, if an average person in a city of 100,000 is willing to pay $80 to reduce her or his death risk by 1/100,000, the statistical value of avoiding one mortality is $8 million. The value of statistical life is also determined from how much more employers pay their workers who have a higher risk of dying on the job.

The cost savings from avoiding a single air-pollution mortality is estimated as the value of statistical life. This value varies by country and year. The cost savings due to reducing air-pollution illness aside from death is then estimated as a percentage (usually 15 percent) of the cost savings from eliminating one air-pollution death. The cost savings due to reducing all nonhealth environmental damage (which does not include global warming, since that cost is determined separately) is also estimated as a percentage (usually 10 percent) of the cost savings from eliminating one air-pollution death.

With these assumptions in mind, a worldwide transition to 100 percent WWS may avoid about $33.8 trillion per year in health costs in 2050.[5] Such cost savings arise due to preventing about five million air-pollution-related deaths, billions more illnesses, and nonhealth environmental damage (aside from climate-change damage) each year. These are the worldwide WWS air-pollution social cost savings.

12.6.3 AVOIDED CLIMATE-CHANGE-DAMAGE COSTS

The economic damage that greenhouse gas emissions impart on the global economy through their impacts on climate is quantified with the **social cost of carbon**. Units of the social cost of carbon are US dollars per metric tonne of CO_2-equivalent emissions. Climate-change-damage costs include the costs due to higher sea levels (coastal infrastructure losses); reduced crop yields for certain crops; more intense hurricanes; more droughts and floods; more wildfires; more air pollution; more migration due to famine; more heat stress and heat stroke; more dengue fever, malaria, lyme disease, and West Nile virus; more fishery and coral reef losses; and greater energy requirements for air-conditioning, among other impacts. Only a portion of these costs will be offset by higher yields of some crops due to higher temperatures and carbon dioxide levels, and lower air- and water-heating requirements in some locations.

The social cost of carbon emissions is likely to increase over time as carbon dioxide and other warming agents accumulate in the atmosphere and temperatures continue to rise. The gas and particle accumulation will accelerate as more people rise out of poverty in developing countries and consume energy faster than their predecessors did. As such, the social cost of carbon is tied to the gross domestic product of a country. Based on data from several studies, an estimate of the social cost of carbon in 2010 was $250 ($125–$600) per tonne of CO_2-equivalent emissions. Accounting for the growth in damage between 2010 and 2050, the projected social cost of carbon in 2050 is $560 ($315–$1,190) per tonne of CO_2-equivalent emissions.[5] Consequently, the climate damage to the world in 2050 due to fossil fuels and bioenergy in a business-as-usual case may be a mean of about $30.9 trillion per year.[5]

12.6.4 SUMMARY OF AVOIDED ENERGY, HEALTH, AND CLIMATE-DAMAGE COSTS

The total social cost of business-as-usual energy is the sum of the private cost, air-pollution cost, and climate cost of energy. Worldwide (among 149 countries), those costs in a business-as-usual case are approximately $16.5, $33.8, and $30.9 trillion per year, respectively, for a total of $81.2 trillion per year.[5] Thus, health and climate costs of emissions exceed private energy costs by a lot, so eliminating emissions reduces social costs enormously.

Using WWS results in zero emissions of air pollutants and climate-damaging pollutants from energy. It also reduces private energy costs from $16.5 to $6.7 trillion per year, or by 59.6 percent. The $6.7 trillion per year WWS cost is the annual energy cost of a worldwide Green New Deal. The **Green New Deal** is the name given to a policy proposal that calls for, among other goals, a country to transition to 100 percent clean, renewable energy for all purposes as fast as possible. The lower private energy cost is due to the 54.4 percent lower annual energy requirements and the 11.4 percent lower cost per unit of energy.[5] WWS also reduces social energy costs from $81.2 to $6.7 trillion per year, or by 91.8 percent. The lower WWS annual social cost arises due to the lower WWS annual private energy cost plus the elimination of health and climate costs associated with business-as-usual fuels.

13

KEEPING THE GRID STABLE WITH 100 PERCENT WWS

One of the greatest concerns facing the implementation of a worldwide 100 percent clean, renewable energy and storage system is whether electricity, heat, cold, and hydrogen will be available when they are needed. In other words, can a 100 percent WWS grid avoid blackouts? The electric grid in a 100 percent WWS world will be very different from that of today. Today, electricity comprises about 20 percent of all end-use energy (or 40 percent of primary energy). In a 100 percent WWS world, electricity will comprise close to 100 percent of all end-use energy, which itself will equal primary energy less transmission and distribution losses. The nonelectricity end-use energy will come from geothermal and solar heat. The sectors that will be electrified (transport, buildings, industry, agriculture/forestry/fishing, and the military) will use more energy-efficient technologies than their fossil-fuel counterparts would. Such technologies include battery-electric vehicles, hydrogen-fuel-cell-electric vehicles, and electric heat pumps, among others. The reduction in energy needs due to the use of more efficient technologies will reduce overall energy demand substantially. Demand will also decrease because no more energy will be used to mine, transport, or process fossil fuels, bioenergy, or uranium for energy. End-use energy efficiency will increase, and policies will encourage less energy use. A future electric grid will also be coupled with electricity, heat, cold, and hydrogen storage. Finally, a future grid will have more long-distance electrical transmission instead of fossil-fuel pipelines. Thus, the main challenge in a future grid will be to match electricity, heat, cold, and hydrogen demand with 100 percent WWS electricity and heat supply plus storage while using demand response. This chapter discusses meeting such demand both on short timescales (seconds to minutes) and long timescales (months to seasons to years).

13.1 VARIABLE VERSUS INTERMITTENT RESOURCES

Wind, waves, and sunlight produce electricity that varies in amount over short and long timescales due to continuous changes in the weather and climate. Thus, these energy sources are **variable** resources. Another word commonly used to describe a variable resource is an **intermittent** resource. However, all energy resources, even those not affected by the weather, are intermittent due to the fact they are shut down during scheduled and unscheduled maintenance. Variable resources are those affected by both the weather and maintenance.

One concern with the use of variable WWS resources is whether they can provide electricity, heat, cold, and hydrogen on demand. Any energy system must respond to changes in energy demand over periods of seconds to years and accommodate unanticipated changes in the availability of electricity and heat generators. It is not possible to control the weather; thus, a sudden change in demand often cannot be met with a variable WWS resource unless either storage, demand response, or long-distance transmission or all three are coupled with electricity generation.

The concern about matching demand with supply, though, applies to all energy sources, not just the variable ones. For example, because geothermal, tidal, coal, and nuclear generators can provide relatively constant (**baseload**) electricity supplies, these generators rarely match energy demand, which varies continuously and significantly. Also, all baseload generators are down for maintenance anywhere from 3 to 50 percent of all hours of a year. As a result, gap-filling resources, such as fossil gas, hydropower, pumped hydropower, and batteries, are needed to meet peaks in demand.

Even nuclear power, which has been designed in some countries to ramp its electricity production up and down slowly, does not match demand. In France, for example, nuclear power ramps up to 100 percent of its power in 20–100 minutes. But this is 40–400 times slower than the ramp rate of hydropower or pumped hydropower storage (100 percent in 15–30 seconds), 60,000–300,000 times slower than the ramp rate of a battery (20 milliseconds), and 4–20 times slower than the ramp rate of an open-cycle fossil-gas turbine (100 percent in five minutes).[37]

13.1.1 RISK OF GRID FAILURE WITH THE CURRENT GRID

Although concern exists about maintaining stability of the current fossil-fuel-dominated electricity grid, the grid generally works, increasing the desire of grid operators to maintain the status quo. But, with increasingly extreme heat waves and cold spells caused, in part, by global warming, the fragility of the current grid is being exposed. Because WWS has been added to the grid in most places, some people have blamed it for blackouts

13.1 VARIABLE VERSUS INTERMITTENT RESOURCES

that have occurred. Close inspection, though, indicates that grid failures in those cases were caused mostly by problems with the grid itself or with fossil-fuel unreliability, not with WWS electricity variability.

Two examples are the August 14–15, 2020, summer blackout in California and the February 14–18, 2021, winter blackout in Texas. Close to 45 percent of California's electricity and 23 percent of Texas' electricity was from WWS in 2020. Despite many people blaming the growth of solar and wind in California for the blackout, the heads of the California Public Utility Commission, California Independent System Operator, and California Energy Commission confirmed that "renewable energy did not cause the rotating outages."[290] Instead, a variety of factors, including an unexpected unavailability of imports from across the west, led to the blackout. In the case of Texas, low temperatures caused fossil-gas, coal, nuclear, and wind electricity generators to fail, with fossil gas being the largest source of electricity and failure.[291] Some frozen wind turbines were shut because none had deicing equipment. A simple solution is to add deicing equipment to wind turbines in Texas. Wind turbines with deicing equipment have operated successfully in cold weather in Iceland, Norway, Sweden, Alaska, Canada, and Russia.

13.1.2 COUNTRIES, STATES, AND REGIONS WITH 100 PERCENT WWS AND A STABLE GRID

Blackouts can occur with a 100 percent WWS grid if the grid is not planned properly. However, proper planning and management can prevent blackouts. In this section, real-world examples of 100 percent WWS grids are discussed.

13.1.2.1 COUNTRIES AND REGIONS WITH 100 PERCENT WWS GRIDS

By 2022, 12 countries had already generated 98.4–100 percent of their electricity from WWS in the annual average (Figure 13.1). Among these, 10 generated 99.5–100 percent of their electricity with WWS. The dominant renewable in all ten countries was hydropower. Iceland and Costa Rica also produced 29.4 and 12.9 percent, respectively, of their electricity from geothermal. Costa Rica and Norway further supplied 10.9 and 10.2 percent, respectively, of their electricity from wind. Some of the countries with high WWS penetration (Paraguay, Bhutan, and Nepal) produced more electricity than they consumed and exported the difference. Paraguay exported to Argentina and Brazil. Bhutan and Nepal both exported to India.

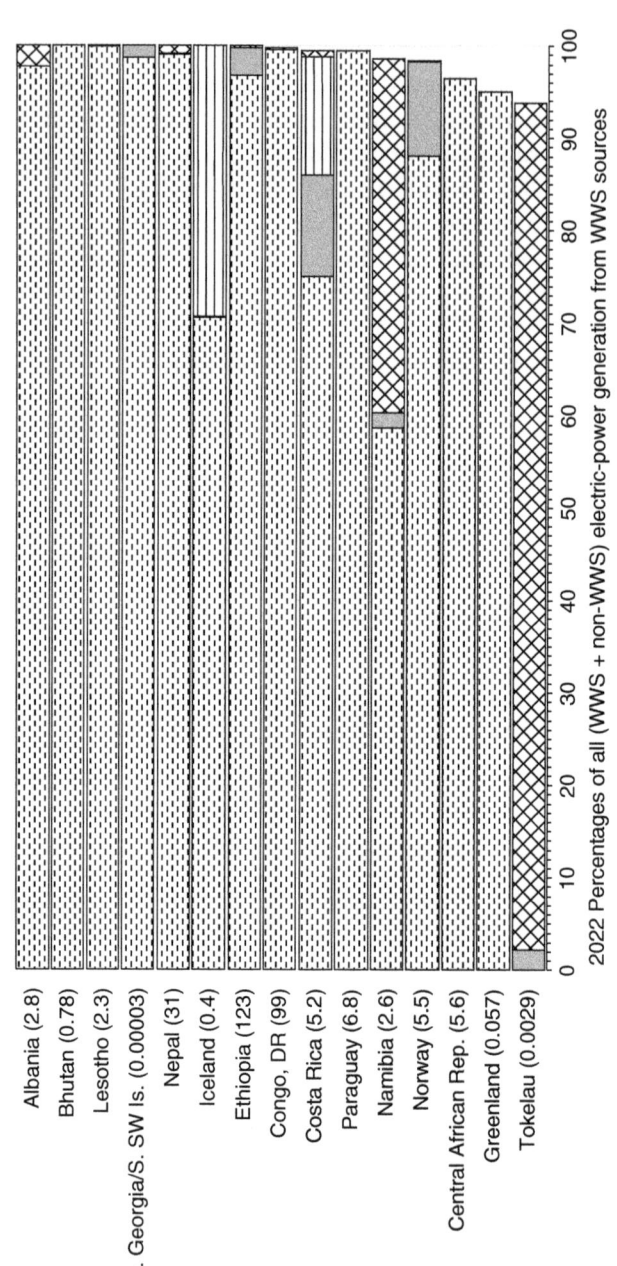

Figure 13.1. 2022 percentages of total electricity generated by wind–water–solar (WWS) sources for the 15 countries or territories with the highest percentage of the total electricity they generated from WWS. Numbers in parentheses are 2022 population (millions). Data from IEA[261] for all countries except Bhutan, Lesotho, S. Georgia/S. Sandwich Islands, Central African Republic, Greenland, and Tokelau, which are from IRENA.[292]

13.1 VARIABLE VERSUS INTERMITTENT RESOURCES

Transition highlight
Overall, 64 countries or territories, representing 15.9 percent of world population, generated 50–100 percent of all electricity they generated with WWS in 2022.[261,292] Whereas electricity production in most places was dominated by hydropower, in Tokelau (an island territory of New Zealand) it was dominated by PV; in Denmark, Luxembourg, Lithuania, and Portugal it was dominated by wind; and in Kenya, by geothermal. El Salvador had a good mix of solar, geothermal, and hydro. Luxembourg had a good mix of wind, solar, and hydro. The world as a whole generated about 27 percent of the electricity it consumed from WWS.

Only a few world regions have achieved high shares of WWS without relying mainly on hydroelectricity. For example, during the first six months of 2019, the Coquimbo region of Chile produced 99.7 percent of its electricity from WWS, with 74.9, 20.8, and 4 percent coming from wind, solar PV, and hydropower, respectively.[293] In 2022, wind, hydro, and solar plus other renewables supplied 88.9, 16.9, and 8.9 percent, respectively, of gross electricity consumption in Scotland. Thus, total WWS production exceeded consumption and the difference was exported to the rest of the United Kingdom.[294] Tokelau is an island territory of New Zealand with 29,000 people spread across three atolls. The territory transitioned from diesel generators to solar PV starting in 2012. In 2022, Tokelau generated 91.6 percent of its electricity from PV and 2.2 percent from wind (Figure 13.1). It also has large banks of battery storage.

During 2024, South Australia generated more than 70 percent of its electricity from wind and solar, with a particularly high share of rooftop PV.[295] During the last week of winter, 2024, WWS supplied, on average, over 100 percent of South Australia's electricity demand.[296] The Australian government will underwrite sufficient additional wind, solar, and storage capacity for South Australia to meet 100 percent of its demand, in the annual average, with just WWS, primarily wind and solar, by 2027.[295] Of further note, from January to September 2024, Germany's WWS electricity production reached 50 percent of total electricity production for the first time (up from 44 percent in 2023). It was also the first time that wind plus solar electricity output (45 percent of the total) exceeded fossil-fuel output (40 percent) in Germany. From January to December 2024, the United Kingdom produced more WWS electricity (37 percent) than fossil-fuel electricity (35 percent) for the first time in its history.

13.1.2.2 US STATES WITH CLOSE TO 100 PERCENT WWS GRIDS

From October 1, 2023 to September 30, 2024, WWS supplied 53.7–109.8 percent of electricity demand in 11 US states.[257] At the top was South

Dakota, which produced the equivalent of 109.8 percent of its demand with just WWS: wind (77.5%), hydro (30.1%), and solar (2.2%). South Dakota produced another 16 percent of the electricity it consumed from fossil gas and 11.2 percent from coal. Thus, South Dakota produced a total of about 137 percent of the electricity it consumed with WWS plus fossils and exported the extra 37 percent. The other 10 states producing over 50 percent of demand with WWS include Montana (86.5%), Iowa (79.4%), Washington State (72.6%), Kansas (70.2%), Oregon (64.2%), Maine (62.1%), New Mexico (59.7%), Wyoming (56.1%), North Dakota (55.1%), and Oklahoma (53.7%). The main electricity source in seven of the eleven states was wind; in the remaining four, it was hydro. In 2024, the second-largest island in the US, Kodiak Island, Alaska, with a population of 15,000 and peak electricity demand of 28 megawatts, became the first US island to run its grid on 100 percent WWS for a year. It did so with 33 megawatts of hydro, 9 megawatts of wind, 3 megawatts of battery storage, and 2 megawatts of flywheel storage.

13.1.2.3 100 PERCENT WWS IN CALIFORNIA

California's WWS generation from October 1, 2023 to September 30, 2024 averaged 47.3 percent of demand, ranking it number 12 among US states in terms of WWS supply as a percentage of demand.[257] However, those data do not illustrate the remarkable progress California, the world's fifth-largest economy at the time, made toward 100 percent WWS from March 7 through June 30, 2024 (late winter, all of spring, and early summer). During that period, WWS supply met 100–162 percent of demand for parts of a record 98 of 116 days on California's main grid, which serves about 82 percent of California's grid demand.[257] Among all 116 days, WWS exceeded 100 percent of demand for an average of 4.84 and a maximum of 10.1 hours per day. Over all hours during the period, WWS supplied 61.3 percent of demand. A multiweek heat wave then enveloped California, causing air-conditioning demand to rise substantially, preventing WWS from meeting 100 percent of demand again until July 27. However, during all of 2024, California met more than 100 percent of main-grid demand with WWS for parts of 132 days.

When WWS supply exceeded demand on California's main grid from March 7 to July 30, 2024, about half the excess electricity was stored in batteries; most of the rest was exported to other western US states. The remainder was curtailed. Most battery electricity was discharged to the grid after sunset and before sunrise. On June 19, 2024, nighttime battery peak

13.1 VARIABLE VERSUS INTERMITTENT RESOURCES

output reached 7.78 gigawatts,[257] equivalent to the peak output of about eight nuclear reactors and to 35.3 percent of average grid demand during the period (22.1 gigawatts). Because batteries were mostly four-hour batteries, they stored up to 31.1 gigawatt-hours of electricity. Battery-electricity supplied an average of 3.9 percent and a peak of 6 percent of California's main-grid daily (24-hour) electricity demand. Thus, batteries supplied up to 12 percent of nighttime demand. The major role of batteries was to shift daytime WWS surpluses to night.

On April 11, 2024, solar PV plus concentrated solar power (CSP) alone provided over 100 percent of demand for the first time in California.[257] After that, through June 30, 2024, solar alone exceeded demand for parts of 22 more days. Grid solar output does not include most distributed, behind-the-meter PV output, which supplied the equivalent of 12.9 percent of California's electricity retail sales from October 1, 2023, to September 30, 2024. Distributed PV is either roof- or ground-mounted PV powering residential, commercial, or industrial buildings directly. Because distributed PV provides electricity for buildings, those buildings need much less daytime grid electricity. Thus, while not contributing much to grid electricity supply, distributed PV reduces grid demand during the day.

During the April 8, 2024, partial solar eclipse, for example, grid demand increased temporarily because distributed PV output decreased, increasing the need for the grid to supply electricity to buildings.[257] At the same time, utility solar supply decreased, but battery output filled in the gap. Not only did no blackout occur during the eclipse, but no blackout occurred during any other day during the 116-day period. Further, during no five-minute interval during the 116 days did WWS supply fall below 21.2 percent of demand. Thus, WWS supplied "baseload" electricity for over a fifth of grid demand.

California's progress in 2024 was due largely to an increase in WWS supply and decrease in demand since 2023. Utility solar, wind, and battery output from March 7 to June 30, 2024, were 31, 8, and 105 percent higher, respectively, on California's main grid than during the same period in 2023, mostly due to nameplate capacity increases of 18, 4, and 73.3 percent, respectively, from June 2023 to June 2024. Overall WWS output increased by 12 percent in 2024 versus 2023. Simultaneously, demand (load) dropped by 1 percent, due to an increase in distributed, behind-the-meter PV. A major benefit of WWS growth was a 40 percent drop in fossil-gas-electricity use between the two periods.[257]

California can eliminate its remaining use of fossil gas and its use of nuclear and biomass in several ways. First, increasing distributed behind-the meter PV and batteries will reduce grid demand. Second, adding more utility PV and batteries will increase WWS supply during day and night. Such growth is already occurring. By May 2025, the peak discharge rate of batteries on California's main grid had increased to above 10 gigawatts. Third, adding offshore wind, which peaks during evenings and summers in California, will increase day and night electricity production. Fourth, shifting more hydro to night from day will provide more nighttime electricity. Fifth, adding enhanced geothermal will increase day and night electricity supply. Finally, using demand response, such as by setting higher electricity rates during times of high electricity demand, will shift more demand from high- to low-demand hours.

13.1.3 EFFECTS OF CLIMATE CHANGE ON THE RELIABILITY OF CONVENTIONAL AND WWS RESOURCES

Weather and climate affect both conventional and WWS energy resources. However, the increased frequency of heat waves due to global warming is a greater danger to the current energy infrastructure than to a WWS infrastructure. The main reason is that coal, fossil gas, oil, biomass, and nuclear electric-power plants depend on cooling water, usually from a river, lake, or the ocean, to operate safely. The effectiveness of the cooling system of such power plants is constrained by the laws of physics, regulations, and access to cooling water.[297] For example, if cooling water temperature rises too high, or if river or lake water levels drop too much, power plants that require cooling must be shut down; otherwise, they risk damage to marine life or themselves.

During an August 2018 heat wave in Europe, for example, high water temperatures caused Sweden to shut two nuclear reactors; France to shut four reactors; and Germany, Switzerland, and Finland to reduce power from several reactors. Germany also shut a coal plant due to high water temperatures, shut another due to low water levels, and reduced output from two fossil-gas plants. Similarly, during a July 2019 heat wave, France and Germany each shut a nuclear reactor, and France reduced output from six more reactors.

High water temperatures (from heat) and low water levels (from drought) are not the only climate-related causes of power-plant shutdowns. Others include wildfires, lightning, hurricanes, and typhoons. Between the 1990s and the 2010s, for example, the frequency of climate-related nuclear plant outages due to all these factors increased by a factor of eight.[298] Such outages are expected to increase in the decades to come.

13.1 VARIABLE VERSUS INTERMITTENT RESOURCES

The only WWS generator that has issues related to cooling water temperatures is CSP. However, CSP plants are cooled more and more by air instead of water, since most CSP plants are located in deserts, where water is scarce. Air cooling reduces the efficiency of CSP output slightly, but not materially, compared with water cooling.

Global warming can affect hydropower output, for both the better and the worse. Water supplies for hydropower depend on rainfall, and global warming changes weather patterns, causing drought in some places and flooding in others. In the global and annual average, though, evaporation must equal precipitation. In other words, the same amount of water that goes into the air must come out. As the Earth warms on average, more water evaporates from the oceans and soils. This means more rainfall plus snowfall must occur in the global average. Thus, although global warming increases droughts in some locations, it increases rainfall to a greater extent in other locations. As such, although hydropower resources will decline in some regions as the Earth warms, hydropower resources should increase in the global average.

The seasonal variation and yearly uncertainty in hydropower output, though, can cause problems, particularly if hydropower is the main source of electricity in a country. For example, although Tajikistan usually supplies more hydropower electricity than it needs and exports the difference, until 2023 it required its citizens to reduce electricity use during winters because of low reservoir levels. Tajikistan addressed this problem by upgrading existing hydropower plants and building new ones.

The increased frequency of drought in some locations due to global warming is increasing the concern about water shortages affecting hydropower output and drinking water supplies. The western United States, for example, which has a lot of hydropower, risks losing a portion of hydropower output due to extended and more frequent drought. However, the western US, and California in particular, import some of their hydropower from Canada. So, the loss of local electricity from hydro can be replaced by increasing imported electricity from hydro. In addition, the west has enormous solar and wind resources. As such, during a drought, more sunlight is available so additional solar electricity production substitutes for reduced hydro availability.

Another impact of high temperatures is on solar photovoltaic output. High temperatures reduce the output of solar PV cells but do not cause the panels to shut down. On the other hand, high temperatures often occur when the sky is cloud free, so more sunlight is available to PV panels during hot periods.

The effects of climate change on wind energy output are more complex. Wind energy output depends on wind speeds 30–250 meters above ground level. Buildings and vegetation slow winds, as does air pollution. For example, measurements from 1974 to 1994 at 79 stations in southeast China indicated wind speeds decreased by 24 percent, resulting in a 60 percent decrease in wind power.[299] This **disappearing wind syndrome** was attributed to the construction of new buildings and the growth of vegetation near meteorological stations. However, heavy air pollution may have also played a role, as it can reduce near-surface wind speeds by up to 8 percent.[300]

The reason that air pollution reduces wind speeds is as follows. Some air pollutants absorb sunlight, heating the air but also preventing the light from reaching the surface, cooling the ground. Remaining air pollutants reflect sunlight, also cooling the ground relative to the air. Thus, all air pollutants cool the ground relative to the air. Warmer air above a cooler ground makes it more difficult for air or winds near the ground to be lifted or for air or winds high above the ground to descend. Since horizontal wind speeds high above the ground are always faster than those near the ground (wind speeds at the ground itself are always zero), it is now more difficult for fast horizontal winds high up to be transported down to the height of wind turbines. Thus, air pollution decreases wind speeds available to turbines. The result is that, as air pollution decreases during a transition to 100 percent WWS, wind power output should increase.

However, wind speeds are also affected by climate change. Global warming may be increasing winds in northern Europe, the southern US Great Plains, and Brazil but decreasing them in southern Europe and the western US.[301]

In sum, global warming has shifted the locations and magnitudes of some wind resources, and air pollution has decreased wind speeds available to wind turbines. Transitioning to WWS, which reverses warming, will undo some of those impacts. Global warming should have little impact on geothermal or tidal electricity since geothermal depends on temperatures well below the Earth's surface and tidal electricity depends on the gravitational attraction between the moon and Earth. Because waves are driven by offshore winds, global warming will affect wave power similarly to how it affects offshore wind power. Finally, higher temperatures due to global warming will increase summer air-conditioning demand and decrease winter heating demand, shifting more demand from winter to summer.

13.2 METHODS OF MEETING ENERGY DEMAND CONTINUOUSLY

Given the minute-by-minute, daily, seasonal, and annual variations in both energy demand and WWS supply, it is critical to develop methods to reliably match demand with WWS supply on all these timescales. With 100 percent WWS, all energy produced will be electricity or heat, with electricity dominating. However, in terms of consumption, a substantial share will be heat used for buildings and industry. Thus, a good portion of electricity will be converted to heat, either with electric heat pumps, electric-induction cookers, or furnaces for industry. A portion of the electricity will also be used for cooling. Another portion will be used to produce hydrogen.

There are several steps needed to design an energy system in which electricity, heat, cold, and hydrogen demand are matched continuously with 100 percent WWS electricity and heat supply, storage, demand response, and transmission. The overall strategy is separated into sets of steps. The first set of steps are for creating a WWS infrastructure. The second set are for matching demand with supply, storage, and demand response within the infrastructure.

13.2.1 STEPS IN CREATING AN INFRASTRUCTURE FOR PROVIDING ALL ENERGY WITH WWS

The first set of steps is to create a WWS infrastructure. This set requires the following:

a. Transition transportation to run on battery electricity and hydrogen-fuel-cell electricity, where the hydrogen is produced from electricity.
b. Transition all building air and water heat so that it is provided with electric heat pumps or district heating. Ensure that district heat is produced by electric heat pumps, solar heat, and/or geothermal heat. Use district heat immediately or store it in water tanks or underground in soil (boreholes), water pits, and aquifers.
c. Transition building air-conditioning to run on electric heat pumps or district cooling. Cold from district cooling should come from heat pumps and be stored in water tanks or ice.
d. Electrify low-to-high-temperature heat needed for industry with electric-arc furnaces, induction furnaces, resistance furnaces and boilers, dielectric heaters, electron-beam heaters, heat pumps, and solar heaters. Use firebricks heated with electric-resistance heating to store low-to-high-temperature heat.

e. Transition all remaining combustion appliances, machines, and processes to electric ones (e.g., electric-induction cookers, electric fireplaces, electric leaf blowers, electric lawnmowers, and electric chain saws).
f. Build wind farms in cold regions because wind output and heat demand both increase with lower temperatures, and wind electricity can help meet that heat demand.
g. Interconnect, through the transmission system, geographically dispersed wind, wave, and solar resources to turn some of their variable supply into more steady supply while reducing overall transmission requirements.
h. Colocate wind and solar together since wind and solar are complementary in nature and colocating them reduces overall transmission requirements.
i. Increase building energy efficiency by using LED lights; installing efficient appliances; reducing air leaks from doors, windows, and skylights; and increasing insulation in walls, floors, and ceilings.
j. Improve passive heating and cooling in buildings by using insulation; thermal-mass materials; ventilated facades; window blinds, awnings, and films; and night ventilation.
k. Provide all electricity with onshore and offshore wind, distributed and utility-scale solar PV, concentrated solar power, geothermal electricity, tidal electricity, wave electricity, and hydroelectricity.
l. Store excess WWS electricity in stationary and vehicle batteries, grid-hydrogen storage, pumped-hydro storage, flywheels, compressed-air storage, gravitational storage with solid masses, and concentrated solar power storage.
m. Store any remaining excess electricity in firebricks for industry, hot and cold storage for buildings, and hydrogen for nongrid purposes.
n. Colocate WWS electricity generation and storage where possible to reduce transmission requirements.
o. Build more short- and long-distance high-voltage AC transmission, long-distance high-voltage DC transmission, and AC distribution lines to interconnect WWS supply and demand centers.

With respect to Step f, building wind farms in cold regions increases the reliability of the overall grid. The reason is that wind energy production in cold regions correlates strongly with heat demand in buildings. This result has been obtained from both data analysis[302] and computer modeling.[303] The correlation is strongest when averaging over large, cold regions, such as Canada, Europe, Russia, and the United States.[303] Even on an hourly timescale, wind energy output is well matched to the demand for heat

13.2 METHODS OF MEETING ENERGY DEMAND CONTINUOUSLY

in buildings. Wind-turbine electricity in cold regions is best used to run ground-, water-, or air-source heat pumps to provide heat for buildings.

With respect to Step g, interconnecting geographically dispersed wind, wave, or solar farms to a common transmission grid smoothens out electricity supply significantly.[304] In fact, interconnecting wind farms over a region as small as a few hundred kilometers in size can eliminate hours of no power summed over all farms.[305] For example, in a 550 kilometer by 700 kilometer region of the Great Plains of the United States, low wind speeds resulted in no output from an isolated wind farm 7.6 percent of the hours during a year between noon and 3 p.m. However, with three interconnected wind farms spread out over the same region, the number of no-power hours decreased to 2.6 percent. With eight wind farms spread out over the same region, the number decreased all the way to zero. In other words, interconnecting geographically dispersed wind farms eliminates times of the year when no wind power output occurs over a region.[305]

Interconnecting wind farms also means that wind, aggregated geographically, acts like a single wind farm with steadier, closer-to-baseload power production. In fact, when 19 wind sites in the midwest, geographically dispersed over an 850 kilometer by 850 kilometer region, were hypothetically interconnected, between 33 and 47 percent of yearly averaged wind power was calculated to be available at the same baseload reliability as a coal-fired power plant.[306] The benefits of interconnecting increased with more and more interconnected sites, although with diminishing returns.

> **Transition highlight**
> One more benefit of interconnecting geographically dispersed wind farms is that it reduces transmission requirements with little power loss. For example, connecting 19 wind farms to a common point and then connecting that point to a far-away city reduces the need for long-distance transmission, for example, by 20 percent with only a 1.6 percent loss of energy.[306]

Colocating a wind and a wave farm has similar benefits as interconnecting geographically dispersed wind farms. Namely, it dampens overall power variations[308] and reduces transmission requirements.[309] Thus, for example, combining wind and wave power at the same location results in lower variability and fewer hours of no power than a wind farm or wave farm acting alone.

With respect to Step h, it is helpful, for grid stability, to build both wind and solar in or near the same location. The reason is that wind and solar are complementary in nature.[303] During the day, when the wind is not blowing, the sun is often shining and vice versa. This occurs because, when surface air pressure is high (such as under a high-pressure system), winds are generally slow. At the same time, descending air in a high-pressure system evaporates clouds, increasing sunlight to the surface. Conversely, when surface air pressure is low (such as under a low-pressure system), winds are generally fast. At the same time, rising air in a low-pressure system increases cloudiness, reducing sunlight to the surface.

The anticorrelation between wind and solar is remarkably consistent worldwide.[303] The implication of this is that colocating wind and solar farms reduces the variability of the overall power output of wind alone or solar alone. Colocating also reduces transmission requirements. For example, when wind and solar are built far from each other, each needs its own transmission line, yet solar does not use its line at night. When wind and solar are colocated, only one line is needed and wind will use the line during the night, minimizing the time during which the line is not used.

Similarly, colocating WWS electricity generation and storage optimizes the use of transmission lines. For example, when utility PV is colocated with battery storage, some of the PV electricity produced during the day can be stored in colocated batteries that return the electricity through the transmission line at night, reducing the peak transmission size needed and maximizing the use of the transmission between day and night.

Transition highlight
Colocating batteries with PV or wind reduces the problems of no sunlight at night and the variability of wind and solar. By the end of 2023, the main type of grid-connected colocated WWS plus battery system in the United States was a PV plus battery system (288 plants with 14.5 gigawatts of PV plus 7.77 gigawatts/24.2 gigawatt-hours of battery storage). There were also 19 colocated wind-battery plants (with 2.98 gigawatts of wind plus 528 megawatts/598 megawatt-hours of battery storage), and 5 colocated PV plus wind plus battery plants (526 megawatts of wind, 76 megawatts of PV, and 69 megawatts/139 megawatt-hours of battery storage).[307] More battery systems colocated with solar PV existed in the US than stand-alone battery systems.[307]

13.2.2 STEPS FOR MATCHING DEMAND WITH SUPPLY, STORAGE, AND DEMAND RESPONSE

The second set of steps for keeping the grid stable involves matching instantaneous electricity, heat, cold, and hydrogen demand with WWS electricity and heat supply, storage, and demand response.

13.2 METHODS OF MEETING ENERGY DEMAND CONTINUOUSLY

One type of **demand response** is the use of financial incentives to shift the time of a flexible demand to a future time, when more energy is available, to satisfy the demand. **Flexible demands** are demands for electricity, heat, cold, or hydrogen that can be shifted forward in time during a day or over several days by demand response.

For instance, charging an electric vehicle is a flexible demand because the time of charging can be shifted to a nonpeak time of grid-electricity use through a financial incentive, such as a lower nighttime electricity rate. In fact, any electricity demand that can be shifted in time is a flexible demand. Examples are electricity used for (a) a wastewater treatment plant, (b) charging a stationary battery, (c) pumping water to an upper reservoir in a pumped hydropower system, (d) storing energy in a flywheel, (e) heating water in a hot-water tank with a heat pump, (f) producing ice for later use, or (g) producing and storing hydrogen for later use.

Similarly, electricity demand for high-temperature industrial heat is a flexible demand if the heating can be scheduled flexibly to take place at a different time of day. For example, a utility may establish an agreement with a manufacturing plant that uses electricity to produce high-temperature heat stored in firebricks that will be used later for an industrial process, to use the electricity during only certain hours of the day in exchange for a better electricity rate. In this way, the utility can shift the time of electricity use to heat the firebricks to a time when more WWS electricity is available, whereas the plant may still draw heat from the firebrick storage continuously 24 hours per day.

Similarly, the demand for electricity to charge battery-electric vehicles is a flexible demand because such vehicles are often charged at night, and it is not critical which nighttime hours the electricity is supplied so long as the needed electricity is supplied during the night. In this case, a utility can use a smart meter to provide electricity for battery-electric vehicle charging when WWS is available and to stop charging when it is not. Utility customers would sign their vehicles up under a plan giving the utility control of the time it supplies the electricity to their vehicles.

Inflexible demands, on the other hand, are electricity, heat, cold, or hydrogen demands (loads) that cannot be shifted in time. Examples are electricity for refrigerators, lighting, and electric-induction cooktops.

Whereas the need for electricity stored in an onboard battery to drive an electric vehicle or to power an electric leaf blower is an inflexible demand, the need for electricity to charge the battery in both cases is a flexible demand. Similarly, whereas the use of low-temperature heat and cold for building heating and cooling are inflexible demands, providing heat

13 KEEPING THE GRID STABLE WITH 100 PERCENT WWS

and cold to the storage (water-tank, soil, water-pit, and aquifer storage) used to serve those demands is a flexible demand. Finally, whereas the use of hydrogen in a hydrogen-fuel-cell-electric vehicle is an inflexible demand, producing the hydrogen with electricity is a flexible demand.

Inflexible demands must be met immediately. However, if the cost of an inflexible demand is high enough, many individuals will forego the demand in order to save money. As such, no demand is truly inflexible. Demands are inflexible only up to a certain cost. Beyond that cost, demands can be shifted in time with demand response. Incentives for demand response can be either a lower price for electricity during times of low demand or a direct payment to the customer in exchange for the customer agreeing not to use electricity during the time of peak demand. The demand that is shifted can be either an electricity, heat, cold, or hydrogen demand.

The detailed steps for meeting demand with supply, storage, and demand response are as follows.

a. Determine the annual end-use energy demand in each energy sector. Size WWS electricity and heat generators to meet the annually averaged demand.

b. Estimate the additional generation, storage, and demand response needed to meet demand continuously rather than annually. Do this by first quantifying (1) electricity and heat demands for low-temperature heating that must be met, (2) electricity demands for cooling and refrigeration that must be met, (3) electricity demands for producing, compressing, and storing hydrogen that must be met, and (4) all other electricity demands (including industrial heat demands) that must be met. Of the demands, identify the flexible ones subject to demand response.

c. Once the generation, storage, and demand response system is designed, carry out the following steps to match instantaneous demand with supply.

d. When the current WWS electricity supply exceeds the current inflexible electricity or heat demand, use the electricity supply to satisfy the demand.

e. When the current WWS solar plus geothermal heat supply exceeds the current inflexible heat demand, use the heat supply to satisfy the demand.

f. Use the remaining instantaneous WWS electricity or heat supply to satisfy as much existing flexible electric or heat demand as possible.

13.2 METHODS OF MEETING ENERGY DEMAND CONTINUOUSLY

g. Use any excess electricity remaining after that to fill electricity, heat, cold, or hydrogen storage. Select which type of storage to fill first.
h. Use any excess geothermal or solar heat after satisfying inflexible and flexible heat demand to fill heat storage.
i. When the current inflexible plus flexible electricity demand exceeds the current WWS electricity supply from the grid, use electricity storage to fill in the gap in supply. Use the electricity to supply inflexible demands first.
j. If electricity storage is depleted and flexible demand remains, use demand response to shift the flexible demand to a future hour.
k. If inflexible plus flexible heat demand in buildings exceeds current supply, use stored heat. If stored heat is exhausted, shift flexible heat demand to a future hour with demand response.
l. If inflexible plus flexible cold demand in buildings exceeds current supply, use stored cold. If stored cold is exhausted, shift flexible cold demand to a future hour with demand response.
m. Provide hydrogen demand with current hydrogen supply, where the current hydrogen is produced by current electricity. If current hydrogen supply is exhausted, use stored hydrogen.
n. If the measures above are insufficient to match energy demand with supply continuously, do one or more of the following: (1) increase the nameplate capacities of some electricity generators; (2) increase electricity, heat, cold, or hydrogen storage nameplate capacity; and/or (3) increase the nameplate capacity of the transmission system; (4) use electricity stored in vehicle batteries to help fill in gaps in supply on the grid; and/or (5) integrate weather forecasts into electricity planning to improve the efficiency of the grid.

In the next sections, some of the steps required for matching demand with supply, storage, and demand response are discussed in more detail.

13.2.3 SPECIFY WWS GENERATOR REQUIREMENTS TO MEET CONTINUOUS DEMAND

The first steps in matching demand with supply, storage, and demand response are to estimate future (e.g., 2050) annually averaged end-use energy demand across all energy sectors, and propose a mix of WWS electricity and heat generators to match that demand. Chapter 12 discussed the methodology for carrying out this step.

Generators sufficient for meeting demand in the annual average are usually not sufficient for meeting demand every minute. Demand varies continuously, as does electricity supply from most WWS resources. Whereas WWS generators provide more than enough electricity during some minutes of a year, they provide less electricity than needed during others. As such, additional methods must be used to match demand with supply. One method is to oversize the nameplate capacity of electricity and heat generators. A second is to use storage. A third is to use demand response.

Oversizing the nameplate capacities of electricity generators increases the average power supply to the grid during the year. With a higher average supply of electricity, more peaks in demand are met before storage is even considered. However, in order to guarantee demand will be met *every* minute of a year without any storage or demand response, it may be necessary to oversize the nameplate capacity of WWS generators by a factor of 5–10 or more. This will result in an exorbitant cost of energy, particularly because the WWS generators will produce 5–10 times more electricity than used, in the annual average. As such, 80–90 percent of the excess electricity will be wasted. When the electricity is wasted, the cost per kilowatt-hour produced goes up, in this case, by a factor of 5–10. Instead, using storage and demand response can limit the need for oversizing generators such that nameplate capacities of generators needed to meet continuous demand are only 10 percent higher, on average among 149 countries, than those needed to meet annually averaged demand.[5]

13.2.4 SPECIFY STORAGE AND DEMAND RESPONSE REQUIREMENTS TO MEET CONTINUOUS DEMAND

No future proposal calls solely for oversizing the nameplate capacity of generators to meet peaks in demand. Future proposals call for combining some additional generation with storage and demand response. In order to determine the ideal amount of storage and demand response needed to meet continuous demand, it is important first to characterize the demand. Specifically, total annually averaged demand should be separated into

1. electricity and heat demand needed for low-temperature air and water heating;
2. electricity demand needed for low-to-high temperature heat for industry;
3. electricity demand needed for cooling and refrigeration;
4. electricity demand needed to produce, compress, and store hydrogen;
5. all other electricity demand (including industrial heat demands).

Each of these electricity and heat demands should then be divided into flexible and inflexible demands. Electricity storage is needed to meet the inflexible electricity demands when current electricity generation is exhausted.

13.2 METHODS OF MEETING ENERGY DEMAND CONTINUOUSLY

Whereas lithium-ion battery storage is still expensive as of 2024, its cost has come down dramatically since 2020. Other forms of electricity storage (CSP storage, pumped-hydropower storage) are already less expensive than lithium-ion batteries, but they are limited to certain geographic regions or face other hurdles that limit the pace at which they will be adopted.

When the cost of stationary battery electricity storage is no longer a barrier, battery storage will be expanded rapidly enough, not only to fill in short-term gaps in electricity supply but also to store large amounts of electricity that can be used for days or weeks if the current WWS electricity supply drops to zero for that length of time. Fortunately, storing battery electricity for days, weeks, or months does not require a battery that provides hundreds or thousands of hours of storage at the peak discharge rate. To the contrary, long-duration electricity storage can be obtained simply by concatenating existing technology batteries that have four, eight, or twelve hours of storage.[310]

For example, suppose a battery stores up to four hours of electricity when the electricity is discharging at the peak discharge rate of 10 kilowatts. This means that the maximum storage capacity of the battery is 40 kilowatt-hours. Suppose also that 10 kilowatts of power must be supplied to meet demands on the grid for 100 hours (4.17 days) straight because no WWS generation is available on those days. The storage capacity needed in this case is then 1,000 kilowatt-hours (10 kilowatts multiplied by 100 hours). One way to provide the necessary storage is to invent a new 100-hour battery that has a peak discharge rate of 10 kilowatts. An easier way, though, is just to concatenate together 25 existing-technology four-hour batteries, each with a peak discharge rate of 10 kilowatts. Once one four-hour battery is depleted, the second one is used, and so on.

There are two main advantages of the second option. The first is that four-hour batteries already exist, whereas 100-hour batteries do not. The second and more important advantage is that having 25 four-hour batteries with a peak discharge rate of 10 kilowatts each allows the peak discharge rate from storage to equal 250 kilowatts over a four-hour period. Such a peak discharge rate may be needed if demand on the grid is that high. The 100-hour battery, on the other hand, can never provide more than 10 kilowatts of power, so can't even home-charge many electric cars at their peak charge rate (20 kilowatts).

Thus, having a short storage time maximizes the flexibility of batteries to meet peaks in demand (kilowatts) and to store electricity for long periods (kilowatt-hours). In summary, concatenating twenty-five 10-kilowatt batteries, each with four hours of storage, allows for either 100 hours of storage at the peak discharge rate of 10 kilowatts or four hours of storage at the peak discharge rate of 250 kilowatts, or anything in between.

Hydropower and batteries are similar in that hydropower can provide electricity at a high discharge rate for a short period or at a low discharge rate for a long period. The main disadvantage of hydropower (unless it is pumped hydropower) is that it recharges only naturally. Pumped hydropower, on the other hand, can be charged whenever excess electricity is available. Also, hydropower is limited by how many new dams can be built and how much water is available behind each dam.

Most other electricity storage types, such as grid-hydrogen storage, gravitational storage, compressed-air storage, and CSP with storage, operate over timescales of minutes to days to a couple of weeks (if the discharge rate is slow).

Low-temperature heat and cold can be stored in soil (borehole storage), water pits, and aquifers over periods of minutes to seasons. The cost of such storage is relatively low, so it can be used seasonally. Firebricks store low-to-high-temperature heat for industry. Per kilowatt-hour of heat stored, they cost less than one-tenth of the cost per kilowatt-hour of electricity stored of a battery. In addition, they avoid the need for furnaces to produce heat since the heat they store is produced within them by electric-resistance heating.

13.2.5 PROCEDURES FOR MATCHING DEMAND WITH SUPPLY IN CASES OF OVER- AND UNDER-GENERATION

Once electricity and heat generator sizes, storage sizes, and demand-response characteristics are defined, a procedure is needed to ensure instantaneous energy demand matches instantaneous supply plus storage plus demand response continuously. Procedures are needed for two main situations. One is when instantaneous generation exceeds demand. The other is when instantaneous demand exceeds generation. Procedures for each case are discussed next.

13.2.5.1 STEPS WHEN INSTANTANEOUS WWS ELECTRICITY OR HEAT SUPPLY EXCEEDS INSTANTANEOUS DEMAND

This section discusses the steps taken when current (instantaneous) WWS electricity or heat supply exceeds the current electricity or heat total demand. Total demand consists of flexible and inflexible demands. Whereas flexible demands may be shifted forward in time with demand response, inflexible demands must be met immediately.

If WWS instantaneous electricity or heat supply exceeds instantaneous inflexible electricity or heat demand, then the supply is used to satisfy that demand. The excess WWS is then used to satisfy as much current

13.2 METHODS OF MEETING ENERGY DEMAND CONTINUOUSLY

flexible electric or heat demand as possible. If any excess electricity exists after inflexible and current flexible demands are met, the remainder is sent to fill electricity, heat, cold, or hydrogen storage. Electricity storage is filled first. Excess WWS electricity is used to charge battery storage. If battery storage is full, remaining electricity is next used to produce hydrogen (for grid electricity, vehicles, or steel or ammonia manufacturing). If hydrogen storage is full, the remaining electricity is used to fill pumped hydroelectric storage, flywheel electricity storage, gravitational mass electricity storage, and compressed-air electricity storage. Any remaining electricity is used to produce (1) low-to-high-temperature heat that is stored in firebricks, (2) chilled water stored in cold-water tanks, (3) ice stored in containers, and (4) low-temperature heat stored in hot-water tanks and/or soil, water pits, and aquifers, respectively. Any residual after that is curtailed.

Another potential source of excess electricity is excess CSP heat. Such heat is first put into CSP heat storage. If such storage is full, remaining CSP heat is used to produce electricity immediately. If CSP electricity is not needed to meet current demand, the excess electricity is used to fill storage in the same order as with excess electricity, as just discussed.

Cold and low-temperature heat storage in water or soil are filled by using excess electricity to produce heat or cold with an electric heat pump. Hydrogen storage is filled by using excess electricity to produce hydrogen with an electrolyzer, then to compress the hydrogen before it is put into a storage tank. Hydropower dam storage is filled naturally with rainfall and runoff.

If any excess geothermal or solar heat exists after it is used to satisfy inflexible and flexible low-temperature heat demands, it is used to fill district heat storage (water-tank and underground heat storage) or heat storage in small hot-water tanks in buildings. Low-to-high-temperature industrial process heat for storage in firebricks is produced from extra electricity with resistance heaters connected to the firebricks[139] or direct resistance heating of the firebricks.[143,144]

13.2.5.2 STEPS WHEN INSTANTANEOUS DEMAND EXCEEDS INSTANTANEOUS WWS ELECTRICITY OR HEAT SUPPLY

When current inflexible plus flexible electricity demand exceeds current WWS electricity supply from the grid, the first step is to use electricity storage (battery, grid-hydrogen, CSP, pumped-hydro, hydropower, flywheel, gravitational-mass, and compressed-air energy storage) to fill in the gap in supply. The electricity is used to supply the inflexible demand first, followed by the flexible demand.

If electricity storage becomes depleted and flexible demand persists, demand response is used to shift the flexible demand to a future minute or hour.

If the inflexible plus flexible heat demand subject to storage exceeds WWS direct heat supply, then stored district heat, if available, is used to satisfy the heat demand. Stored hot water in building water tanks is used to satisfy building hot-water demand. If stored heat becomes exhausted, then the remaining inflexible heat demand must be met immediately with electricity. The remaining flexible heat demand is shifted forward in time with demand response.

Similarly, if the inflexible plus flexible cold demand subject to storage exceeds the cold energy available in ice or water storage, the remaining inflexible cold demand must be met immediately with electricity. The remaining flexible cold demand is shifted forward in time with demand response.

If the industrial-process heat demand subject to firebrick storage exceeds the heat stored in firebricks, then half the excess becomes an inflexible demand, which must be met immediately with electricity. The rest becomes a flexible demand, which can be met with current electricity or shifted forward in time with demand response. If a demand shifted forward is not met after the maximum number of demand-response hours, it is turned into an inflexible demand.

Finally, if the current hydrogen demand depletes hydrogen storage, the remaining hydrogen demand becomes an inflexible electrical demand that must be met immediately with current electricity.

13.2.6 MEASURES NEEDED WHEN INSTANTANEOUS DEMAND CANNOT BE MET WITH CURRENT SUPPLY OR STORAGE

If the measures just described are insufficient for meeting power demand with supply continuously, the system must be modified to ensure reliability. The standard for reliability is a **loss of load expectation** of no more than one day (24 hours) in ten years. Loss of load arises due to the failure to match instantaneous demand with instantaneous supply or storage. It can arise due to a severe lull in the supply of both wind and sunlight over a long period, scheduled or unscheduled maintenance of many energy generators at the same time, an unexpected heavy demand due to a heat wave or a cold spell, transmission-line congestion, or a transmission-line outage.

If supply does not match demand consistently, it will be necessary to do one or more of the following: (a) increase the nameplate capacity of

13.2 METHODS OF MEETING ENERGY DEMAND CONTINUOUSLY

the electric-power generators, (b) increase the electricity, heat, cold, or hydrogen nameplate storage capacity available, and/or (c) increase the nameplate capacity of the transmission system to allow more renewables that are far away and subject to different weather patterns to help fill in gaps in supply. Additional options are to (d) use electricity stored in vehicle batteries to help fill in gaps in supply on the grid, and/or (e) integrate weather forecasts into system operations to improve efficiency of the grid. Some of these measures also help to increase the efficiency of the grid, even when supply matches demand continuously. These techniques are discussed next.

13.2.6.1 OVERSIZING WIND, WATER, AND SOLAR GENERATION TO HELP MEET DEMAND

Oversizing the nameplate capacity of WWS installations can help to increase the number of times during a year that available WWS electricity supply exceeds inflexible demand. This solution, therefore, reduces the number of outages that occur during a year. Whereas oversizing helps to meet hours of peak demand, it also results in excess supply during many other hours. Ideally, the excess supply is put into electricity storage or used to produce heat, cold, and/or hydrogen, which are either stored or used immediately. Using excess WWS in these ways avoids WWS curtailment, reducing system cost.

13.2.6.2 OVERSIZING STORAGE TO HELP MEET PEAKS IN DEMAND

Oversizing storage is a second way to improve matching power demand with supply and storage on the grid. Because wind, solar, and wave power are variable WWS resources, they often overproduce electricity. Existing storage may not be enough to hold the excess electricity. Purchasing more storage can help to allow the excess electricity to be used during additional hours when WWS electricity is not available. The same applies to heat, cold, and hydrogen storage.

13.2.6.3 INCREASING TRANSMISSION NAMEPLATE CAPACITY TO HELP MEET DEMAND

Interconnecting geographically dispersed variable renewable resources, such as wind, solar, and wave power, smoothens the aggregate power supply from these generators. Similarly, interconnecting wind and solar, which are complementary in nature, smoothens out overall supply. As such, an additional solution to keeping the grid stable is to import far-away variable renewable electricity through an upgraded transmission system. The long-distance transmission lines should be high-voltage direct current (HVDC) lines to reduce line losses, thus reduce system costs. Whereas each country will ideally produce its own energy, interconnecting renewables among nearby countries reduces overall costs as well.[311] The reason is that some countries have better wind and hydropower resources than others, whereas others have better solar resources. Interconnecting nearby countries to take advantage of the plentiful resources in each country reduces times that no power is available and allows storage to be replenished more regularly. Interconnecting also allows aggregate transmission requirements to be reduced with little loss in annual energy output.

13.2.6.4 HELPING TO BALANCE DEMAND WITH VEHICLE-TO-GRID

Another method to help match demand with supply is to store electric energy in battery-electric vehicles, then to withdraw such energy when it is needed for the grid. This concept is referred to as **vehicle-to-grid**.[312] With vehicle-to-grid, a utility operator enters into a contract with an individual battery-electric-vehicle owner to allow electricity transfers back to the grid any time during an agreed-upon period in exchange for a lower electricity price.

Vehicle-to-grid is expected to wear down batteries faster by increasing the number of charge and discharge cycles the battery goes through during a year. However, only a small percentage of vehicles need to participate in a vehicle-to-grid plan to help meet demand with supply. One study suggests only 3.2 percent of vehicles in the US are needed to smoothen out US electricity demand when 50 percent of demand is supplied by wind.[313] Through computer modeling, other studies found that using vehicle-to-grid would help to match demand with supply and storage hourly in the eastern United States and in the Aland Islands (Baltic Sea), respectively.[314,315] An alternative to vehicle-to-grid is simply to avoid charging battery-electric vehicles when electric power is in short supply and charge them when too much WWS electricity is available. Such an action is a form of demand response.

13.2.6.5 USING WEATHER FORECASTS TO PLAN FOR AND REDUCE BACKUP REQUIREMENTS

Forecasting the weather (winds, sunlight, waves, tides, and precipitation) gives grid operators more time to plan ahead for a backup energy supply. Good forecast accuracy can also allow reserves to be shut down more

13.2 METHODS OF MEETING ENERGY DEMAND CONTINUOUSLY

frequently, increasing the overall grid efficiency.[316] Forecasting is done usually with a numerical weather-prediction model, the best of which can produce minute-by-minute predictions one to four days in advance with good accuracy. Forecasting is also done using statistical analyses of local measurements. The use of forecasting reduces uncertainty and makes planning more dependable, thus reducing the impacts of variability, reducing costs.

13.2.7 ANCILLARY SERVICES: LOAD FOLLOWING, REGULATION, RESERVES, AND VOLTAGE CONTROL

An electric-power system needs to meet two major requirements. One is to match current demand with current generation plus storage. The second is to manage power flows between and among individual transmission facilities. This section discusses both issues in light of practical grid operations.

Meeting instantaneous demand is difficult because demand and supply continuously vary in time. Demand can vary over seconds to minutes due to the random turning on and off of millions of individual inflexible demands. Demand can also vary seasonally due to changes in heating and cooling demand during the year. WWS electricity and heat production also varies minute by minute to season by season.

Ancillary services are services performed by equipment and people that are, according to the US Federal Energy Regulatory Commission (FERC), "necessary to support the transmission of electric power from seller to purchaser given the obligations of control areas and transmitting utilities within those control areas to maintain reliable operations of the interconnected transmission system."[258]

The main ancillary services include load following, regulation, spinning reserves, supplemental reserves, replacement control, and voltage control. These are discussed, in turn.

13.2.7.1 LOAD FOLLOWING

Load following is the use of power generators or storage to follow the rate of change (megawatts per minute) of the demand for electricity, averaged over a 5–15-minute period. Load following can also be accomplished by shifting part of the demand forward in time (demand response).

> **Transition highlight**
> Load following can be accomplished by rapidly reducing the power consumption of large industrial customers. Some industrial demands that can be reduced quickly include air liquefaction; induction and ladle

metallurgy; water pumping with variable speed drive; and production by electrolysis of aluminum, chlor-alkali, potassium hydroxide, magnesium, sodium chlorate, and copper.[317] Indeed, as the US National Research Council states, "The ability of industry to cut peak electric loads is a motivator for utilities to incentivize demand response (shifting loads to off-peak periods) in industry ... In combination with peak-load pricing for electricity, energy efficiency and demand response can be a lucrative enterprise for industrial customers."[318]

Load following can similarly be accomplished by reducing power consumption in residences and commercial buildings. For example, demand response can help push the time of battery charging for electric vehicles to late at night. It can also push the time of heating water in a hot-water tank or of running a dishwasher, washing machine, or dryer to a time of low overall electricity demand on the grid. Similarly, demand response can push the time of running a wastewater treatment plant or hydrogen production by electrolysis to a time of low demand.

Load-following generators or storage devices can have a relatively modest ramp rate. Thus, in a WWS world, hydropower, a portion of CSP output, and a portion of solar PV output may be used to follow demand, as can CSP storage or any electricity storage. In fact, any WWS generator can load follow, since if a generator projects that it will have electricity available over a coming period (e.g., because of wind or solar forecasting), it can compete in a real-time electricity market to sell the electricity for load following.

13.2.7.2 REGULATION

Regulation is the automatic (through electronic grid controls) use of pre-contracted power generators, storage, or demand reduction to fill in small gaps between the actual demand and the demand that results from load following. Over a 5–15-minute period, the actual demand varies every second due to hundreds to millions of people turning demands on or off. Load following meets only the average change in demand over a period, not the minute-by-minute change in demand. Minute-by-minute regulation can be met with generators, storage, and demand-reduction methods that have ramp rates 5–10 times the minimum ramp rates required for load following.

With WWS, regulation will be performed mostly by hydropower, pumped-hydro storage, batteries, grid-hydrogen storage, flywheel storage, compressed-air storage, and gravitational storage with solid masses. The first two ramp to 100 percent power within 15–30 seconds. Batteries and

13.2 METHODS OF MEETING ENERGY DEMAND CONTINUOUSLY

grid-hydrogen storage ramp up in 20 milliseconds. The rest, in less than a second. The best electricity producers for regulation are generally fast-responding storage technologies that can repeatedly operate through many cycles without degrading. Batteries, for example, respond quickly, have outputs that can be controlled precisely, and can operate through many cycles. As such, they are ideal for regulation.

When intermittent WWS generators, such as wind, solar, or wave power, are aggregated together over a large geographic region, their overall variability decreases.[306] This reduces the regulation requirement for, and thus the cost of, wind, solar, and wave power. In other words, aggregation makes it easier for battery storage to meet the difference between instantaneous fluctuation in demand and WWS generation.

Because of the fast response time needed for regulation services, such services are not obtained from market signals, which respond too slowly. Instead, generators, storage, and demand reduction used for regulation are contracted ahead of time.

Frequency Regulation

Frequency regulation helps to maintain the AC frequency on the grid. The AC frequency on the grid is required to be close to 60 hertz in the US and some other countries, and 50 hertz in Europe and remaining countries. The deviation allowed from this mean frequency in the US is 59.7–60.3 hertz. In most of Europe, it is 49.8–50.2 hertz. In some places in Eastern Europe, it is broader, 47–53 hertz. Frequency is maintained within a narrow range because too high or too low a frequency can destroy electrical equipment on the grid and end-use electrical appliances and devices.

When the frequency on the grid falls above the range allowed, averaged over a minute, grid operators reduce the frequency down into the permissible range. They do this by reducing electricity added to the grid from generation or storage or by increasing demand on the grid. In other words, a frequency that is too high can be reduced by slightly reducing the quantity of electricity added to the grid. Conversely, adding more electricity to the grid can push a low frequency into the normal range. The increase or decrease of generation, storage, or demand to regulate frequency on the grid is called **frequency regulation**.

Transition highlight
Traditionally, grid operators have increased frequency by increasing generation from gas, coal, oil, nuclear, or hydroelectric plants. In a WWS world,

operators will increase frequency by increasing hydroelectric, geothermal, wind, and PV output as well as output from batteries, pumped-hydro storage, CSP storage, flywheels, compressed-air storage, and gravitational storage with solid masses. Conversely, grid operators will reduce frequencies that are too high by reducing generation and storage or turning on artificial demands. In the case of wind turbines, an operator can increase generation by running some turbines in partial-load mode such that, when a decrease in grid frequency occurs, the operator increases turbine output by varying blade pitch. Alternatively, the operator may control short releases of electricity from wind turbines to the grid.[319] Similarly, some PV plants can be run at less than full output such that, when a decrease in grid frequency occurs, output is increased. Inverters can also be optimized to provide frequency control to the grid.[320]

Spinning, Supplemental, and Replacement Reserves and Voltage Control

Three ancillary services – spinning reserves, supplemental reserves, and replacement reserves – are used to supply demand in the event that a large power generator or transmission line goes down.

Spinning reserves are electricity generators or storage media, online and connected to the grid, that can increase output immediately in response to a major generator or transmission outage. Spinning reserves are required to reach full output within 10 minutes.

Supplemental reserves are the same as spinning reserves, but they are not required to respond immediately. They can be offline but, when turned online, they must reach full output within 10 minutes.

Replacement reserves are the same as supplemental reserves, but they must reach full output within 30 minutes. They are also used to replenish spinning and supplemental reserves to their precontingency status.

Voltage control is the injection or absorption of reactive power to maintain transmission-system voltages within required ranges. Such control is needed on the timescale of seconds. Whereas **real power** is the energy per unit of time used to run a motor or a heat pump, **reactive power** is imaginary power that does not do any useful work but simply moves back and forth within power lines. It is a byproduct of a circuit that has capacitors or inductors. Reactive power not only smoothens voltage on the transmission grid by supplying or absorbing it, but it also provides needed power on a timescale of seconds to avoid blackouts. Transformers, motors, and generators all need reactive power to produce a magnetic flux.

13.3 STUDIES ON MEETING DEMAND WITH 100 PERCENT WWS

Several countries now provide 100 percent of the electricity that they produce with WWS (Figure 13.1). However, today, electricity is only 20 percent of total end-use energy. In the future, all countries, will ideally provide 100 percent of their electricity and heat with WWS. Because no country has an all-sector 100 percent WWS energy system today, computer modeling of grid stability with a 100 percent WWS system is needed to guide its build-out. In this section, past modeling of 100 percent systems is first described. Then, results from a modeling study that simulates matching all-sector energy demand with WWS supply, storage, and demand response for most of the world is discussed.

13.3.1 PREVIOUS STUDIES OF MATCHING DEMAND WITH OR NEAR 100 PERCENT WWS

Many scientific studies have supported the proposal of matching electricity demand with 100 percent or nearly 100 percent WWS supply in one or more energy sectors. Whereas several of these studies have examined matching demand with supply in the annual average, dozens of others have treated matching demand with supply at a time resolution ranging from 30 seconds to a few hours. Both types of studies are summarized below.

In 1975, Sorensen examined the supply of wind and solar in Denmark and suggested that enough in combination might be available to supply all energy for building heat, transportation, industry, and electricity in the annual average by 2050.[321] In 1976, Lovins wrote an article called, "Energy Strategy: The Road Not Taken," describing how the United States might be able to transition entirely to renewable energy and efficiency, a transition he called the "soft energy path."[322] In 1996, Sorensen then estimated the annually averaged biomass, biofuels, and biogas energy and WWS (wind, hydro, and solar) energy needed in combination to provide all world energy.[323]

In 2001, Jacobson and Masters estimated the number of wind turbines needed in the annual average to replace 60 percent of coal in the United States to satisfy the 1997 Kyoto Protocol.[324] In 2005, Jacobson et al. modeled the health and climate benefits of transitioning 100 percent of US on-road vehicles to hydrogen-fuel-cell-electric vehicles, where the hydrogen was produced by either electrolysis (with the electricity from wind), steam-methane reforming, or coal gasification.[325] Wind-electrolysis was found to be the cleanest among the options for producing hydrogen. Also in 2005, Archer and Jacobson used data from 7,753 surface weather stations and 446 stations measuring weather in the lower atmosphere worldwide to map and quantify the land and coastal-wind-energy potential for every inhabited continent.[326] They concluded enough wind was available in high-wind-speed locations (where the mean annual wind speed exceeded 6.9 meters per second at an 80-meter hub height) to power the world for all purposes many times over.

In 2005 and 2007, respectively, Czisch and Czisch and Geibel developed a least-cost optimization model that simulated the electric-power grid over Europe, North Africa, and the Middle East divided into 19 regions interconnected by HVDC transmission.[327,328] They found that a renewable electricity supply of wind, solar photovoltaics, concentrated solar power, bioenergy, and hydroelectric power could keep the grid stable at low cost over a year. In 2006, Lund examined optimized mixes of wind, PV, and wave energy that could meet 100 percent of electricity demand in Denmark.[329] He concluded 50 percent wind, 20 percent photovoltaics, and 30 percent wave may be optimal.

In 2009, Hoste et al. found that by combining baseload geothermal, intermittent wind, intermittent solar, and gap-filling hydroelectric, and assuming an expanded transmission network, California could meet 100 percent of its hour-by-hour electricity demand during April and July 2020.[330] Thus, bundling intermittent wind and solar with baseload geothermal and gap-filling hydropower may increase the ability of all resources to meet demand.

Soon after, Jacobson evaluated the effects of 9 electricity sources and 12 options for powering vehicles on 11 impact categories (climate, air pollution, land use, water supply, grid reliability, wildlife, energy security, and more).[178] The study concluded that WWS resources may be the best in terms of minimizing such impacts and should be used to advance solutions to global warming, air pollution, and energy security. Later in 2009, Jacobson and Delucchi estimated the WWS resources needed to provide 100 percent of the world's 2030 end-use power demand for all purposes in the annual average.[331] The main barriers were stated to be social and political, not technical or economic. Material requirements presented some challenges but not limitations. The study considered grid stability and land needs. Results were obtained without nuclear power, fossil fuels with carbon capture, direct air capture, biomass, biofuels, or biogas. As such, the study was the first to propose a worldwide all-sector WWS energy system that eliminated energy-related air pollution, climate-relevant emissions, and energy insecurity simultaneously.

In 2011, Jacobson and Delucchi then developed more detailed roadmaps for the world and the US to meet annually averaged power demand

13.3 STUDIES ON MEETING DEMAND WITH 100 PERCENT WWS

with WWS supply.[101,332] Their group subsequently developed WWS roadmaps for New York,[333] California,[334] Washington State,[335] the 50 US states,[190,336] 139 countries,[284,181] 143 countries,[337] 145 countries,[4,65,96] 149 countries,[5,43] 53 towns and cities in North America,[338] and 74 metropolitan areas worldwide.[339]

In 2009, Lund and Mathiesen modeled the hour-by-hour matching of electricity demand with wind, wave, solar, and biomass supply in Denmark for the year 2050 and determined that such a system can remain reliable.[340] In 2010, Mason et al. modeled the electricity system in New Zealand assuming 22–25 percent wind, 12–14 percent geothermal, 1 percent biomass, 53–60 percent baseload hydro, 0–12 percent added hydro for filling gaps in supply, and assumptions about demand response.[341] They found that matching demand with supply with this system (near 100 percent WWS) was feasible. In 2011, Hart and Jacobson modeled the California electric-power grid under 2005 to 2006 conditions as if the grid were transitioned to renewables.[316] They found that matching power demand with 99 percent of electricity supplied from noncarbon sources was possible, even without considering demand response or large-scale storage. Gaps in supply were filled primarily by hydropower.

Between 2011 and 2024, hundreds of additional studies on 100 percent renewable energy systems were carried out by at least 42 independent groups based in multiple countries. These include, among others, studies led by Connolly,[342] Mathiesen,[343,344] Hart,[345,346] Elliston,[347] Rasmussen,[348] Steinke,[349] Becker,[350] Frew,[351] Bogdanov,[352,353] Child,[354] Aghahosseini,[355] Blakers,[356] Barbosa,[357] Lu,[358] Gulagi,[359] Esteban,[360] Zapata,[361] Sadiqa,[362] Barasa,[363] Caldera,[364] Liu,[365] Teske,[366] Hansen,[367] Oyewo,[368] Marczinkowski,[369] Li,[370] Alves,[371] Kiwan,[372] Zozmann,[373] Cole,[374] Pombo,[375] Marocco,[376] Oyewo,[377] Icaza,[378] Kuriyama,[379] Shaikh,[380] Breyer,[381] Brown,[382] and Diesendorf.[383] All the papers support the hypothesis that 100 percent renewable energy systems, either in just the electric-power sector or among all energy sectors, are doable. They found that matching 100 percent or near 100 percent time-dependent power demand with time-dependent supply and storage in one or more energy sectors is feasible at low cost. One study from the US National Renewable Energy Laboratory found that a 100 percent renewable and stable US electricity grid might cost less than five cents per kilowatt-hour, which is less than the cost of electricity from a new fossil-gas plant.[374] Breyer et al. provide a detailed history of the literature on 100 percent renewable energy systems through 2022.[381] In addition, two reviews of several 100 percent renewables studies similarly found that the methods used in such studies are rigorous.[382,383]

13 KEEPING THE GRID STABLE WITH 100 PERCENT WWS

13.3.2 MATCHING DEMAND WITH WWS SUPPLY, STORAGE, AND DEMAND RESPONSE: CASE STUDY

Most studies on 100 percent renewables discuss meeting demand in specific parts of the world. An important question is whether demand can be met among all energy sectors in all world regions with 100 percent WWS. This section summarizes results from one computer modeling study that simulated matching demand with 100 percent WWS supply and storage in 149 countries divided into 29 world regions.[5] The 149 countries emit about 99.75 percent of world fossil-fuel carbon dioxide. The regions include 13 multicountry regions (East Africa, North Africa, Southern Africa, West Africa, Central America, Central Asia, China region, Europe, India region, the Middle East, Northwest South America, Southeast South America, and Southeast Asia) and 16 individual countries or pairs of countries (Australia, Canada, Cuba, Haiti–Dominican Republic, Israel, Iceland, Jamaica, Japan, Madagascar, Mauritius, New Zealand, the Philippines, Russia–Georgia, South Korea, Taiwan, and the US).

Electricity grids were assumed to be interconnected perfectly within each country and among all countries in each region. Winds and solar fields, and building heat and cold demands were predicted in each country every 30 seconds for three years with a global weather-prediction model. Such a model predicts the weather worldwide. Modeled wind speeds and solar output were used to estimate electric-power generation from onshore and offshore wind turbines, wave devices, solar PV panels on rooftops, utility-scale solar PV panels, CSP plants, and solar thermal heat collectors in each country every 30 seconds for the three years. The time-dependent electricity production was used as input into a grid model, which matches power demand with supply, storage, and demand response in each region or country. Additional inputs into the grid model include geothermal and hydropower electricity generation; geothermal heat generation; electricity, heat, cold, and hydrogen storage sizes; transmission losses; time-dependent electricity, heat, cold, and hydrogen energy demands; and characteristics of demand response.

Eleven types of storage were treated: Two were for heat storage in buildings (hot-water-tank storage and underground thermal-energy storage), two were for cold storage (cold-water-tank storage and ice storage), five were for electric-power storage (batteries, grid-hydrogen storage, pumped-hydropower storage, conventional-hydropower storage, and CSP storage), and two were for vehicle transport (batteries and vehicle hydrogen storage).

13.3 STUDIES ON MEETING DEMAND WITH 100 PERCENT WWS

Demand response was treated by allowing flexible demands, if not met immediately, to be shifted forward 30 seconds at a time for up to eight hours. If a flexible demand was not satisfied by then, it was converted into an inflexible demand that had to be satisfied immediately. Electric heat pumps provided building air cooling and air and water heating, including for washing machines, clothes dryers, and dishwashers. Cooking was done with electric-induction cooktops. Some building heating and cooling was subject to district heating. Simulations were run from 2050 to 2052.

Results suggest energy demand matched energy supply plus changes in energy storage minus losses of energy every 30 seconds for all three years in all 29 world regions encompassing all 149 countries. The 2050 WWS mean private energy cost per unit of all-energy-consumed among all 29 regions was 8.8 cents per kilowatt-hour, in 2020 US dollars. The private energy cost includes the costs of new electricity and heat generators, short- and long-distance transmission, distribution, electricity storage, heat storage for buildings and industry, cold storage for buildings and industry, hydrogen production/compression/storage, and heat pumps for district heating and cooling. Private energy costs for individual regions ranged from 6.4 cents per kilowatt-hour (Canada) to 15.7 cents per kilowatt-hour (Haiti–Dominican Republic). The largest part of private energy cost was for generation (which includes capital, operation, maintenance, and decommissioning costs), followed by transmission and distribution, battery electricity storage, hydrogen production, and thermal-energy storage, respectively.

The 2050 WWS energy cost per unit of energy was relatively low for big regions, such as Canada, Russia, Africa, China, Europe, United States, and for small countries with good WWS resources, such as Iceland and New Zealand. The large geographical areas of the big regions allow for the effective aggregation of wind and solar, resulting in less intermittency of the combined resources versus wind or solar alone. These regions also have a good balance of solar and wind, which are complementary in nature seasonally. Further, they have substantial existing hydro resources that can provide peaking power. Iceland has a lot of hydropower, geothermal electricity and heat, and wind.

Costs were highest in small countries with high population densities (Cuba, Israel, Mauritius, South Korea, and Taiwan). These countries have little room for onshore wind or utility PV, the lowest-cost generators of WWS electricity. Nevertheless, WWS annual private energy costs in all five regions were still 40–62 percent lower than with business-as-usual, indicating that a transition to WWS reduces energy costs even under the least favorable circumstances.[5]

13 KEEPING THE GRID STABLE WITH 100 PERCENT WWS

Transition highlight
In island countries, such as in the Caribbean, the price paid for business-as-usual electricity is a mean of about 33 cents per kilowatt-hour. This price reflects, among other factors, the cost of transporting fuel to the islands and price hikes due to frequent supply shortages. Such high costs should not occur when WWS electricity, which is produced locally, is combined with storage. On the islands of Haiti–Dominican Republic, Cuba, and Jamaica, for example, a price of 33 cents per kilowatt-hour is 2.1–3.5 times the estimated electricity cost with 100 percent WWS (15.7, 9.3, and 10.7 cents per kilowatt-hour, respectively). In addition, energy requirements in those three regions with WWS are 60.4, 44, and 57.7 percent lower, respectively, than with business-as-usual, so the annual private energy cost people will pay with WWS is one-fifth to one-eighth of that which they pay currently.

Since a 100 percent WWS system relies a lot on electricity storage, one big question is, Will battery prices drop sufficiently to make a transition to WWS seamless? If battery prices do drop, then that eliminates the major roadblock to a rapid transition, which is the concern about avoiding blackouts on the grid due to low availability of WWS electricity. In 2021, lithium-ion battery cell and pack prices were an average of $101 and $132 per kilowatt-hour of storage, respectively.[384] By 2024, those prices had dropped to $56 and $73 per kilowatt-hour, respectively.[385] A battery cell is an individual container that stores electricity in the form of chemical energy. For example, the cylindrical battery that goes into a flashlight is a single cell. A battery pack is a package that contains battery cells, software, and a heating and cooling system. Battery packs are installed in an electric car to deliver electricity to the car or on the wall of a home to deliver electricity to the home. Battery packs (including cells) cost about 30 percent more than the battery cells alone. Based on the battery-cost data and trends, the 2035 battery cell and pack costs are projected to be less than $25 and $33 per kilowatt-hour of storage, respectively. The 149-country study discussed here assumed a battery pack cost of $60 per kilowatt-hour of storage.[5] As such, if battery costs do indeed drop to $33 per kilowatt-hour of storage, then a transition to 100 percent WWS should be even faster, more feasible, and more cost-effective than discussed here.

The overall **upfront capital cost** of transitioning the 149 countries, to be spent through 2050, was $58.2 trillion, in 2020 US dollars. This is the approximate capital cost of a worldwide Green New Deal. Individual regional ranges were $1.4 billion for Iceland to $15 trillion for China. The US cost was $5.7 trillion. Europe's was $5.1 trillion.[5]

A more relevant metric is the **annual energy cost**. This is the cost per unit of energy multiplied by the amount of energy consumed per year.

13.4 ESTIMATING FOOTPRINT AND SPACING AREAS OF WWS GENERATORS

Since a WWS infrastructure consumes 54.4 percent less energy per year in 2050 than a business-as-usual infrastructure,[5] the annual energy cost must be much lower with WWS than with business-as-usual. In fact, transitioning from business-as-usual to WWS decreased the annual private energy cost by 59.6 percent, from $16.5 to $6.7 trillion per year, among all 149 countries.

The social cost of energy is the private cost of energy plus the health and climate costs of energy. In a 100 percent WWS world, WWS technology production, WWS electricity and heat generation, WWS storage, and WWS transmission result in zero emissions, so all health and climate costs associated with energy are eliminated. As such, WWS energy private cost equals WWS energy social cost. The health and climate costs of zero emissions are zero. From the study, transitioning to WWS reduced annual social energy costs by 91.8 percent (from $81.2 to $6.7 trillion per year).[5]

Transition highlight
The social cost of ending world anthropogenic carbon dioxide emissions is only about 8 percent that of continuing the emissions. More specifically, the 2050 energy cost of transitioning world energy to 100 percent WWS is ~$121 per tonne of carbon dioxide eliminated. On the other hand, the energy, health, climate, and total social costs of not transitioning to WWS in a business-as-usual world are $299, $611, $558, and $1,468, respectively, per tonne of carbon dioxide emitted. Since WWS energy cost equals WWS social cost (because WWS has no health or climate cost associated with energy), the ratio of the social cost of implementing WWS to the social cost of not implementing WWS is about 8 percent.[5]

The large annual private-energy-cost savings resulting from moving from business-as-usual to WWS means the capital cost of a WWS system can be paid back quickly. For example, averaged over 149 countries, the savings result in a capital-cost payback time of 5.9 years. The larger annual social-energy-cost savings result in a capital-cost payback time of less than a year. In sum, the capital cost of WWS pays for itself with energy, health, and climate cost savings rapidly, and even the amount paid back is through energy sales rather than subsidy. Whereas private-energy-cost savings can be seen by consumers, social-energy-cost savings are seen by a country as a whole.

13.4 ESTIMATING FOOTPRINT AND SPACING AREAS OF WWS GENERATORS

A question about an all-renewable energy infrastructure is, How much land does it require? To answer this question, it is important first to differentiate between land footprint and spacing. Footprint (Section 11.7.1) is the area on the top surface of soil or water needed for each energy device. It does not include areas for underground structures. Spacing is the area between some devices, such as wind turbines, wave devices, and tidal turbines, needed to minimize interference of the wake of one turbine or device with the flow of wind, waves, or tides to downwind turbines or devices. Spacing area can be used for multiple purposes, including rangeland, farmland, ranching land, forest land, solar panels, open space, and open water. In fact, adding a wind farm to agricultural land reduces the land available for agriculture by only 4 percent.

Footprint is the most relevant metric for determining land use because that is the land area that cannot be used for any purpose aside from energy generation. Spacing area can be used for nonenergy or other energy purposes and for growing vegetation. In a 100 percent WWS system, new land footprint arises only for new solar PV plants, CSP plants, onshore wind turbines, geothermal plants, solar thermal plants, and hydroelectric plants. Offshore wind, wave, and tidal generators are in the water, so they do not take up new land, and rooftop PV does not take up new land. New land spacing area arises only for new onshore wind turbines.

The footprint area of a wind farm is relatively trivial. It is primarily the area of the cement around the bases of all the turbine towers in the farm that are visible above the ground. Cement underground is not part of the footprint since it doesn't prevent vegetation from growing on the soil above it. Paved roads associated with wind farms are also considered footprint, but unpaved roads are not since such roads can return to their natural state over time.

The total new land area for footprint required to power 149 countries with 100 percent WWS is about 0.13 percent of the 149-country land area.[5] Almost all of that is for utility PV and CSP. The spacing area for onshore wind to power these countries is about 0.38 percent of the 149-country land area. Since WWS uses no fuel to run, WWS has no footprint associated with mining fuels to operate. However, WWS, fossil fuels, and bioenergy all require one-time mining for raw materials to build their infrastructures.

Together, the new land footprint and spacing areas for 100 percent WWS across all energy sectors sum to 0.51 percent of the 149-country land area. This is equivalent to about 1.47 times the land area of California for virtually all world energy. Most of this land area is multipurpose spacing land. For example, solar PV panels can be installed on some of the spacing area between wind turbines. In comparison, about 37 percent of the world's land is used for agriculture and about 3 percent is urban land. As illustrated in the following example, replacing fossil fuels with 100 percent WWS may reduce land requirements for energy substantially in some countries.

13.5 JOB CREATION AND LOSS DUE TO A TRANSITION

Transition highlight
In the United States in 2024, about 30.2 million acres (122,200 square kilometers) of land were used to grow corn to produce ethanol fuel. This area equals 1.24 percent of US land. The ethanol produced from this land provided only 4 percent of the energy used by US transportation. The fossil-fuel industry occupied another 1.3 percent of US land.[10] Thus, the total area used for energy (corn ethanol plus fossil fuels) was about 2.54 percent of US land. On the other hand, converting the US for all purposes to 100 percent WWS by 2050 is estimated to require only about 1.06 percent of US land (0.16 percent for footprint due to utility solar PV and CSP, and 0.9 percent for spacing between onshore wind turbines) for new WWS generators.[5] Thus, such a conversion would free up at least 1.48 percent of US land that would no longer be used by the ethanol or fossil-fuel industries.

13.5 JOB CREATION AND LOSS DUE TO A TRANSITION

A transition to WWS destroys and creates jobs. Construction and operation jobs are lost in the fossil fuel, bioenergy, and nuclear industries. Such jobs are mainly in mining, transporting, and processing of fuels and building and running power plants. Transitioning also reduces jobs in building internal combustion engines, fossil-gas water and air heaters; fossil-gas stoves, turbines, and storage facilities; coal plants; pipelines; fueling stations; facilities, and refineries.

However, a transition creates jobs in the manufacture and installation of solar PV panels, CSP plants, wind turbines, geothermal plants, tidal devices, and wave devices. It also creates jobs in the electricity, heat, cold, and hydrogen storage industries. For example, jobs are needed to produce batteries, pumped-hydro storage, compressed-air storage, gravitational storage, CSP storage, borehole storage, water-pit storage, aquifer storage, ice storage, water-tank storage, and hydrogen storage. More jobs are needed building transmission lines with WWS. Jobs are also needed to install battery-charging stations and to build battery-electric and hydrogen-fuel-cell-electric vehicles, heat pumps, induction cooktops, electric lawn mowers, arc furnaces, induction furnaces, and resistance furnaces and boilers. Jobs are also needed to weatherize homes, make homes more energy efficient, and retrofit homes with electric appliances.

Construction and operation jobs include direct, indirect, and induced jobs. **Direct jobs** are jobs for project development, onsite construction, onsite operation, and onsite maintenance of the electricity-generating facility. **Indirect jobs** are revenue and supply chain jobs. Such jobs include suppliers of construction materials and components; analysts and attorneys who assess project feasibility and negotiate agreements; workers at banks financing the project; equipment manufacturers; and manufacturers of blades and replacement parts. The number of indirect manufacturing jobs is included in the number of construction jobs. **Induced jobs** result from the reinvestment and spending of earnings from direct and indirect jobs. They include jobs resulting from increased business at local restaurants, hotels, and retail stores, and for childcare providers, for example.

Construction jobs (direct, indirect, and induced) are full-time (40 hours per week) jobs but temporary since they last only the duration of the construction project. If each WWS construction project lasts one year but the entire WWS infrastructure is built over 30 years, then the number of long-term (30-year), full-time construction jobs is simply the number of temporary (one-year) construction jobs needed to build the entire WWS infrastructure over 30 years, divided by 30 years. Operation jobs are considered long term in that they last as long as the energy or storage facility lasts. They are needed to manage, operate, and maintain an energy-generation or storage facility. In a 100 percent WWS system, long-term jobs are effectively indefinite because, once a plant is decommissioned, another one must be built to replace it. The new plant requires additional construction jobs as well.

Transition highlight
Whereas a transition to WWS may reduce the world workforce by almost 1 percent due to job losses in the fossil-fuel, bioenergy, and nuclear industries, it will more than make up for those losses with job gains in WWS generation, storage, and transmission. In fact, a transition may increase the number of long-term, full-time jobs created over the number of jobs lost by about 22.9 million among 149 countries. This is the sum of 19.9 million permanent construction jobs and 28.3 million operation jobs produced minus the number of jobs lost (25.3 million).[5]

In sum, a worldwide transition to 100 percent WWS while keeping the grid stable is expected to save consumers money by reducing energy, health, and climate costs; minimize land use; and create many more long-term, full-time jobs than are lost. As such, there appear to be few downsides to a transition.

14

TIMELINE AND POLICIES NEEDED TO TRANSITION

The solution to global warming, air pollution, and energy security requires not only a technical and economic roadmap but also popular support, political will, and a rapid rollout of the solution. In fact, the main barriers to transitioning to 100 percent clean, renewable energy are neither technical nor economic; instead, they are social and political. People need to be confident that a solution is possible, to understand what changes they can make in their own lives to solve the problems, to make such changes, and to support policymakers in passing laws to speed a transition. Policymakers, themselves, need to understand the urgency, and therefore take bold steps. One of the most important factors leading to a change is education about what is needed and how fast it is needed. This chapter discusses the necessary timeline for a transition, obstacles in the way of meeting the timeline, actions that individuals can take, and policies that are needed to overcome the obstacles to reach the 100 percent goal on time.

14.1 TRANSITION TIMELINE

A critical step in implementing a transition to 100 percent WWS is to develop a transition timeline. The perfect timeline is one in which a 100 percent transition occurs in all energy sectors immediately. That won't happen. Jacobson and Delucchi postulated that transitioning all world energy by 2030 is technically and economically feasible but noted that, for social and political reasons, a completion date two decades later, such as 2050 may be more practical.[331] Since then, most 100 percent WWS roadmaps developed have proposed an 80 percent transition by 2030 and 100 percent by no later than 2050. Because technologies are advancing and WWS costs are dropping more quickly than previously thought, newer studies[5,43,65] propose an 80 percent transition by 2030 and 100 percent ideally by 2035–2040, but certainly no later than 2050. For new and existing electricity generation and new buildings, a transition by 2030 is possible. For existing buildings, industry, and transportation, a partial transition by 2030

14 TIMELINE AND POLICIES NEEDED TO TRANSITION

but a complete transition by 2035 is more feasible. This is particularly the case because some technologies, such as long-distance hydrogen-fuel-cell-electric aircraft, may not be commercially available until 2030 to 2035.

Whereas new WWS infrastructure will be installed upon the natural retirement of the current infrastructure, policies are needed to push the remaining existing infrastructure to retire early to allow the conversion to WWS at the rapid pace necessary to solve air pollution, climate, and energy-security problems.

14.1.1 TRANSITION TIMELINES FOR INDIVIDUAL TECHNOLOGIES

Some technologies can transition faster than others. Below are proposed timelines for transitioning technologies in different sectors, either naturally or through aggressive policies.

Development of super grids and smart grids: As soon as possible, countries should expand the transmission-and-distribution systems to supply more geographically dispersed WWS energy to demand centers. The grid should be managed with modern **smart grid** communication technology that detects and reacts to local changes in supply and use by invoking generation, storage, and demand response.

Power plants: By 2026, no more construction of new coal, nuclear, fossil-gas, oil, or biomass electricity-generating plants should occur; all new power plants built should produce WWS electricity. By 2030, all existing non-WWS electricity-generating plants should be retired and replaced with WWS plants.

Heating, drying, and cooking in the residential and commercial sectors: By 2026, all new devices, machines, and equipment for heating, drying, and cooking should be powered by electricity, direct heat, and/or district heating. By 2030, all existing heating, drying, and cooking equipment should be retired and replaced by electric products.

Marine vessels: By 2026–2030, all new boats (speed boats, yachts, barges, dredgers, transfer vessels, towing vessels, fishing boats, patrol vessels, ferries, military boats, etc.) and ships (container ships, cruise ships, military ships, etc.) should be battery-electric or hydrogen-fuel-cell-electric, and all new and existing port operations should be electrified. Policies are needed to incentivize the early retirement of nonelectric ships that do not naturally retire before 2030. By 2035, all existing nonelectric marine vessels should be retired and replaced by electric versions.

14.1 TRANSITION TIMELINE

Industrial heat: By 2026, as much new medium-to-high-temperature industrial-process heat as possible should come from firebricks heated by electric-resistance heating. Remaining medium-to-high-temperature heat should come from electric-arc furnaces; electric-induction furnaces; electric-resistance furnaces, kilns, and boilers; electric crackers; electron-beam heaters; dielectric heaters; and solar heaters. Low-temperature heat should come from electric heat pumps and solar heaters. By 2030–2035, all existing combustion-based heating should be replaced as above.

Rail and bus transport: By 2026, all new buses should be battery-electric and new trains should be battery-electric or hydrogen-fuel-cell-electric. By 2030–2035, all existing rail and bus transport should be replaced by electric vehicles.

Nonroad vehicles: By 2026, all new nonroad land-based vehicles (agricultural machines, construction vehicles, military vehicles) should be battery-electric or hydrogen-fuel-cell-electric. By 2030–2035, all existing nonroad vehicles should be replaced by electric vehicles.

Heavy-duty truck transport: By 2026–2030, all new heavy-duty trucks and buses should be battery-electric or hydrogen-fuel-cell-electric. By 2030–2035, all heavy-duty truck transport should be replaced by electric versions.

Light-duty on-road transport: By 2026, all new light-duty on-road vehicles should be battery-electric. By 2030–2035, all existing on-road vehicles should be transitioned.

Short-haul aircraft: By 2027, all new small, short-haul aircraft should be battery-electric. By 2030–2035, all existing short-haul aircraft should be transitioned.

Long-haul aircraft: By 2030–2035, all remaining new aircraft should be hydrogen-fuel-cell-electric or battery-electric, and existing combustion aircraft should be retired and replaced.

Whereas much new WWS infrastructure can be installed upon the natural retirement of conventional infrastructure, new policies are needed to force remaining existing infrastructure to retire early in order to allow the complete conversion to WWS to occur by 2030–2035.

Some suggest that retiring fossil plants early will create **stranded assets**, incurring a large cost. However, averaged worldwide, the annual social cost of a WWS energy system is less than 10 percent that of a fossil-fuel and bioenergy system.[4] As such, replacing a fossil-fuel-based plant before the end of its life saves substantial health and climate damage and money.

Transitioning also increases the number of jobs available. As such, society benefits from stopping the operation of existing fossil-fuel plants and replacing them with new WWS plants as soon as possible.

During the transition, fossil fuels, bioenergy, and existing WWS technologies are needed to produce the new WWS infrastructure. As such, the time-dependent transition to WWS infrastructure may result in a temporary increase in emissions before such emissions are eliminated. However, as more and more WWS is built, the infrastructure used to build WWS will consist more and more of WWS technologies.

14.1.2 HOW THE PROPOSED TIMELINE MAY IMPACT CARBON DIOXIDE LEVELS AND TEMPERATURES

The WWS strategy proposed here is to transition the world to 80 percent WWS by 2030 and 100 percent by no later than 2050, but ideally by 2035–2040, while eliminating nonenergy emissions at the same pace.

An important question then is, How long will carbon dioxide levels take to recover once carbon dioxide emissions have been halted? Carbon dioxide levels in the air in 1750, before the Industrial Revolution, were about 276 parts per million. In 2024, they were up to a peak of about 428 parts per million. Global computer simulations suggest that eliminating 80 percent of all human-created carbon dioxide emissions by 2030 and 100 percent by 2050 may reduce carbon dioxide levels in the air down to about 350 parts per million by 2100, which is a reasonably safe level, albeit higher than the 1750 level.[10] On the other hand, holding constant or increasing carbon dioxide emissions going forward will increase carbon dioxide levels in the air substantially compared with today.

How would globally averaged near-surface air temperatures respond if we eliminated emissions? Global air-temperature rise is roughly proportional to cumulative carbon dioxide emissions, regardless of the timing of those emissions. Therefore, the more emissions accumulate, the greater the temperature rise. The less the cumulative emissions, the lower the rise.

Between 1850 and 2019, the world emitted an estimated 2,390 billion tonnes of carbon dioxide from fossil-fuel combustion, cement manufacturing, and land-use change.[7] This resulted in the average global surface temperature between 2011 and 2020 rising by about 1.09 degrees Celsius compared with the 1850–1900 period.[7] However, in 2023, warming increased to 1.36–1.48 degrees Celsius above the 1850–1900 mean.[8,9] Temperatures in 2024 rose to more than 1.5 degrees Celsius above the mean.

14.2 OBSTACLES TO OVERCOME FOR A TRANSITION

Based on 2024 temperature data, the world may be beyond 1.5 degrees Celsius average global warming. On the other hand, average global warming must be measured over a period longer than one year, so average warming may still be lower than 1.5 degrees Celsius. Based on 2020 estimates, to limit average global warming to no more than 1.5 degrees Celsius above the 1850–1900 mean temperature (with a 50 percent probability), we could allow no more than another 500 billion tonnes of carbon dioxide emissions from all human sources after 2020. To limit warming to no more than 2 degrees Celsius, we could allow no more than 850 billion tonnes of carbon dioxide emissions.[7]

From 2020 to 2023, global carbon dioxide emissions from fossil-fuel burning and cement production were about 146.5 billion tonnes, and those from land-use change were about 23.6 billion tonnes, for a total of 170 billion tonnes. If carbon dioxide emissions stay constant at 2020–2023 average emission levels going forward, enough carbon dioxide will accumulate in the air for the Earth to warm, on average, by 1.5 degrees Celsius relative to 1850–1900 levels by 2031 and 2 degrees Celsius by 2039.

On the other hand, if emissions of carbon dioxide from all energy and nonenergy sources are decreased by 80 percent from 2023 levels by 2030 and by 100 percent by 2050, cumulative emissions from 2020–2050 of only 418 billion tonnes of carbon dioxide will occur by 2050. This will avoid 1.5 degrees Celsius global warming compared with the 1850–1900 period. This is the strategy proposed here: to transition the world to 80 percent WWS by 2030 and to 100 percent by no later than 2050, but ideally by 2035 to 2040, and eliminate nonenergy emissions at the same pace. To accomplish this goal, aggressive policies are needed.

14.2 OBSTACLES TO OVERCOME FOR A TRANSITION

While technically and economically feasible, a transition of the entire world's energy infrastructure to clean, renewable energy in all sectors is a daunting task that faces significant social and political hurdles. Below, several of these challenges are discussed.

14.2.1 VESTED INTERESTS IN THE CURRENT ENERGY INFRASTRUCTURE

Possibly the greatest challenge to overcome in transitioning the world to 100 percent clean, renewable energy is the challenge of repelling vested interests. Fossil-fuel companies have accumulated enormous wealth and political influence since the start of the Industrial Revolution. As a result, they have been able to implement legislation that has given them an entrenched financial benefit in many countries through tax-code breaks, direct grants, and subsidies. In addition, most of the existing energy infrastructure is a fossil-fuel infrastructure, and most people's day-to-day lives depend on that infrastructure (through transportation; building heating, cooling, and refrigeration; lighting, etc.). Further, over 25 million jobs worldwide depend on the fossil-fuel infrastructure.

Given the great financial resources available to the fossil-fuel industry and the dependence of many people's jobs and livelihoods on the current infrastructure, great care is needed to ensure that a transition to WWS brings along the same or a better standard of living, the same or more jobs, and the same or better comfort to as many people as possible. This means encouraging shareholders of fossil-fuel companies to move their investments to clean, renewable energy and storage and setting policies to retrain workers and to ensure that WWS electricity supply is at least as reliable as the current supply of electricity on the grid.

Movement is already occurring in all three areas. Investors are moving capital resources into WWS. The WWS infrastructure is creating more jobs per unit of energy than the fossil infrastructure is. The WWS system is also proving to be even more reliable and less expensive than the fossil-based system where WWS has replaced the fossil system.

> **Transition highlight**
> In an early example of the benefit of a transition to WWS, in 2017, a large (100-megawatt, 129-megawatt-hour) battery system was installed in South Australia to fill in gaps in electric-power supply for a 317-megawatt wind farm. After one year, the battery saved $40 million (Australian dollars) in grid stabilization costs. This was due to the rapid speed (20 milliseconds) at which batteries can react to shortages.[387]
>
> In another example, during July 2019, NextEra energy determined that investing in a combined wind (250 megawatt), solar PV (250 megawatt), and battery storage (200 megawatt, 800 megawatt-hour) system was less expensive than investing in a fossil-gas peaker plant.[388]
>
> In a third example, in 2024, 94.1 percent of all new US electric-power generation and storage is expected to be from WWS: 11.3 percent from wind, 58.9 percent from solar, and 23.9 percent from batteries.[389] Similarly, in 2024, China added 357 gigawatts of solar PV plus wind nameplate capacity to its grid (6.9 gigawatts per week), which is the equivalent to the peak power of about 357 nuclear reactors and the annual energy output of about 100 nuclear plants.[390]

14.2 OBSTACLES TO OVERCOME FOR A TRANSITION

14.2.2 ZONING ISSUES (NIMBYISM)

The second obstacle that needs to be overcome to implement a 100 percent WWS system is the **not-in-my-backyard** syndrome (**NIMBYism**). NIMBYism is an objection to the siting of something that a person thinks is unpleasant or dangerous in their neighborhood while not objecting to the development of the same thing somewhere else. Classic examples of NIMBYism are the siting of a landfill or a hazardous waste site near a neighborhood. However, NIMBYism extends to most every type of infrastructure development. People generally do not want to see additional buildings or facilities, including energy facilities, near them.

Whereas the installation of more onshore and offshore wind and solar farms and transmission lines faces opposition in many locations, it faces less opposition than the building of more coal and fossil-gas plants, nuclear plants, oil and gas wells, refineries, and pipelines. The reason is that most people realize that WWS does not bring air pollution or catastrophic risk along with it, whereas fossil technologies bring both. Nuclear reactors bring the risk of catastrophic failure and exposure to radiation above background levels. A second reason is that, whereas no one wants to add anything to the landscape, the addition of WWS often reduces land use relative to a fossil-fuel infrastructure. Similarly, whereas land used for a coal or nuclear reactors cannot be used for another purpose at the same time, land occupied by a wind farm can be used for multiple purposes simultaneously: agriculture, animal grazing, open space, or even housing solar PV and batteries.

In addition, many people are becoming accustomed to rooftop PV, so they object to it less than in the past. Some types of PV are also integrated into building design, so it is difficult to discern whether the PV is even present. Most wind farms are located away from buildings because winds are fastest where no obstacles exist on the ground. With the advent of floating offshore wind turbines and solar arrays, visual objections to siting offshore wind and offshore solar are virtually eliminated because people can't see such turbines or PV arrays.

Probably the most difficult WWS infrastructure to site is above-ground transmission. Siting transmission lines and pipelines today already results in NIMBYism, for good reason. Above-ground transmission lines and pipelines are not pretty. In addition, sparks from above-ground transmission lines can trigger wildfires. Wildfires around the world caused by transmission-line malfunctions have resulted in enormous damage, including loss of life. Transmission-line-sparked wildfires have occurred not only in California, Maui, and Texas but also in Australia and Spain.[391]

Wildfires triggered by transmission-line sparks were a problem long before WWS, but the issue still needs to be addressed. The best solution is to bury the lines underground. To that end, in 2021, the northern California utility Pacific Gas & Electric began planning to bury, over several years, 16,000 kilometers of existing transmission lines to reduce fire hazard. Although underground transmission lines are more expensive upfront than overhead lines, burying lines may cost less over the long term because it eliminates devastating fire damage due to transmission lines.[392] Burying transmission lines also largely solves their NIMBYism problem.

Another partial solution to reducing reliance on transmission lines is to install more rooftop PV and batteries in fire-prone regions and to use more storage, in general, instead of transmission. A third partial solution is to improve safety measures by clearing more brush and learning from the causes of previous fires.

Adding new overhead or underground transmission lines due to 100 percent WWS will enable the elimination of all oil and fossil-gas pipelines. Such pipelines leak and/or rupture, sometimes causing explosions that result in damage and death. Particularly dangerous are leaky fossil-gas distribution pipes, which exist almost everywhere in densely populated cities.

Finally, many new transmission lines will be built on the same pathways as existing transmission lines, reducing the need for new land.

Because of the delays caused by zoning requirements, the installation of many new transmission pathways for a 100 percent WWS system may be limited in some countries and states. Fortunately, though, an alternative to transmission is more energy storage. Because storage costs are declining rapidly, the future WWS system may be dominated by storage over transmission.

14.2.3 COUNTRIES ENGAGED IN CONFLICT

A third obstacle to a large-scale WWS buildout is the difficulty in building new energy infrastructure in countries suffering from conflict, such as civil war, terrorism, or war with another country. In 2024, six countries or regions of the world had conflicts with more than 10,000 deaths per year (Myanmar, Gaza, Niger–Burkina Faso–Mali, Mexico, Ukraine, and Sudan). Another 15 had conflicts with between 1,000 and 10,000 deaths per year (Colombia–Venezuela, Afghanistan, Somalia–Kenya, Nigeria, Iraq, Pakistan, Democratic Republic of the Congo–Rwanda–Burundi–Uganda, Brazil, Sudan–South Sudan, Nigeria–Cameroon–Niger–Chad, Syria, Yemen–Saudi Arabia, United Arab Emirates, Cameroon–Nigeria, Ethiopia, and Haiti). Dozens of additional countries suffered less than 1,000 casualties during the year from conflict.

Because millions of people live in countries with ongoing conflict, bringing distributed, clean, renewable energy to these countries is even more pressing. Oil pipelines and transmission lines are often targets for destruction or theft, so local microgrids may be the best way to provide energy safely to communities in countries engaged in conflict until the conflict is resolved. Major problems in a war-torn country are famine and poverty. Microgrids, if set up and maintained properly, can help produce food, water, and energy together, mitigating both problems. One way to help alleviate some of the difficulty in setting up a microgrid in a war-torn country is for other countries to provide aid in the form of clean, renewable WWS microgrid technology.

Some conflicts between countries arise over energy. For example, one country that provides fossil gas to another may withhold the gas to extract concessions, or use money from the sale of the gas to fund a war against another country. Alternatively, one country may need more energy, so it invades a neighbor to take control of its oil- and fossil-gas wells or coal mines. Transitioning a country entirely to clean, renewable energy, where the WWS resources are obtained from within the country, will make the country more energy independent, reducing reasons for conflict. However, many countries, even if they produce most of their own energy in the annual average, will benefit from trading energy with their neighbors to reduce the cost of matching power demand with supply. The reason is that, whereas the wind may not blow in one country at a given time, it is more likely to blow somewhere among several countries. The same applies to solar.

14.2.4 COUNTRIES WITH SUBSTANTIAL POVERTY

Countries with a large segment of their population in poverty will benefit from transitioning to 100 percent WWS as soon as possible. Millions of people die from indoor plus outdoor air pollution each year. Of these about 20 percent are children under the age of five. Almost all of the indoor air-pollution deaths (about 2.6 million per year in 2016) are due to the burning of biomass and coal in developing countries for home heating and cooking.

The main step in eliminating indoor air-pollution death is to eliminate indoor burning of fuel for cooking and heating. This can be accomplished most readily with the use of electric-induction stoves and electric heat pumps, respectively. For remote communities, the electricity may be obtained from a microgrid that combines solar PV or wind with battery storage. Such an infrastructure costs money, and impoverished communities usually do not have access to such money. However, costs of WWS generation and storage continue to decline. In addition, national governments can help financially. International aid will help with this solution too.

Impoverished countries will also benefit from a large-scale transition to WWS. Their end-use energy requirements will go down by an average of 54.4 percent; their annual cost of energy will consequently drop by about 60 percent.[4] Their annual social costs of energy will decrease by nearly 90 percent.

However, the main barrier to a transition to WWS that these countries face is the capital cost of an investment. WWS generators have no fuel cost, but they have an upfront capital cost. Although that capital cost pays for itself relatively quickly, raising upfront capital is not a trivial matter. To that end, wealthier countries need to work with more impoverished countries to help finance the upfront cost of a transition.

14.2.5 TRANSITIONING LONG-DISTANCE AIRCRAFT AND LONG-DISTANCE SHIPS

Whereas 95–97 percent of the technologies needed for a transition to 100 percent WWS are currently commercialized, the rest are not yet available. The two most obvious technologies not yet commercialized are long-distance aircraft and long-distance ships. About 15 percent of worldwide commercial aircraft flights by number and 46 percent of such flights by distance are mid-haul flights (three to six hours) or long-haul flights (longer than six hours).[79] The best WWS solution for such flights may be the development of hydrogen-fuel-cell-electric aircraft. However, for such aircraft to work cost-effectively, fuel-cell sizes and efficiencies need to improve. A transition to long-distance WWS aircraft may not occur until 2030–2035 because of the improvements and testing needed to commercialize hydrogen-fuel-cell-electric aircraft for mid- and long-distance travel. For short-haul flights (less than three hours), aircraft will be primarily battery-electric.

Long-distance ships are also proposed to be mostly hydrogen-fuel-cell-electric although some may be battery-electric.[79] Long-distance ships can stop during their journey to refuel (if they are hydrogen-fuel-cell-electric) or recharge (if they are battery-electric). In addition, ships have less constraint on mass than do aircraft, so ships are easier to design. As such, all new ships should be electric by between 2026 and 2030.

14.2 OBSTACLES TO OVERCOME FOR A TRANSITION

14.2.6 COMPETITION AMONG SOLUTIONS

Another obstacle facing a rapid transition is the competition among proposed solutions to the problems of air pollution, global warming, and energy security. Given the severity of the problems facing the world and the short time available to fix them, competition among energy plans can result in poor technologies being implemented, thereby slowing down the solution. The main competitors for solutions to date have been a WWS solution and an **all-of-the-above** solution. An all-of-the-above solution includes WWS but also includes nuclear, fossil fuels with carbon capture, bioenergy with or without carbon capture, direct air capture, hydrogen from fossil gas with or without carbon capture, and nonhydrogen electro-fuels, among other technologies. However, those non-WWS technologies are opportunity costs (Chapter 8). They result in either more air pollution, more greenhouse gas emissions, more energy-security risk, more land-use risk, higher costs, or longer planning-to-operation times, or all of these, relative to WWS. As such, spending money on non-WWS technologies results in less benefit and a longer delay before enough WWS can be implemented to eliminate air pollution, global warming, and energy insecurity. The best way to overcome this obstacle is to educate the public and policymakers about it and guide them to the WWS solution.

14.2.7 SLOW PROGRESS

One more obstacle is the fact that, despite substantial effort to date, progress toward 100 percent WWS has been slow, on average, worldwide. Whereas several countries have reached 100 percent WWS in their electricity sector (Figure 13.1), the energy produced in these countries represents a small portion of world energy. In addition, electricity is only 20 percent of end-use energy. Part of the issue is that, although WWS and electrification is growing, it is barely exceeding increases in demand as the world population increases and less-developed countries become more developed and energy-thirsty. Another part of the issue is that transportation, buildings, and industry have been transitioning more slowly than electricity.

Although the concern about slow progress is valid, the good news is the solution exists: Technically and economically, every country can transition to 100 percent WWS for all energy purposes while reducing annual energy costs, air-pollution costs, and climate costs.[5,43] On average, countries will gain far more jobs than are lost worldwide. Only modest amounts of land will be needed for a transition. In some cases, less land will be needed than is currently occupied by the fossil-fuel and bioenergy industries. The key is to deploy, deploy, deploy, as fast as possible. The slow progress is really an indication of the other barriers listed above that are getting in the way. Part of the solution is to dismantle those barriers one by one, and another part is to keep our eye on the ball – focus on electrification of almost everything and on providing the electricity with 100 percent WWS.

14.3 WHAT CAN INDIVIDUALS DO?

The solution to air pollution, global warming, and energy insecurity requires actions by individuals, businesses, nonprofits, and policymakers. This section discusses some of the actions that individuals can take to reduce energy use and to change the type and source of the energy they consume.

Most individuals consume energy in all the main energy sectors (electricity, transportation, building heating and cooling, and industry) that produces carbon dioxide and air pollutants. We consume electricity for lights, computers, refrigerators, and food production. We drive to work or for leisure. We desire buildings of a certain temperature for comfort. We consume products and food created by industry. Because we consume, we can control our emissions both by changing our habits and choices and by changing the type and source of energy we use.

Because each person is in a different situation in terms of both the energy they consume and their financial and practical ability to transition their energy, each recommendation given here may apply to some people but not others.

14.3.1 NEW HOME CONSTRUCTION

People building a new home from scratch can plan their home not only to minimize energy use but also to produce their own WWS energy. The first step in planning such a home is to ensure that it uses only one energy carrier, electricity, rather than both electricity and fossil gas. There is no reason to have two carriers of energy in a building since every appliance that runs on fossil gas has an electrical equivalent that is the same or better in quality. To the contrary, adding fossil gas to a home or another building adds unnecessary cost, including for a gas-hookup fee charged by the utility, for ditches to bury fossil-gas pipes, for fossil-gas pipes themselves, for a fossil-gas meter, for vents to provide air and ventilation for fossil-gas combustion, for carbon monoxide monitors, and for inspection of all of the above. These items sum to tens of thousands of dollars of unnecessary

14.3 WHAT CAN INDIVIDUALS DO?

cost per home. Fossil-gas use in buildings also releases health-affecting air pollutants. In-home health problems from fossil-gas fumes and combustion products are eliminated by eliminating fossil-gas use in the home.

The appliances in a building that run on fossil gas often include an air heater, a water heater, a cooktop, a clothes dryer, and a pool heater. For buildings on district heating loops, air heating and water heating (and rarely, air cooling) are supplied from centralized boilers and chillers, so in-house appliances are not needed for those tasks.

For buildings not on district heating loops, a fossil-gas air heater, water heater, clothes dryer, and pool heater can be avoided by using an electric heat pump version of each. A fossil-gas cooktop can be avoided by using an electric-induction cooktop. Because heat-pump air heaters all automatically run in reverse as air-conditioners, the use of a heat pump for air heating and air-conditioning eliminates the need for a separate air-conditioner, saving additional money. Because heat pumps consume one-fourth of the energy of fossil-gas heaters, the installation and use of a heat pump saves a lot of money over time. Finally, the use of a heat-pump air heater/air-conditioner that has an indoor unit in each room of the house and an outdoor unit to exchange air between the inside and the outside eliminates the need for air ducts throughout the house, saving more money.

The third step in new-home planning is to minimize energy use by maximizing energy efficiency. One way to make a building energy efficient is to install thick insulation within walls, below the bottom floor, between floors, above the ceiling, around water pipes, and in and around windows. Concrete, which helps to keep the temperature relatively stable inside a building, can also be used as floor or wall thermal-mass material. Triple-glazed windows minimize heat and cold loss through the windows. The greater the insulation, the less the energy needed to heat or cool a home, reducing energy consumption and extending heat-pump life. Similarly, smart-glass windows reduce cooling and heating needs. They are translucent during summer, blocking incoming sunlight, and transparent during winter, maximizing sunlight into your home.

A second way to minimize energy use is to install light-emitting diode (LED) lights everywhere. A third way is to use only energy-efficient electric appliances (refrigerator, dishwasher, washing machine, television, vacuum, etc.).

The fourth step in new home planning is to install rooftop solar PV panels, garage batteries, an inverter to control both, and an electric-car-charging port. In some windy locations, a backyard wind turbine is ideal. The solar PV or wind system should be sized to meet at least 100 percent of your home's annually averaged electricity demand, including for any battery-electric vehicles you might use. Ideally, the PV system will be sized larger than this to account for any future growth in demand and to help the grid when it is not providing enough electricity to meet everyone's demand. One or two batteries is helpful to store electricity during times of low demand, such as during the morning, and to discharge electricity during times of peak demand, such as after sunset. Such batteries help to relieve stress on the grid and save you money.

The fifth step is to use sustainable building materials for construction. One option is to use prefabricated, largely recycled steel for the structure. Since the steel beams are produced with precision in a factory and are mostly from recycled material, using them minimizes waste. Wood homes typically result in wood waste during construction. Concrete can also be recycled and used as aggregate, either as a sub-base material or mixed as an aggregate with new concrete. Some sustainable building materials include composite roofing shingles and bamboo. Composite roofing materials require much less repair and replacement than do asphalt shingles, tile, and wood shake. Bamboo regrows in 3 years compared with 25 years for most trees.

14.3.2 HOME RETROFITS

Hundreds of millions of buildings already exist worldwide, and the turnover is only 1 percent per year. Given the substantial energy use by people in existing buildings, retrofitting such buildings so that they are all electric is essential for solving the air pollution, climate, and energy-security problems we face.

Solutions for existing buildings are usually not technically challenging. However, unlike with new buildings, where one can save a lot of money by going to 100 percent WWS, retrofitting existing buildings usually saves less.

One of the lowest-cost ways to reduce energy use in your home, and thus reduce energy cost, is to weatherize your home by sealing cracks in windows, doors, and skylights and adding insulation around water pipes. Adding or changing the insulation below the floor or in the attic also helps. Even insulating an attached garage can minimize heat loss.

A second step is to change lights to LED lights and to change appliances to electric, energy-efficient appliances. Certainly, if a fossil-gas water heater needs replacing, it should be replaced with a heat-pump water heater. Most fossil-gas air heaters in buildings are centralized units that provide heat sent through ducts to the rest of the building. When or before the heater breaks down, it should be replaced with a centralized heat-pump air heater

14.3 WHAT CAN INDIVIDUALS DO?

and air-conditioner that also uses ducts. Similarly, a fossil-gas clothes dryer should be replaced with a heat-pump dryer. A fossil-gas cooktop should be replaced with an electric-induction cooktop. Gradually, all appliances in a building should be changed from fossil gas to electric.

A third step is to install rooftop PV or a backyard wind turbine, one or two batteries, an inverter, and an electric-car-charging port, just as for a new building.

Whereas some of these measures are low cost and others can be taken when an existing appliance breaks down, the rest require new investment. Although some people can afford such investment, many cannot. Thus, policies are needed to encourage and help finance retrofits. Otherwise, only a few of the needed retrofits will be performed.

14.3.3 RENTING

Many people rent apartments, condominiums, or town houses, so they have little control over either their source of electricity or the appliances that they use. However, renters do have some control. First, renters pay electricity bills to a utility. Many utilities today, particularly community-choice-aggregation utilities, but also others, offer to sell customers 100 percent WWS electricity. Renters can subscribe to such programs. Second, even if an apartment has a fossil-gas cooktop, it is possible to purchase an individual induction cooktop burner for $40–$100. These just plug into the wall and replace the fossil-gas cooktop. Third, weatherizing doors and windows in a rental can save a lot of money, as can changing lights to LED lights. If the building is on a district heating and cooling system, then fossil gas is not used for any other purpose in the rental, and you are done. If your rental has a fossil-gas air or water heater, then it would be up to the landlord to change these. The first step in doing this is to request the landlord to invest in electric heat-pump air and water heating, especially when existing units break down. The second step is to ask your local policymaker to incentivize or require such changes.

14.3.4 TRANSPORTATION

People travel primarily by either foot, bicycle, car, streetcar, bus, train, ship, or airplane. Many own or lease cars. If a new car is needed, the next one should be battery-electric. Whereas only expensive, shorter-range battery-electric cars were available until a few years ago, a variety of lower-cost, longer-range models are now available almost everywhere. Many people can also opt for battery-electric bicycles or scooters or public transportation. Traveling by streetcar, bus, train, ship, or plane usually results in less energy consumed per unit of distance traveled than traveling by car.

14.3.5 CONSUMER CHOICES

Individuals make conscious or unconscious choices every day about how much energy they consume and, therefore, how much pollution they emit. Consumers decide what to eat, where to travel, whether to use a computer or television, and what goods to buy. Based on today's fossil-fuel use worldwide, each product purchased or action taken results in a certain **carbon footprint**, which is the lifecycle CO_2-equivalent emissions of the product or action. However, as we go to 100 percent WWS and simultaneously eliminate nonenergy emissions, the carbon footprint of actions and products heads to zero, which is good news. In the meantime, though, certain actions can reduce air pollution and carbon emissions. Such actions include telecommuting for work instead of driving to and from work, holding more meetings online instead of in person, minimizing air travel, using more public transportation, biking or walking instead of driving, switching off lights when they are not needed, dimming the screen on a computer or phone, reducing meat consumption, and purchasing fewer nonessential goods.

> **Transition highlight**
> In an extreme example of how consumer choices can affect energy use, on July 11, 2021, Richard Branson and three others flew into space 85.3 kilometers above the surface of the Earth as the world's first space tourists. However, the 1.5-hour, 11,260-kilometer flight resulted in 135,000 kilograms of carbon dioxide emissions per person along with other rocket-fuel exhaust emissions.[393] The average person in the US emits 3.4 kilograms of carbon dioxide per 1.5 hours, so the emissions per person for that 1.5-hour flight were 40,000 times those of the average person. A better solution is not to travel to space for tourism, thereby avoiding wasteful energy use.

14.4 POLICIES

The policies necessary to transform to 100 percent WWS differ by country, depending largely on the willingness of the government and people in each country to institute a rapid change. This section first defines several types of policy options that each country can consider implementing. Policy options for different energy sectors are then discussed in more detail.

14.4 POLICIES

14.4.1 POLICY OPTIONS FOR A TRANSITION

Below are some policy options that have been used in the past. The list is by no means complete.

Renewable portfolio standards, also called renewable electricity standards, are policy mechanisms requiring a certain fraction of electricity generation to come from specified renewable energy sources by a certain date. Thus, for example, a mandate of 80 percent WWS by 2030 and 100 percent no later than 2035 is a renewable portfolio standard. To date, 100 percent renewable portfolio standards have been enacted in many countries, states, cities, and towns.

Financial incentives and laws for increasing energy efficiency and reducing energy use are policy methods to reduce the demand for energy. For example, a law requiring the use of low-energy-consuming LED light bulbs instead of high-energy-consuming incandescent ones reduces energy demand. Demand reduction reduces the pressure on energy supply, which makes it easier for WWS electricity or heat supply to match demand.

Laws requiring demand response are helpful because they force utilities to incentivize customers to shift the time of their electricity use from a time of peak electricity demand to a time of lower demand during the day or night. Demand response is usually accomplished in one of two ways. The first way is for utilities to increase the price of electricity during times of peak electricity use. The higher price of electricity incentivizes customers to use less electricity during those times of day. The second way is for utilities to pay customers to use less electricity during times of high electricity demand. In both cases, the customer can react manually to the change in rate or payment by reducing or stopping energy consumption during times of peak demand. Alternatively, the customer can agree to have internet- or radio-controlled switches installed on air-conditioners or other devices to automatically reduce energy use during times of peak demand.

Feed-in tariffs are subsidies to cover the difference between electricity-generation cost (ideally including grid-connection cost) and wholesale electricity prices. Feed-in tariffs have been an effective tool for stimulating the market for renewable energy. To encourage innovation and the large-scale implementation of WWS, which will itself lower costs, feed-in tariffs should be reduced gradually. Otherwise, technology developers have little incentive to improve.

Output subsidies are payments by government to energy producers per unit of energy produced. For clean, renewable energy producers, the justification of such subsidies is to correct the market because fossil-fuel and bioenergy producers are not paying for the pollution they emit, which has health and climate costs for society. In other words, the subsidies attempt to address the **tragedy of the commons**, which arises because air is not privately owned. As such, air historically has been polluted without polluters paying the health, climate, and environmental costs of its pollution.

Investment subsidies are direct or indirect payments by governments to energy producers for research and development. Historically, such subsides have been given mostly to conventional fuel producers through legislation and clauses in tax codes, such as deductions and credits for specified activities. Conventional generators have benefited historically from such subsidies by not paying externality costs of the pollution that their energy creates. Investment subsidies are now available to WWS energy sources in many countries.

One type of investment subsidy is a **loan guarantee**, whereby the government guarantees a loan to a company for building a facility. Without a guarantee, many large energy projects will not be approved for a construction loan. Loan guarantees have been historically provided for conventional fuels and will benefit WWS as well.

On a smaller scale, **municipal financing** of residential energy-efficiency retrofits and solar installations help to overcome the financial barrier of the high upfront cost to individual homeowners. **Purchase incentives and rebates** can also help stimulate the market for electric vehicles.

A potential policy tool that has not been used widely to date is a **revenue-neutral carbon tax or pollution tax**. This is a tax on polluting energy sources, with the revenue transferred directly to nonpolluting energy sources. In this way, no net tax is collected, so the cost to the public is zero in theory. If the tax is a revenue-neutral carbon tax, it may not address air pollution. For example, a biomass or coal-with-carbon-capture electricity plant can claim they are low carbon (which itself is not accurate), thus avoiding much of the tax, yet still emit substantial health-affecting air pollutants along with carbon. A second problem is that heavy polluters can choose to pay the tax and withstand a lower profit margin or raise their price yet still pollute.

A related tool is straight **pollution tax**, such as a **carbon tax**. Such a tax is not so popular since it is perceived as a cost to the public since the money collected from the tax doesn't necessarily go back to reducing energy prices. Instead, polluting companies merely increase the price they charge to customers in order to pay the tax.

Another noneconomic policy is a **mandatory emission limit** (usually a vehicle tailpipe or stationary stack exhaust-emission limit). This is a **command-and-control** policy option implemented widely under the US Clean Air Act Amendments and in many other countries to reduce vehicle emissions. By tightening emission standards, including for carbon dioxide, policymakers can force the adoption of cleaner vehicles. Such emission limits have also been set

14.4 POLICIES

for other pollution sources. A disadvantage of emission limits (unless they are zero) is that they allow the fossil-fuel industry and its upstream mining of fossil fuels to persist while the industry pursues incremental reductions in tailpipe or stack emissions. However, a mandate of absolutely zero tailpipe pollution emissions eliminates new fossil-fuel-powered vehicles, because only battery-electric and hydrogen-fuel-cell-electric vehicles can meet a zero-emission limit. Thus, such a policy leads to the phase-out of internal-combustion-engine vehicles.

Related to mandatory emission limits is **cap and trade**. Under this policy mechanism, mandatory emission limits lower than current emission levels are set for an entire industry, and pollution permits are issued corresponding to the total emissions allowed. Polluters in the industry can then buy and sell pollution permits among each other. The net result is lower overall emissions and a payment by the polluters for the remaining emissions. The problem with this mechanism is similar to that with emission limits. Unless the cap is zero, pollution will persist long into the future.

Lastly, making a modification to a building code is a popular way of eliminating the use of fossil gas in new and existing residential and commercial buildings, ensuring buildings have sufficient battery-electric vehicle-charging infrastructure, ensuring buildings provide enough of their own electricity with rooftop PV and batteries, and improving building energy efficiency.

14.4.2 POLICY OPTIONS BY SECTOR

Current energy markets, institutions, and policies have been developed to support the production and use of fossil fuels, bioenergy fuels, nuclear power, and clean, renewable energy. New policies are needed to ensure that a 100 percent clean, renewable energy system develops quickly and broadly in each country and that dirty and dangerous energy systems are not promoted. Below, several policy mechanisms are proposed for each energy sector to accomplish these goals. For each sector, the policy options are listed roughly in order of proposed priority.

14.4.2.1 ENERGY-EFFICIENCY AND BUILDING-ENERGY MEASURES

Expand energy-efficiency standards.

- Incentivize the conversion from fossil-gas water and air heaters to electric heat-pump heaters.
- Promote, through incentives, rebates, and municipal financing, energy-efficiency measures in buildings.

- Incorporate "green building standards" in building codes to reduce energy use in building design, construction, and use.
- Incentivize landlord investment in energy efficiency and reduced energy use in buildings.
- Create a green building tax credit program for the corporate sector.

14.4.2.2 ENERGY-SUPPLY MEASURES

- Increase the percent of WWS energy supply required under renewable portfolio standards.
- Extend or create WWS production tax credits (a tax credit for every kilowatt-hour of electricity produced).
- Invest in job retraining from business-as-usual energy to WWS energy.
- Incentivize the expansion of building electricity storage.
- Identify and preapprove regions where WWS generation and transmission can be added. Streamline the permit approval process for large-scale WWS power generators and transmission.
- Work with local government to streamline zoning and to award permits for new electricity generation and storage within existing planning efforts to reduce the cost and uncertainty of projects and to expedite their build-out.
- Streamline the awarding of permits for installing behind-the-meter solar and wind.
- Lock in fossil-fuel and nuclear plants to retire under enforceable commitments.
- Implement taxes on air pollution and carbon emissions by current utilities to encourage their phase-out.
- Incentivize home and community battery storage that accompanies rooftop PV.

14.4.2.3 UTILITY PLANNING AND INCENTIVE STRUCTURES

- Incentivize district heating and cooling and community seasonal heat and cold storage.
- Incentivize utility-scale grid electric-power storage.
- Require utilities to use demand–response management to reduce the need for electricity storage on the grid.
- Incentivize the use of excess WWS electricity to produce and store hydrogen, low-to-high temperature heat for industry, low-temperature heat for buildings, and cold for buildings to help manage the grid.

14.4 POLICIES

- Develop programs to use electric-vehicle batteries after their useful life in vehicles for stationary storage.
- Implement **net metering**, whereby rooftop solar owners can sell electricity back to the grid to offset a part or all of the cost of the electricity that they buy from the grid.

14.4.2.4 TRANSPORTATION MEASURES

- Mandate battery-electric vehicles for government transport (mail delivery, service vehicles, military-base vehicles) and use incentives and rebates to encourage battery-electric vehicles for commercial and personal use.
- Promote more public transport.
- Increase biking and walking infrastructure, such as bike lanes, sidewalks, crosswalks, timed walk signals, etc.
- Set up time-of-use electricity rates to encourage vehicle charging during off-peak hours.
- Use incentives or mandates to stimulate the growth of private fleets of battery-electric buses.
- Incentivize battery-electric ferries, riverboats, tugboats, speed boats, dredgers, and other short-range watercraft.
- Adopt zero-emission standards for all vehicles, with 100 percent of new production required to be zero-emission of every air pollutant. Incentivize the transition by gasoline and diesel superusers fastest, since the top 10 percent of gasoline users, for example, consume more gasoline than the bottom 60 percent.[394]
- Ease the awarding of permits for installing electric-charging stations in public parking lots, hotels, suburban metro stations, on streets, and in residential and commercial garages.
- Incentivize the electrification of freight rail and shift freight from trucks to rail.

14.4.2.5 INDUSTRIAL-SECTOR MEASURES

- Provide financial incentives to electrify low-to-high temperature manufacturing and firebrick storage for industry.
- Provide financial incentives to use WWS electricity to produce heat for industry.
- Encourage industry to take part in demand–response measures to help match grid power demand with supply.
- Encourage steel, concrete, and silicon manufacturing plants to produce those products without emitting carbon dioxide chemically.

The set of all measures listed above is a limited set. Yet, many are necessary to speed a transition, which would otherwise drag on long past 2050. Each town, city, state, province, or country must select its own policies based on what works best for it. Yet, the cost reductions that have already occurred combined with economies of scale due to further WWS expansion are a cause for optimism. If effective policies are put in place, an all-sector transition to 100 percent WWS, ideally by 2035 to 2040, but no later than 2050, is possible.

14.4.3 US INFLATION REDUCTION ACT

One example of a policy measure that has been successful in several respects but unsuccessful in others is the 2022 US Inflation Reduction Act. On the plus side, this act called for subsidies for new distributed and centralized wind, solar, and geothermal electricity generators; new battery-electric vehicles and hydrogen-fuel-cell-electric vehicles; more insulation of floors, walls, and ceilings; more weatherstripping of doors and windows; new electric heat pumps for air heating and air-conditioning; new electric heat-pump water heaters; new electric heat-pump clothes dryers; new electric-induction cooktops; new home and grid-scale batteries; and new electrolyzers, hydrogen storage, and fuel cells for green hydrogen, among other WWS technologies.

In 2023, the Inflation Reduction Act spurred the purchase, in the United States, of 752,000 new rooftop PV systems, 139,000 solar-water-heating systems, 81,000 geothermal heat pumps, 49,000 battery-electric storage systems, 42,000 small-scale wind-energy systems, and 36,000 hydrogen-fuel-cell-electric storage systems.[395] Thus, about 1.1 million households received subsidies in the form of tax credits for WWS electricity and heat generation and storage.

In addition, the act spurred, in 2023 alone, 699,000 new home-insulation projects, 694,000 window and skylight weatherstripping projects, 488,000 central air-conditioner insulation projects, 660,000 exterior door weatherstripping projects, 293,000 water-heater insulation projects, 283,000 water-boiler insulation projects, 268,000 electric heat-pump for air heating and air-conditioning purchases, 104,000 electric heat-pump water-heater purchases, 93,000 electric panel and circuit upgrades, and 37,000 home energy audits. A total of 2.3 million homeowners received tax credits for these projects.

Unfortunately, the act also funded technologies identified in Chapter 8 as not useful. Such technologies include carbon capture, synthetic direct air capture, blue hydrogen, nonhydrogen electro-fuels, bioenergy, and small and large nuclear reactors. These unhelpful technologies account for about 40 percent of the funding for energy under the act.

15

MY JOURNEY

Since the early 2000s, substantial progress has been made toward raising awareness about the need for a rapid transition to WWS. A movement toward 100 percent clean, renewable energy was created. Popular support for such a transition has grown, as have the numbers of laws and commitments furthering that goal. Such commitments have been made by cities, states, provinces, countries, international businesses, nonprofits, community groups, policymakers, and individuals. This chapter discusses my personal journey to develop and implement 100 percent WWS roadmaps, and how these roadmaps and collaborations with other scientists, cultural influencers, business leaders, and community leaders have helped to shape the 100 percent WWS movement.

15.1 FIRST EXPOSURE TO SEVERE AIR POLLUTION

During the summer of 1978, as a 13-year-old tennis player, I traveled to San Diego from my home in northern California to play in a tennis tournament. What struck me was the air pollution, which was visible on and off the freeways. I could see it, taste it, and smell it. The pollution hung like a morbid pall. It was one thing to sit inactively in a car in the middle of this pollution. It was another to run through it for hours while playing tennis. During play, my throat and lungs became irritated, and my eyes became scratchy. Taking deep breaths while lunging for tennis balls was a chore. If this soup of pollution was hurting me after only a few minutes, I imagined the damage it caused people who lived in it on a daily basis. Indeed, living in such pollution is equivalent to smoking two to three packs of cigarettes per day.

I took additional trips to Los Angeles and San Diego during the next few years. After only the second trip, I realized that the smog was not just a one-time event. After the third trip, I concluded this pollution was a way of life in these cities. I then began to ask myself, "Why should anyone live like this?" and "Isn't there a solution to this problem?" I decided then, that when I grew up, I wanted to understand and try to solve this avoidable air-pollution problem, which affects so many people. I knew what I wanted to do for my career.

Later in my teens, I learned from popular magazine articles about the emissions of greenhouse gases since the Industrial Revolution creating a blanket over the Earth, trapping heat and increasing globally averaged temperatures. I also learned about acid rain and its impact on forests and lakes. I realized that these problems were intimately connected to the air-pollution problem. The main sources of the chemicals that cause air pollution, fossil-fuel and bioenergy combustion, also produced chemicals that cause global warming and acid rain. If combustion was the problem, then eliminating combustion must be the solution.

Yet, at that time, I was busy trying to finish high school and play competitive tennis. I had neither the time nor the skills to help solve the problem. I did figure though that if I kept studying and learned as much math and science (and later, at university, engineering and economics) as possible, I would equip myself as best I could to solve these problems.

15.2 HUNGRY FOR KNOWLEDGE

When I started my undergraduate degree program at Stanford University during the autumn of 1983, it was not obvious to me what courses to take or major to declare related to my goal of solving pollution, climate, and acid rain problems. Indeed, there was no major available to study air pollution, climate, or acid rain. There was not even a general environmental major. The closest degree (not even an undergraduate degree) was a master of science (MS) degree in Environmental Engineering, which centers around the study of groundwater pollution.

Because it was not evident to me what undergraduate major to focus on, I took five engineering courses, each in a different engineering major, during the spring and autumn of 1984. These included courses in material science, electrical engineering, aeronautics and astronautics, statics, and environmental science and technology. The last course broadly covered water pollution, urban air pollution, acid rain, global climate, and energy. Professor Gil Masters was the instructor. His engaging teaching style consistently won him teaching awards. This course was by far the most interesting to me among the five. It was taught through the Civil Engineering Department. I decided then to major in Civil Engineering although there were hardly any other relevant courses available in the department or the university.

One additional relevant course I did find was on small-scale energy systems. It was also taught by Professor Masters. In that course, I learned about the efficiencies and engineering characteristics of solar panels, wind turbines, and other types of renewable energy systems. I took that course during the winter of 1985, which was in the middle of my sophomore year. Although I was interested in the renewable energy information in that course, I wanted to understand air-pollution, climate, and acid-rain problems before focusing on solutions. As such, the information in that course stayed dormant inside me until the year 2000 when I dusted off my notes from it to come up with an aha moment.

In the meantime, I had taken a basic economics course during my freshman year. Since I had several credits from high school that I could transfer to Stanford, I had room to consider a second major. Although I was not so excited about taking economics courses, I felt it was important to do so because I knew that, if I wanted to understand and solve large-scale air-pollution, climate, and acid-rain problems, I should understand costs, financing, and economics. As a result, I selected economics as a second major.

During the rest of my undergraduate career, I built up skills in engineering and economics. However, I was disappointed by the fact that there were simply no other courses that I could take that focused on my interests in air pollution, climate, and acid rain. Toward graduation, I thought deeply about what my next step would be. I learned about a program at Stanford, called the co-term program, that would allow me to complete an MS degree concurrently with my undergraduate degrees. I decided to apply to the co-term MS degree program in Environmental Engineering through the Department of Civil Engineering. This program focused on groundwater pollution, which was still not my main interest. However, it was the closest I could come to a program that moved me toward my goal. Since the program took only one more year, I used the time as an opportunity to broaden my knowledge about the environment and obtain some research skills.

15.3 LESSONS FOR LIFE

After completing my undergraduate degrees and MS degree at the end of the winter quarter in 1988, I contemplated my next move. I knew I wanted to enter a doctor of philosophy (PhD) degree program to study the atmosphere. However, I had also been playing tennis continuously for about twelve years, including four years on the Stanford University tennis team, and I felt that I should try to go onto the professional circuit for a period. I allocated myself a year and a half, at which point, during the autumn of 1989, I would enter a PhD program. While I always hoped to break through the tennis ranks and rise to the top, I was under no illusion that I was good enough to do that, so I planned my obsolescence ahead of time. Otherwise, like many of my tennis-player friends, I could have stayed out on the tennis circuit for years traveling and playing but eventually having to come back to real life. I also felt that the longer I procrastinated before trying to understand the problems that I was interested in solving, the longer I would need to come to a solution. I felt passionate about finding a solution. As such, I compromised by giving myself a limited time to play.

Halfway through the one-and-a-half years of travel, I developed a bone fracture in my right knee that resulted in a piece of bone breaking loose and floating around my knee joint. This required surgery to screw the bone chip back in and suture up my meniscus, which the floating bone chip had shredded. During the surgery, the doctor mistakenly sutured my right peroneal nerve down. That nerve runs down from the knee to the foot. By compressing the nerve, he paralyzed my foot, giving me a drop foot. I could push down but not pull my foot back up. He realized this the next morning and took me back in for another surgery to undo the suture. Even though the nerve was not cut, it was compressed so severely that my foot remained paralyzed for two years and did not recover fully for seven years. Feeling gradually came back to my leg and foot at the rate of one inch per month down my leg, starting at my knee. Needless to say, this was the end of my competitive tennis career although I tried to use a prosthetic to lift up my foot after I stepped down on it. This allowed me to run without tripping and play tennis with a drop foot. However, while I could play singles and doubles, the loss of speed made it almost impossible to win singles matches.

I never regret playing tennis. It taught me several lessons that I still use today.

One is to focus, regardless of all the distractions around you. Always keep your eye on the ball. Don't let outside noise or movement disturb you. I have applied this in my research countless times. There are many distractions that can take focus away from research on solutions.

Second, there is no substitute for practice and hard work. The harder one trains and practices tennis, the less chance that a loss will be due to not being fit or not having trained sufficiently. The more one studies and practices a research subject, the deeper one's knowledge of the subject becomes and the less chance there is of making an error.

15.4 MODELING REGIONAL POLLUTION AND THE WEATHER

Third is switching gears. While playing tennis through high school and university, I simultaneously took heavy course loads. This was difficult, but it taught me to be efficient in my current work, where I need to teach, research, advise students, write proposals, respond to emails, review applications, attend committee meetings, and attend conferences and seminars. Even though switching gears to work on different tasks has been stressful at times, I learned how to focus on each task, based on years of training.

Fourth is being accurate and truthful. There is nothing more important than trying to be as accurate as possible in research and while reviewing other people's work. In most tennis tournaments growing up, players would call their own lines. Thus, if someone cheated, it would be known quickly. As such, most players tried to be honest despite the temptation to get an extra point here or there. Honesty is important in science as well. It is incredibly important to be accurate not only in one's own work but also when describing another's work. Too often, scientists criticize other studies without fully understanding them either because they don't want to spend the time to understand them, or in some cases, because they simply want to make other researchers look bad to make their own work look better in comparison. These scientists would have benefited from learning tennis etiquette.

15.4 MODELING REGIONAL POLLUTION AND THE WEATHER

During my hiatus from education to play tennis, I looked at PhD programs at the University of Washington and the University of California, Los Angeles (UCLA), both of which had strong atmospheric science programs. I ended up going to UCLA because I found an advisor there, Professor Richard Turco, who had a project available on the main topic I wanted to study, urban air pollution. In addition, I felt that if I wanted to understand and solve air-pollution problems, I needed to be in a living laboratory. Los Angeles was just the place. It was and still is ground zero for air pollution in the United States.

I started my PhD at UCLA during the autumn of 1989 and completed my studies in June of 1994 with an MS degree and PhD degree in Atmospheric Sciences. There, I learned how to build physical equations describing phenomena in the atmosphere, how to solve the equations, how to write computer programs to represent the equations and their solutions, and how to compare model results with data. My project was to build an urban air-pollution–weather prediction model, apply it to study air pollution in Los Angeles, and compare model results with data. I loved my project from beginning to end.

An **air-pollution computer model** is really a four-dimensional (three dimensions in space and one in time) mathematical representation of atmospheric processes, including gas processes, aerosol-particle processes, transport processes, solar and heat radiation processes, ground-surface processes, and ocean processes, among others. At the time, air-pollution models were decoupled from weather-prediction models. In other words, the sources of wind speed and direction data to move gases and particles from one place to another in the air-pollution model were either observations or an electronic database, where the data were produced by running a separate weather-prediction model with the results transferred to an electronic file. No regional air-pollution model at the time had a built-in weather-prediction model, where the winds drove pollution movement *and* the pollutants themselves fed back to modify the weather.

In 1990, I began my computer-modeling research by writing computer algorithms to simulate gas chemistry, and aerosol-particle-physics and -chemistry processes. Some of the codes I eventually developed for the air-pollution model included three algorithms that simulated gas chemistry. Others simulated aerosol-particle physics and chemistry, considering the size and composition of the particles.

I then coupled these gas and aerosol-particle algorithms with existing transport and radiation algorithms developed by Dr. Owen Toon and my advisor, Professor Turco. Dr. Toon was a colleague of my advisor. I worked during the summer of 1990 in Dr. Toon's lab at NASA Ames Research Center in Mountain View, California.

The resulting combination of gas, aerosol-particle, radiation, and transport algorithms consituted an air-pollution model. I set up the model to predict gas and particle concentrations in space and time in the Los Angeles basin. However, the model was applicable anywhere that sufficient data were available. At the time, though, the only source of winds for the air-pollution model was an electronic file with interpolated observations. This was not ideal, because the observations were far apart in space and not very frequent in time. Fortunately, in our research group at UCLA, another graduate student, Rong Lu, was building a regional **weather-prediction model**. Such a model predicts winds, temperatures, pressures, and humidity.

In 1993, I coupled Rong's weather-prediction model with my air-pollution model in such a way that heating rates from the radiation algorithms in the air-pollution model fed back to the weather model to change temperatures in the weather model. Heating rates are calculated by considering the transfer of sunlight and heat through the atmosphere and the interactions of the sunlight and heat with gases, aerosol particles, and cloud particles in the air. Simultaneously, I set up the "coupled model" so that winds from the weather-prediction model now moved gases and particles around in the air-pollution model. Also, temperatures and air pressures from the weather-prediction model now fed back to affect gas and aerosol-particle chemical reaction rates and physical interactions in the air-pollution model. Little did I know at the time that this was the first air-pollution–weather-prediction model coupled with feedback for gases and aerosol particles to be developed worldwide that could facilitate the study of urban or regional air pollution. I gave a copy of the resulting model back to Rong for him to use as well, which he did. Sadly, Rong passed away from a brain tumor just a few years after he graduated.

I named the coupled model GATOR-MMTD (Gas, Aerosol, TranspOrt, Radiation-Mesoscale Meteorological and Tracer Dispersion model).[396] I subsequently used the model to simulate gas, aerosol, radiative, and meteorological parameters and to compare results with hourly data in different locations in the Los Angeles basin.[397,398,399]

15.5 MODELING GLOBAL POLLUTION AND CLIMATE

In early 1994, before graduating from UCLA, I was fortunate to land a job as an assistant professor at Stanford University in what became the Department of Civil and Environmental Engineering. While still at UCLA, I wrote a large research proposal to the US Environmental Protection Agency to expand my regional air-pollution model to the global scale in order to study global climate and air pollution. I included on the proposal my advisor and Professor Akio Arakawa, primary developer of the UCLA General Circulation Model (GCM). The UCLA General Circulation Model predicted winds, temperatures, air pressures, cloud cover, and moisture on a global scale, but it lacked detailed treatment of gases, particles, clouds, or radiation. The goal of the project was for me to couple, with feedback, the UCLA GCM to a globalized version of my air-pollution model. I would perform the same type of interactive coupling as I did between the regional pollution and weather models. To my shock and that of everyone else on our proposal team, the proposal was funded. The hard part, work on the project, followed.

Shortly after the project was funded, I left for my new job at Stanford. I then worked on the project while also teaching and advising students. I first stretched the air-pollution portion of the GATOR-MMTD model to the global scale. Due to the intense treatments of air-pollution chemistry, particle physics, and radiation, this resulted in a detailed global air-pollution model. I then coupled this global air-pollution model, with feedback, to the UCLA GCM weather-prediction model. The result, in 1995, was the first coupled global air-quality–weather-climate model worldwide to treat the feedback of gases and size- and composition-resolved aerosol particles to weather and climate, and vice versa.[400,401]

A regional air-pollution model requires inputs at its horizontal boundaries. Such inputs include wind speeds and directions, temperatures, pressures, gas concentrations, and aerosol-particle concentrations. These values ideally come from a global model. At the time, however, boundary inflows into regional models came primarily from interpolated data. Since I now had both global and regional coupled air-quality, weather, and climate models, I realized I could improve the regional model by coupling the global model to it and feeding in boundary conditions from the global model to the regional model, not only horizontally into the side boundaries but also vertically into the top boundary of the regional model. I did this coupling and ensured atmospheric processes on all scales were solved consistently. The first version, completed in 1998, was a nested global-through-regional model.

A **nested model** starts at the global scale to produce meteorological variables (winds, temperatures, pressures, etc.) and gas and aerosol-particle concentrations. These variables and concentrations are then fed into one or more smaller, more finely resolved domains placed anywhere within the global domain. Within each smaller domain may lie one or more even smaller, more finely resolved domains that receive boundary conditions from the next-larger domain. In this way, air pollution and/or weather can be modeled anywhere in the world at high resolution while receiving boundary conditions that vary continuously in time and space. I eventually called the nested model GATOR-GCMOM (Gas, Aerosol, TranspOrt, Radiation-General Circulation, Mesoscale, and Ocean Model).[402,403,404] It was the first model worldwide to nest gases, aerosol particles, cloud particles, radiation, and meteorology from the global to local scale. Since then, many models have made efforts to nest pollution, weather, and climate in a similar manner.

15.6 BLACK CARBON, THE KYOTO PROTOCOL, AND WIND VERSUS COAL

I have used the global and nested models over the years to study several phenomena, including the impacts of aerosol particles on ultraviolet radiation,

15.6 BLACK CARBON, THE KYOTO PROTOCOL, AND WIND VERSUS COAL

temperatures, and ozone in urban areas[405,406] and the effects of black carbon on climate. One hypothesis I proposed was that black carbon, the main component of soot, may be the second leading cause of global warming after carbon dioxide in terms of direct radiative forcing.[11] I further hypothesized that controlling black-carbon emissions may be the fastest method of slowing global warming.[12,13] The contention that black carbon was the second-leading cause of warming was supported by others, including in a major review article.[14]

Up to the year 2000, I focused on understanding air-pollution and climate problems. In 2000, I began looking into solutions. In particular, I became concerned about whether the United States would ratify the Kyoto Protocol. The **Kyoto Protocol** was an international climate agreement adopted by the United Nations Framework Convention on Climate Change on December 11, 1997. The protocol called for developed countries to reduce greenhouse gas emissions. Between the adoption of the protocol and 2024, 191 countries and the European Union ratified it. The only party to the protocol not to ratify it was the United States. Canada renounced its ratification in 2012.

In 2000, I pondered what was needed for the US to satisfy its requirement under the Kyoto Protocol, which was to reduce greenhouse gas emissions to 7 percent below 1990 levels. I first calculated this could be accomplished by reducing about 59 percent of 1999 US coal emissions. I then wondered how many wind turbines were needed to replace such coal. I dusted off my notes from the 1984 environmental science and technology course I took from Professor Masters. I remembered an equation he derived that could help answer this question. The equation gave the capacity factor of almost any wind turbine given three parameters: mean wind speed, turbine nameplate capacity, and turbine blade diameter. The equation didn't exist anywhere except in his course notes, which I had luckily saved.

After thinking about this a bit, I went to Professor Masters to discuss my goal. He was encouraging and agreed to help me on a paper. I then used the equation along with information about a recently developed efficient 1.5-megawatt wind turbine to calculate the number of such turbines that might be needed to replace enough coal for the US to satisfy the Kyoto Protocol. This analysis resulted in a very short (three-quarters of one page) paper in *Science Magazine* called "Exploiting wind versus coal."[265] Despite its brevity, the paper received an enormous amount of pushback, primarily from coal supporters. One person kept sending negative comments about the paper to hundreds of people on a blind email list. He defended himself by saying that he was just a concerned citizen. However, an investigation by a reporter revealed that this concerned citizen was Glenn Schleede, a senior vice president of the National Coal Association from 1976 to 1980. Despite the negative feedback, I was energized to investigate more the potential of wind energy to meet national energy needs in the US and, ultimately, worldwide.

The US signed the Kyoto Protocol on November 12, 1998. During May 2001, shortly after **George W. Bush** became US President, he had to decide whether to submit the Kyoto Protocol for ratification to the US Senate. At that time, I was working on a draft of the paper called "Control of fossil-fuel particulate black carbon and organic matter, possibly the most effective method of slowing global warming."[12] The main conclusion of the paper was "Reductions in black carbon plus organic matter emissions from fossil-fuel sources will not only slow global warming, but also improve health." During May 2001, I sent a draft of the paper to some people I knew at the US Environmental Protection Agency to provide comments. Coincidentally, about the same time, the White House asked the agency if they were aware of any new papers on causes of climate change. My paper was mentioned. On May 18, 2001, the White House then requested the agency to ask my permission for the agency to send the draft paper to the White House. I agreed but asked that the paper not be cited or quoted, as it was not yet published. The White House agreed, and the paper was sent.

Sure enough, on June 11, 2001, President Bush gave a speech in which he explained why he was not sending the Kyoto Protocol to the Senate for ratification. In his speech, he said, "The Kyoto Protocol was fatally flawed in two fundamental ways." First, the protocol would have "a negative economic impact, with layoffs of workers and price increases for consumers." Second, "Kyoto also failed to address two major pollutants that have an impact on warming, black soot and tropospheric ozone. Both are proven health hazards. Reducing both would not only address climate change, but also dramatically improve people's health."[407]

The first reason for President Bush's rejection of the Kyoto Protocol was not true, because a transition to renewable energy creates more jobs than are lost and reduces costs to consumers. While the second reason was true, it was not a reason to pull out of the Kyoto Protocol. I was flattered that the President thought my conclusions were important enough to mention them almost verbatim. However, I was not happy that he used them to draw an incorrect inference, namely, that the Kyoto Protocol should not be ratified. The correct implication of the conclusions was that the Kyoto Protocol should have been ratified but that it could have been improved by including black carbon (and ozone). I do understand though, that, even if the President did not use this reason, he would have found another reason to prevent ratification. Nevertheless, I learned a lesson here. It is extremely important for scientists to ensure their results are used by policymakers in the way that they are meant to be used. This is not easy since, once a paper is published, it is impossible to ensure that everyone applies the results properly. However, scientists can take steps, such as by disseminating follow-up statements or opinion-editorials, to clarify the implications of a paper.

15.7 ANALYZING WIND AND OTHER WWS TECHNOLOGIES

The feedback I received from the 2001 paper on wind versus coal motivated me to examine wind energy in more detail. I asked a doctoral-degree student, Cristina Archer, who was knowledgeable about meteorology, if she wanted to work on a wind-mapping project for the United States. She agreed to work on this project along with her main project, which was to analyze a wind circulating around the Monterey Bay, the Santa Cruz eddy, that she discovered while visiting the Santa Cruz beach. Over the next few years, she developed two of the world's first wind maps from data alone at 80 meters above ground level. One map was for the United States, and the other, for the world as a whole.[305,326] During her studies, she also found that interconnecting geographically dispersed wind farms could turn completely intermittent wind power into partial baseload power.[305,306]

Another student, Mike Dvorak, then performed high-resolution computer modeling of wind resources offshore of California and offshore of the US east coast.[408,409,410] These studies uncovered not only substantial offshore-wind resources in the US but also that the times of peak offshore wind speeds often coincide with the times of peak energy demand.

Meanwhile, I was using the GATOR-GCMOM model to compare impacts on air pollution and climate of different energy technologies. I compared the effects of diesel versus gasoline vehicles on air pollution;[411] the effects of gasoline vehicles versus hydrogen-fuel-cell-electric vehicles and versus gasoline-battery-electric-hybrid vehicles on US air quality and climate;[325] the effects of hydrogen-fuel-cell-electric vehicles versus gasoline vehicles on the global ozone layer and climate;[99] and the effects of ethanol vehicles versus gasoline vehicles on air-pollution mortality.[211,212,213]

After collaborating on the wind analyses and comparing the impacts of different fuels and electricity sources, I began, in 2008, to think that it would be useful to review systematically different proposed solutions to global warming, air pollution, and energy security. At the time, several technologies were being proposed as solutions to replace fossil fuels for electricity. Others were being proposed as solutions for transportation. Aside from WWS, major proposed technologies for electricity generation included nuclear and coal with carbon capture. For transportation, ethanol-powered vehicles were being proposed along with battery-electric vehicles and hydrogen-fuel-cell-electric vehicles.

Given our previous work on evaluating some of these technologies, I felt ready to perform such a review. The study, published in 2009, compared 9 electricity-generating options and 12 options for powering vehicles as proposed solutions to global warming, air pollution, and energy security.[178] It considered impacts of each option on 11 impact categories: CO_2-equivalent emissions, air-pollution deaths, footprint area, spacing area, water consumption, resource abundance, effects on wildlife, thermal pollution, water chemical pollution/radioactive waste, risk of energy supply disruption, and normal operating reliability.

With respect to reliability, in 2008, I asked another graduate student, Graeme Hoste, to see if it were possible to match California's hourly electricity demand with only wind, solar, geothermal, and hydroelectricity supply. Geothermal would provide constant electricity each hour; solar and wind would provide variable electricity; and hydroelectricity would fill in gaps in supply. Graeme completed a report on this topic.[330] He found that, in theory, California could meet 100 percent of its monthly averaged hour-by-hour power demand during the two months tested, April and July 2020. The upshot was that when treated as a bundle with geothermal and hydroelectric, wind and solar are more reliable for meeting peak demand than they are when treated individually. This result, which was borne out in hundreds of subsequent independent studies, was used in my 2009 review paper.

The overall conclusion of the review paper was that the best electricity-generating technologies for minimizing the 11 impacts were onshore and offshore wind, solar PV, concentrated solar, geothermal, tidal, wave, and hydro. These technologies were referred to as **wind, water, and solar** (**WWS**) technologies. The paper also proposed the use of battery-electric vehicles and hydrogen-fuel-cell-electric vehicles for transportation.

15.8 100 PERCENT WIND–WATER–SOLAR AND THE TED DEBATE

The review paper garnered substantial interest by the climate and energy communities. Shortly after the paper was published, I was approached by *Scientific American* to consider writing a follow-up paper about the feasibility of powering the world with the best technologies identified in the review.

15.8 100 PERCENT WIND–WATER–SOLAR AND THE TED DEBATE

While pondering this, I asked a colleague, Dr. Mark Delucchi, who was a research scientist at the Institute of Transportation Studies at University of California, Davis, whether he would be interested in partnering with me on such a study. He agreed. Together, we analyzed the technical and economic feasibility of transitioning the world's all-purpose energy to 100 percent WWS. We analyzed the change in energy demand upon a conversion to WWS, the numbers of WWS devices needed, the land areas required for footprint and spacing, the materials needed, reliability, and costs. The study concluded that, with aggressive policies, a transition to 100 percent WWS by 2030 was technically and economically possible, but for social and political reasons, a more likely transition end-goal was around 2050.

The paper was published in November 2009.[331] It was immediately attacked as pie in the sky and an impossible dream. Yet, within 16 years, the 100 percent goal outlined in the paper had advanced significantly, and a movement had sprung up around it. Since then, many countries and US states have either reached or almost reached 100 percent WWS in their electric-power sector, and many 100 percent laws and commitments have been put in place by cities, states, countries, and businesses. WWS costs have also come down substantially, and the public is overwhelmingly supportive of the goal. Nevertheless, the solution is still a long way from being fully realized.

As a result of the review paper and the *Scientific American* paper, I was asked to take part in a debate at a TED (Technology, Entertainment, Design) conference in Long Beach, California, on February 11, 2010. The debate was with Stewart Brand, former editor of the *Whole Earth Catalog*, turned nuclear advocate. The debate was on nuclear power versus renewables.[412] I had given lots of talks but had not experienced this debate format before, where we each had six minutes to lay out a case and a few minutes each for rebuttals.

At the beginning of the debate, the moderator, Mr. Chris Anderson, asked the audience of about 2,000 if they favored or opposed nuclear energy. To my surprise, about 75 percent favored it. Given the strong support for Mr. Brand's position from the get-go and the facts that he was an experienced speaker at TED events, had such a charming personality, and was the first to speak, I thought I was doomed. The only advantage I had was that the data I was about to show were based substantially on new, raw research that our group had performed and that he had not seen. Such data included data from wind-mapping studies and comparisons among different energy sources. He was using information exclusively from third-party sources, and it was mostly dated.

After his six minutes of speaking, I felt more in a hole because he just presented lots of information with a compelling style. I felt I needed to get my words out correctly from the beginning; otherwise, I would lose the audience. I focused on my first few words and managed to start strong. I then reeled off statistics and my own graphs. While I was presenting my six minutes of information, I could see out of the corner of my eye the concern growing on his face. At the same time, I could feel the audience resonating with my arguments and statistics. In fact, a few times, the audience started cheering me on. A vote after the debate indicated that I had switched about 10 percent of the audience, or 200 people, such that in the end, 65 instead of 75 percent favored nuclear.

This debate, coupled with the review paper that found that nuclear was better than some technologies but not so good as WWS technologies, rendered me a target for nuclear advocates. Most never understood or did not want to understand that my goal has always been focused on solving air-pollution, climate, and energy-security problems with the technologies that result in the fastest solution, the lowest cost, and the greatest benefit. Nuclear electricity, while beneficial in some respects and on some timescales, is not so good as WWS electricity. Thus, it isn't that I am *against* nuclear power. Instead, from a scientific point of view, nuclear is not so good as other technologies, and I needed to state that honestly, as I have been trained to do. The fact that nuclear has problems has consistently been borne out over the years since the 2009 review paper and 2010 debate.

Subsequent to the *Scientific American* paper, Dr. Delucchi and I wrote a more detailed version of the global 100 percent WWS roadmap and also a roadmap for the US as a whole. These were published in the journal *Energy Policy*.[101,332] Concurrently, I engaged another doctoral student, Elaine Hart, to develop an optimization model to simulate the matching of electricity demand with geothermal, wind, solar, and hydroelectricity supply in California. She completed that work, confirming in more detail what Graeme Hoste had found.[316,345,346] Another doctoral student, Bethany Frew, followed Elaine's work with an optimization model study that found a combination of renewables would allow reliable electricity grids across the US.[351] Yet another doctoral student, Eric Stoutenburg, examined the impact of combining wind and wave power to increase grid reliability.[308,309]

15.9 THE SOLUTIONS PROJECT

In the midst of the flurry of research activity, I was invited to a dinner at the Axis Café and Gallery in San Francisco on July 10, 2011, to discuss the

15.9 THE SOLUTIONS PROJECT

economic viability of renewable energy alternatives for the state of New York. At the time, I had no idea that this meeting would catalyze a series of events that would turn a scientific theory into a popular mass movement to transition the world to clean, renewable energy and storage. Nor did I envision at the time that this movement would result in country, state, and city laws and proposed laws, including the Green New Deal, along with business commitments.

The dinner was hosted by my now good friend, Marco Krapels. At the time, Marco worked in the banking business. He had also started a non-profit, Empowered by Light, whose goal was to bring solar power to remote communities worldwide. Marco had invited several others to this dinner, from both the local area and from New York. In particular, he invited Mark Ruffalo and Josh Fox. Mark is an actor and activist who had been asked by the governor of New York, Andrew Cuomo, to participate in a New York renewable energy task force. Josh directs documentaries and was coming off the success of his 2010 documentary *Gasland*, which exposed the problem of fossil-gas fracking to a worldwide audience. The movie received an Academy Award nomination.

I had been invited to the dinner because Marco and several others had seen our *Scientific American* paper about powering the world for all purposes with clean, renewable energy. My contribution to the dinner would be to offer ideas about what New York could do to obtain its own energy from something other than fracked fossil gas. At the time, fracking was not legal in New York, but the governor was under significant pressure to legalize it. Fracking was legal in nearby Pennsylvania, and hundreds of thousands of wells had already been drilled there, causing damage and upheaval in many communities.

At the dinner, Marco, Josh, Mark, and I bonded together. It was as if we had known each other for years. We were all passionate about finding a sustainable solution to New York's energy problem and eliminating air-pollution and climate problems in general. The interesting thing was that we all had different backgrounds. I was a scientist. Marco was a businessperson. Mark and Josh were cultural heroes – artists and entertainers. Later, this combination of science–business–culture would prove to be a powerful one, much stronger than any of the individual parts.

At the dinner, we discussed the fracking issue in New York and whether there was an alternative energy solution to it. I speculated that New York could be powered entirely by clean, renewable energy for all purposes. This prompted the question as to whether I would be interested in developing a WWS plan for New York State. My immediate impulse was that this would be great to do, but I knew how much work it would take, so I wavered. Put on the spot, I told them I would be willing to write a one-paragraph summary but that it would be better for someone else to take that summary and turn it into a real roadmap given the effort involved.

Josh Fox then suggested I speak to two Cornell University professors, Tony Ingraffea and Robert Howarth, to get their perspective on a New York plan. Tony was a professor in civil and environmental engineering who had worked a lot on methane leaks from cement casings in fossil-gas fracking wells. Robert was a professor in ecology and evolutionary biology who had just published a seminal paper with Tony on methane leakage rates from the fracking of shale rock.[184] I spoke with them and Josh Fox by phone almost two weeks later. We discussed, among other topics, the different renewable energy resources available in New York.

After the call, I sent Tony and Robert a wind-resource analysis of New York that my doctoral student, Mike Dvorak, had put together. Josh Fox, Robert Howarth, Tony Ingraffea, and I then scheduled a follow-up call with a larger group, adding Gianluca Signorelli, a work associate of Marco Krapels; Stan Scobies, a retired researcher in New York familiar with the energy landscape; and Marcia Calicchia, a policy expert and researcher at Cornell.

On the follow-up call at 4 p.m. Pacific time on the afternoon of September 13, 2011, I was asked again if I could write a 100 percent renewable energy plan for New York. Feeling the stress of a lot of other work commitments, I wavered again. I repeated that I could only write a brief one-paragraph summary and then help guide someone else do the rest. Later that night, when I started writing the paragraph, I began to think more deeply about a state plan for New York, and something in me clicked. I thought to myself, "If we really want to solve the problems of air pollution, global warming, and energy security, we need granular state- and country-level plans." I also realized that a New York plan would be a natural extension of the global and US roadmaps that Mark Delucchi and I had previously developed.

My inspiration took over. I worked into the night. I found wind-, solar-, and hydroelectric-resource data and air-pollution-mortality data for New York State. I then did an energy-, air-pollution-, and climate-cost analysis of fossil fuels versus WWS for the state. Finally, I identified a set of policy mechanisms that could be proposed to implement 100 percent WWS in the state.

In the morning, I woke up from my trance and happily emailed my paragraph-turned-14-page single-spaced draft 100 percent New York energy roadmap to the group. They were as shocked as I was. Stan quipped, "I figure about three more inspiring conference calls ought to have the whole thing done." We now had a starting point to change the New York energy infrastructure.

The first draft catalyzed a flurry of activity and edits among several people in the group that we had just formed. Ultimately, the New York roadmap went through 40 drafts before it was completed and published in the journal *Energy Policy* 18 months later, on March 13, 2013.[333]

At the time, though, our immediate goal was to develop a draft paper that was complete enough to present to New York Governor Andrew Cuomo and his staff. I worked feverishly on this paper, engaging students and incorporating comments by several in the group. The group as a whole also started holding regular phone conversations. Marco, Mark, Josh, and I also began talking more, taking part in several middle-of-the-night phone conversations. We became close as we shared a common passion and goal to solve major problems.

The oral communication and writings from our group had become substantial enough that we decided we needed a name for the group. We wanted the name to represent something positive because we felt it was better to be for than against something. Since this all-volunteer group was focused on solutions for New York's energy future, we settled on **The Solutions Project**. On December 15, 2011, Marcia sent the group its first brochure, which summarized The Solutions Project as a collaboration among "scientists, renewable energy industry pioneers, experts in renewable energy financing and investment, mission-oriented investors, businesses, labor organizations, inner city community groups, farmers, environmentalists, and many celebrities and cultural figures." Its initial main goals, laid out in a brochure published on February 25, 2012, were to

- raise mass awareness about the viability of renewable energy,
- raise awareness about the 100 percent WWS roadmap we were developing for New York, and
- leverage private capital to demonstrate the success and viability of some specific renewable energy projects.

On January 20, 2012, The Solutions Project, still behind the scenes, began recruiting members for an advisory council. Ultimately, the advisory council consisted of scientists, business leaders, and celebrities. Some of the celebrities that agreed to take part included Deepak Chopra, Leonardo DiCaprio, Jesse Eisenberg, Woody Harrelson, Ethan Hawke, Scarlett Johansson, Sean Lennon, Julianne Moore, Leilani Munter, Elon Musk,

Edward Norton, Yoko Ono, Robert Redford, Eileen Rockefeller, Antonio St. Lorenzo, Wendy Schmidt, Martha Stewart, Channing Tatum, Michelle Williams, and Deborah Winger. The purpose of the Advisory Council was not only to provide feedback but also to help The Solutions Project meet its main goal, which was to bring awareness to the world about the potential of a 100 percent transition to clean, renewable WWS energy.

During early March 2012, a new integral member came on board The Solutions Project. Jon Wank had been working in advertising and learned about The Solutions Project from Marco Krapels. Jon's passion for making a difference resulted in him quitting his job and volunteering to work with us on media and social engagement. It was Jon's ingenuity that led to the development of state, country, and city infographics. Each infographic summarizes a 100 percent roadmap for a different locale. The infographics have been used worldwide and are still available online.[413]

15.10 EFFECTS OF NEW YORK STATE ROADMAP ON POLICY

The original task of The Solutions Project was to bring a science-based 100 percent New York energy roadmap to the governor of New York. The hope was that the governor might see that a WWS plan is the best solution for the state's future and to translate the plan into law. Such an action would also mean that hydraulic fracturing was not needed. To that end, The Solutions Project began to engage other groups, including the National Resources Defense Council (NRDC), which had a major presence in New York. After some discussion, NRDC committed, on January 2, 2012, to support The Solutions Project goal of bringing 100 percent WWS to New York. This was an important first step.

The Solutions Project held its first all-hands meeting at Cornell University on May 21, 2012. It was there that many of those involved in numerous phone calls could finally meet each other and focus on a path forward. Such a path included the idea of getting information out to large numbers of people. Some of the ideas that came forth were to hold a concert, go on a speaking tour, hold rallies, and develop media content. Marcia introduced finger puppets that we could wave at each other during contentious discussions to lighten the mood.

Shortly after that meeting, on June 20, 2012, Mark Ruffalo, Marco Krapels, and I met at Stanford University, California, where we spoke with students, some of whom were helping to develop the New York energy-transition roadmap. We then went over to Google in Mountain View, where the three of us gave a joint talk to a packed house.[414] We offered our

15.10 EFFECTS OF NEW YORK STATE ROADMAP ON POLICY

experiences through the lenses of culture (Mark Ruffalo), business (Marco Krapels), and science (me), about how it is possible and necessary to transition from fossil fuels to 100 percent WWS.

We then met with the Google energy and sustainability team, directed by Rick Needham. Google had been making investments in renewable energy for several years. In 2007, they installed a 1.6-megawatt PV array on their Mountain View buildings. In 2010, they purchased two wind farms in North Dakota and contracted for additional wind in Iowa. Our meeting with them was a meeting of like minds. We discussed the importance of companies working with scientists, cultural and community leaders, and policymakers to spur a transition. Ultimately, in 2017, Google became the first company in the world to provide at least 100 percent of its annually averaged electricity needs with WWS.

On the same day as the Google visit, we visited Facebook, also in Mountain View, and their sustainability and energy-efficiency team, led by Bill Weihl. We discussed the need for Facebook and other businesses to participate in a 100 percent WWS transition.

The talk at Google was received so well that the three of us (Mark, Marco, and I) were asked to give another joint talk, this time at a conference on Nantucket Island, Massachusetts, on October 6, 2012. There, policymakers from across political parties and media were gathered. Some included Secretary of State John Kerry, President of Americans for Tax Reform Grover Norquist, presidential advisor David Gergen, US Treasury Secretary Larry Summers, and MSNBC commentator Chris Matthews.

The talk[415] led to an invitation to Chris Matthews' home in Washington, DC, for a dinner. On February 27, 2013, Mark, Marco, and I presented The Solutions Project vision at the dinner to US Senator Kirsten Gillibrand, Secretary Kerry, some members of the US House of Representatives, and some members of President Obama's staff. Whereas these meetings didn't lead to direct policies at the time, they provided new information to policymakers, increased the familiarity of policymakers with what we were doing, and allowed us to learn about the concerns of others.

In a parallel effort to raise awareness on a large scale, during August 2012, Jon Wank created an animated cartoon, called *Tommy and the Professor*. Mark Ruffalo came up with this idea at our May 12, 2012 retreat at Cornell. He and I were brainstorming outside, and he suggested a brilliantly crazy idea to produce a cartoon where he was an annoying kid, and I was a pedantic professor trying to teach him about renewable energy. Months later, Jon Wank brought this cartoon idea to life. Tommy (voiced by Mark Ruffalo) was the annoying student who hung out with a girl (voiced by Zoe Saldana).

15 MY JOURNEY

After seeing that the Professor's house was the only one on the block with its lights on after a blackout, the two came to a class on renewable energy taught by the Professor (voiced by me). There, the kids learned about energy from wind, water, and sunlight. After the lecture, Tommy had an epiphany about what 100 percent WWS meant and wanted to inform the world about it.

In the meantime, during the rest of 2012, several students at Stanford plus Tony Ingraffea, Bob Howarth, Stan Scobies, economist Dr. Jannette Barth, and I continued to revise the New York State energy roadmap. I also engaged Dr. Mark Delucchi to work on the New York paper. He has been a coauthor on most roadmaps since then.

On November 17, 2012, Mark Ruffalo, Marco Krapels, Josh Fox, Jon Wank, and I brought the latest version of the New York State roadmap with us to a meeting with NRDC in New York City. The goal of the meeting was to discuss how they and The Solutions Project could bring 100 percent WWS to New York State. On February 5, 2013, NRDC and The Solutions Project wrote a joint letter to the governor of New York informing him about our roadmap. The letter, signed by Frances Beinecke (President of NRDC), Mark Ruffalo, and me, stated in part, "We are writing today to let you know about a clean energy framework for New York that has been developed by scientists, financial specialists, business leaders, and policy experts under the Solutions Project and reviewed and updated for practical implementation in New York State by the Natural Resources Defense Council." The letter then went on to propose several first steps that the governor could take to reach 100 percent WWS. These included increasing the installation of large-scale wind and behind-the-meter solar, scaling up energy efficiency, removing barriers to the adoption of electric vehicles, creating a **green bank** (which leverages public and private financing to advance clean energy projects), and expanding the existing solar rooftop program.

Soon after (during March 2013) our New York roadmap paper was published and received attention in the press, particularly in New York. The roadmap spread like wildfire throughout New York. As a result, I was asked to give a talk at an antifracking rally about the potential of New York to go to 100 percent WWS. The rally was on the doorstep of the governor's office in Albany. I spoke on June 17, 2013, to a crowd of thousands. This was the first and only rally I ever spoke at. It was gratifying because the crowd was on board with transitioning to 100 percent WWS. Our roadmap for New York gave people hope that there was an alternative solution to fossil gas.

Ultimately, on December 17, 2014, Governor Andrew Cuomo banned fracking in New York due to concerns about its health risks. A decision on fracking had been held in abeyance for six years while health effects data were gathered. Grass-roots organizations in the state ensured the health risks were considered. The ban was facilitated by the fact that an alternative to fracking now existed. That alternative, supported strongly by the grass-roots organizations, was WWS.

> **Transition highlight**
> With the fracking issue behind him, Governor Cuomo submitted to the New York Public Service Commission a proposal for New York to obtain half of its electricity from renewable sources by 2030. On August 1, 2016, the Public Service Commission approved the proposal, giving rise to a mandatory and enforceable 50 percent WWS law by 2030 for the state of New York, called the Clean Energy Standard. The law required half of the state's electricity to come from onshore and offshore wind, solar, and hydroelectric power. Our New York roadmap called for 80 percent WWS in all energy sectors by 2030 (and 100 percent by 2050), but the Clean Energy Standard requirement of 50 percent WWS in the electric-power sector was a good first step. It was the first piece of proposed legislation for which the 100 percent WWS roadmaps served as a scientific basis.

On July 18, 2019, Governor Cuomo went further by signing a law requiring that New York reach 70 percent WWS by 2030 and up to 100 percent by 2040 in the electric-power sector. The law also required sufficient emission reductions in other sectors to occur that overall greenhouse gas emissions in the state must decrease 85 percent by 2050. It was nice to see that science, business, culture, and community could come together to motivate the passage of a law mandating a clean, renewable energy solution.

15.11 EFFECT OF CALIFORNIA ROADMAP ON CITY POLICIES

In the meantime, by September 2012, I wanted to develop a 100 percent all-sector WWS roadmap for California. Aside from the facts that California is the most populous US state and its residents have an affinity for renewable energy and efficiency, our research group had already analyzed the electricity grid and offshore wind resources in California. I formalized a Solutions Project student research group at Stanford and engaged over 20 new students to help develop the California roadmap. I also engaged several of the scientists who had helped with the New York roadmap, on the California plan.

We finished an early draft of the California roadmap on January 8, 2013. I sent a copy that day to the Sierra Club nonprofit environmental organization to review it for practical implementation. Jodie Van Horn, at the Sierra Club, became the point person, and she obtained internal suggestions for improvement. This was the first of many iterations of the California roadmap over the next 18 months.

In early 2013, The Solutions Project was still an all-volunteer group. However, the group was considering becoming a formal nonprofit. Before deciding to take this leap, we considered other options. Given that we were now interacting more with the Sierra Club, it was logical to discuss an option to merge with them as well. On February 6, 2013, Mark Ruffalo approached Michael Brune, the Executive Director of the Sierra Club, about the possibility of The Solutions Project and Sierra Club becoming "one diverse coalition around science, business, and culture advancing 100 percent renewables."

The Sierra Club was very interested and proposed to incorporate The Solutions Project as a campaign inside of the Sierra Club and to move resources from their *Beyond Coal Campaign* to a *100 Percent WWS Campaign* led by The Solutions Project. The Sierra Club was interested because The Solutions Project offered raw science, the media power of celebrity, and creativity. The offer was tempting to The Solutions Project because it negated the need for us to go through the pains of forming a nonprofit. However, it would also reduce the autonomy of The Solutions Project, whose success to date was based on being free to pursue ideas without gaining the consensus of a large board.

After lengthy deliberations, including a meeting in New York City on April 20, 2013, The Solutions Project chose to become its own nonprofit. Its focus remained, for the time being, on educating the public and policy-makers about the 100 percent roadmaps. The Sierra Club saw this as an important mission as well but shifted to focus on what it does best, grass-roots campaigning. In this case, the campaign centered around getting cities across America to commit to 100 percent WWS. This effort, called the "Ready for 100 Campaign," started in earnest in late 2015.

On October 29, 2015, Jodie Van Horn from the Sierra Club contacted me again. We talked about the Sierra Club's pending cities campaign, the goal of which was to get 100 US cities to commit to 100 percent WWS within three years. She asked if I could prepare 100 percent WWS roadmaps for some targeted cities. On the one hand, I knew that each roadmap took a lot of work. On the other hand, I knew they were important for analyzing the ability of cities to transition. I agreed but told her it would take some time. Indeed, the first roadmap paper we finally completed for 53 towns and cities in North America was not published until two and a half years later, on June 30, 2018.[338]

Transition highlight
The Sierra Club used the 100 percent WWS state roadmaps and resulting infographics[413] to help provide confidence to communities and town and city leaders that 100 percent WWS was possible. When the town and city roadmaps we were developing finally dribbled in, they became helpful as well. However, by the time the roadmaps were published in 2018, the Sierra Club had already obtained 100 percent commitments from over 100 US cities, meeting their goal of at least 100 cities within three years. By April 2022, when the Sierra Club ended the cities campaign, this number had increased to over 200 cities, towns, and counties.[416] The 100 percent movement had really accelerated with the help of many groups working together.

The Solutions Project finally became an independent nonprofit in mid-2013. Much of the rest of the year was spent hiring an Executive Director, Sarah Hope, and staff; setting up the organization; and preparing a launch event. The launch took place on June 19, 2015, in New York City. Leo DiCaprio joined The Solutions Project team at the event.

During May 2013, while still updating the California roadmap, I embarked on an individual roadmap for Washington State. Washington was chosen due to the substantial renewable resources it has (wind and existing hydroelectric in particular). I thought that Washington might be a state that could reach 100 percent quickly with respect to electric power. During the autumn of 2013, I recruited a new set of over 20 students to work more intensely on the roadmap. The Washington State roadmap paper was ultimately published in 2016.[335]

15.12 THE LETTERMAN SHOW

On September 25, 2013, a television producer, Mike Buczkiewicz, invited me to appear on the *Late Show with David Letterman*. Buczkiewicz said that he invited me because he saw my interview in the documentary *Gasland Part II* by Josh Fox. Josh had interviewed me for his documentary on July 12, 2011, the day after we first met in San Francisco. His film is about the damage done by fossil-gas fracking. My short interview in his documentary was about how it was possible to get off fossil gas, coal, and oil and move completely to renewables. Mr. Buczkiewicz liked the positive message I related. He told me that when he showed the clip to Mr. Letterman, the latter asked, "Why haven't we had that guy on before?" Mr. Letterman himself had a serious concern about the future of humanity in the face of climate change, so he felt hopeful that a solution might exist.

15 MY JOURNEY

I was humbled about the thought of being on the *Late Show*. Only a handful of scientists had appeared on a late-night show on a major network. One who I deeply admired was the late Professor Stephen Schneider, who was a climate scientist and ironically had an office a few doors down from me at Stanford University. He had appeared on *The Tonight Show Starring Johnny Carson* in 1977 to talk about climate science. Ever since I heard about Stephen Schneider doing this, I thought that being able to speak on such a stage about science would be almost impossible, but amazing if it could happen. When I was offered an appearance on the show, I felt a weight and responsibility. I needed to communicate clearly and accurately, given that Stephen was such an effective communicator, and I had nothing like his skills.

I appeared on the show on the evening of October 9, 2013.[417] I spent the whole day training in New York City with Mark Ruffalo and others. Mark came with me backstage to the *Green Room* to help prepare me further. I thought I had everything under control. My time slot was right after Lucy Liu and before that of the musical guest, the Weeknd. I ended up speaking for 11 minutes of the show. This was a long interview for the *Late Show*.

As I was walking toward the stage with about 30 seconds to go before appearing live, the *Late Show* band music was playing, and the lights were blinding. Even though I had been teaching for over two decades and had given hundreds of talks and had prepared all day, I panicked. I was overthinking what I wanted to say, and all of a sudden, I had too much running through my head. Then, the thought "I'm going to be speaking to millions of people" crept into my head. With seconds ticking by, all I could think about was, "This is going to be a disaster." I kept walking toward the stage in a panic, as if I were walking to my death.

Then, with about four seconds to go, I remembered one of my tennis lessons: Keep your eye on the ball. I thought to myself, "Ok, relax. Take a breath. Just focus on one thing to say. Say something about why I want to change the energy system." Because my concern about air pollution is the reason that I started this career, I immediately knew I should just focus on this topic. I decided to say something about how many people die from air pollution. As morbid as that sounds, I felt that few people were aware of this fact, and it was important. I believed that if I said that, everything else would flow.

When I finally sat down and the music died down, I was still jittery but felt more under control. I also remembered what the producer had told me, "Look toward Dave and keep your hands planted." Mr. Letterman made me feel at ease and threw me a softball statement to respond to, "Tonight, you have something positive that you can present to all Americans." My response was, "So Dave, we're developing science-based plans to eliminate global warming, air pollution, including the 2.5 to 4 million deaths that occur worldwide each year ..." His immediate response was, "Due to air pollution? That many people are dying due to air pollution?" Once he asked that, I knew I had made the right decision to focus on that topic, because it grabbed his attention and likely the attention of those watching. After that, I relaxed, and everything flowed for the rest of the interview.

The response to my being on the show was overwhelmingly positive. The New York governor's staff were watching, as were people interested in energy and the environment across America. Our Solutions Project student group at Stanford University had organized its own viewing party, which humbled me further. The show brought 100 percent clean, renewable energy to the large-scale public sphere. The concept of 100 percent WWS was no longer just a niche idea. It had begun to capture the public's imagination and was accelerating into a movement.

Soon after the show, a new star entered The Solutions Project scene. On November 26, 2013, Marco Krapels hosted another dinner in San Francisco, where he introduced me and others to Leilani Munter. Leilani was a racecar driver who was passionate about clean, renewable energy. She enthusiastically embraced the 100 percent goal and wanted to bring it to 80 million US racecar fans who normally would not care about energy. Over the years, since that meeting, she has been a spokesperson for electric racing and electric-powered racing stadiums. She also graciously volunteered her time at events and offered to speak out whenever possible about 100 percent WWS.

15.13 IMPACT OF CALIFORNIA ROADMAP ON CALIFORNIA LAW

The California roadmap paper was ultimately published on July 22, 2014.[334] Once it became public, the press began writing articles about it, and some articles reached the office of California Governor Jerry Brown. On August 22, 2014, Mr. Brown's senior advisor on energy and environmental issues, Cliff Rechtschaffen, emailed me stating,"We read with interest about your recent paper on a 2050 renewables strategy for California. I'm wondering if you would be interested in coming up to brief a group of advisors and policymakers who work on climate and energy issues in the administration." Needless to say, I was ecstatic at the thought because the whole goal of our work was to perform science in order to inform policymakers about solving air-pollution and climate problems. Air pollution alone in California caused at the time about 13,000 deaths per year. This was also a chance to inform the staff of the governor of the fifth-largest economy in the world about a pressing global and local issue.

On October 27, 2014, Marco Krapels and I trekked to Sacramento to present our 100 percent WWS roadmap for California. The group we presented to included Mr. Rechtschaffen, Mr. Brown's senior policy advisor Ken Alex, California Energy Commissioner David Hochschild, and others. I laid out the 100 percent roadmap for California in detail. The roadmap called for an 80 percent transition of all energy sectors by 2030 and 100 percent by 2050. Marco and I were peppered with lots of good questions. We left feeling like we gave it our best shot but had no idea at the time what the impact might be. We did know that nothing would happen right away, if at all, because Mr. Brown was running for reelection, and the election was coming up in 11 days.

Mr. Brown was reelected. During his inauguration speech, on January 5, 2015, he pleasantly surprised us by proposing several laws that followed logically from our discussion and our California roadmap. He proposed that at least 50 percent of all electricity in the state of California should come from WWS sources by 2030. He also proposed reducing petroleum use in vehicles purchased in the state by 50 percent by 2030, which meant electrifying 50 percent of the vehicle fleet by 2030. We had proposed 80 percent by 2030. He finally proposed doubling the efficiency of existing buildings by 2030. We had proposed providing all heating, cooling, and electricity in buildings with WWS and reducing energy use through energy-efficiency measures. Whereas the oil industry gutted Mr. Brown's proposal to reduce petroleum use, California did pass the 50 percent WWS portfolio standard and the energy-efficiency standard through California SB 350.

In another surprise, on September 19, 2016, Governor Brown signed California SB 1383, which required a 50 percent reduction in black-carbon emissions below 2013 levels by 2030. After two decades of work on black carbon, my scientific results were finally paying off with specific, enforceable legislation.

Between 2016 and 2018, the 100 percent movement swelled nationally. The Solutions Project was central to this expansion. They, along with the Sierra Club and a few other key groups mobilized almost 100 nonprofits in a group later called the **100 Percent Network**, around the 100 WWS percent goal. The main purposes of The Solutions Project had shifted slightly, not only to raise awareness that 100 percent WWS is possible but also to bring 100 percent WWS to 100 percent of the people.

To amplify the latter point, The Solutions Project created a subgroup called **100.org**. The purpose of 100.org was to provide small grants for small organizations or individuals who had creative ideas about how to engage people to transition to 100 percent WWS, especially people who might not otherwise prioritize a transition. The group 100.org also provided awards to honor transition leaders and people with innovative ideas. It further sought an **equitable transition** to 100 percent WWS. This involved, for example, proposing that at least a certain fraction of new jobs during a transition would be reserved for disadvantaged communities. Through The Solutions Project and 100.org, communities in many inner cities suddenly became part of the 100 percent movement.

Due to the groundswell of public support for 100 percent WWS in California following the passage of SB 350, State Senator Kevin DeLeon proposed a follow-up law, **SB 100**. The law called for 60 percent of all grid electricity in the state to come from eligible renewables by 2030 and the remaining 40 percent to come from eligible renewables, large hydro, or other zero-carbon technologies not yet invented, by 2045.

In California, eligible renewables include several WWS technologies – wind, solar, geothermal, and small (run-of-the-river) hydro – but not large (conventional) hydro. However, because California has so much existing large hydro and imports more from Oregon, Washington State, Idaho, and Canada, and because state lawmakers wanted to maintain large hydro as part of their 100 percent goal, it was necessary to create a separate category from eligible renewables to put large hydro in. So, the last 40 percent of SB 100 allows for eligible renewables plus "zero-carbon resources," which means primarily large hydro. The California Senate states that other possible zero-carbon technologies include "new technologies we may not know about today."[418] Notwithstanding nonexistent technologies, California's SB 100 indicates one way the state can meet 100 percent of electricity supply is with 100 percent WWS technologies.

> **Transition highlight**
> SB 100 was signed into law on September 10, 2018. This was a landmark legislation mandating the transition to effectively 100 percent WWS in the state's electric-power section. The passage of this law was the kind of result hoped for when we embarked on the first 100 WWS percent roadmap in 2009.

15.14 STATE AND COUNTRY ROADMAPS AND PARIS CONFERENCE

While developing Washington State and California roadmaps during October 2013, I came to realize that at the rate we were creating roadmaps one state at a time, it could take decades to finish all 50 states. I then decided to automatize the process. I expanded the Washington State roadmap spreadsheets to include all 50 states. This began my journey of developing 50 individual state roadmaps. These were published 20 months later, in June, 2015.[190] Once we finished the first draft of the 50-state paper during February 2014, Jon Wank created infographics for each state that were posted on a clickable map online.[413] These maps were invaluable because they were simple and informative.

A ripe opportunity soon arose for us to use the infographics. On August 27, 2014, Leilani Munter, Marco Krapels, Mark Ruffalo, Brandon Hurlbut (a Solutions Project board member), Andres Lopez (a Solutions Project staff member), and I went to the White House to meet with Vice President Joe Biden. The purpose of our visit was to provide information to the Vice President about our 100 percent roadmaps. He was taking a leading role in issues related to the US renewable energy infrastructure. Ahead of the meeting, we provided his staff with an infographic of our 100 percent WWS roadmap for Delaware, the state Mr. Biden had represented for 36 years in the US Senate. We were told that we had only 30 minutes to speak with him, so we each prepared short remarks.

Mr. Biden came in 15 minutes late and apologized for being tardy. The first thing he said, before we could put in a word, was that he could stay only a few minutes but that he fully supported our 100 percent WWS roadmaps. He cautioned, though, that the political landscape was such that it would be difficult for the federal government at the time to take any action on the roadmaps. We later chuckled, because Mr. Biden ended up staying with us for an hour and talked almost the entire time. He was endearing and easy to listen to. Halfway through, he said he really had to go and started to leave. I asked a parting question, and he came back in the room and spoke for another 30 minutes. For me, the best thing that came out of it was the feeling he was on board with solving the problem in a big way.

On September 21, 2014, New York held a climate march, attended by about 400,000 people. I was not able to attend, but The Solutions Project had a large and central presence there. Leonardo DiCaprio joined Mark Ruffalo, Marco Krapels, Jon Wank, Leilani Munter, and Brandon Hurlbut from The Solutions Project. The United Nations building itself was even lit up with the sign, "Solutions Exist."

Mr. DiCaprio, who was good friends with Mark Ruffalo, was passionate about what The Solutions Project stood for. He wanted to play a key role in helping to disseminate research results on our behalf. To that end, on September 23, 2014, he spoke in front of the United Nations General Assembly after being designated as a United Nations Messenger of Peace.

15.14 STATE AND COUNTRY ROADMAPS AND PARIS CONFERENCE

He spoke about climate change and about the solutions we had developed. He stated in front of the world body, "New research shows that by 2050 clean, renewable energy could supply 100 percent of the world's energy needs using existing technologies, and it would create millions of jobs."

A month later, on October 29, 2014, Leo DiCaprio came to Stanford University to interview me and several students working with me on state roadmaps. He was producing a documentary called *Before the Flood*. He and director Fisher Stevens spent the day filming with us. Marco Krapels was there as well. We all saw firsthand the passion that the students had for trying to solve the climate, pollution, and energy problems we face. The documentary ended up taking a different direction so did not focus on solutions, but some of the interviews were used as a postscript on the DVD version of the film. This event motivated our students for months after.

During 2015, I was not only completing the 50-state roadmaps but had also embarked on an even more ambitious project, which was to develop roadmaps for most countries of the world. Although we had developed roadmaps for the world as a whole in 2009 and 2011, individual countries could not practically implement a world plan, so country roadmaps were needed. The work on country roadmaps began during August 2014. Our group at Stanford first put together a sample roadmap for Ukraine. We expanded this in earnest to 139 countries by March 2015. This was the number of countries for which we could find raw energy data from the International Energy Agency. Work on the country roadmaps continued all year but accelerated feverishly to meet a deadline we set for ourselves, the end of November 2015, in order to have the roadmaps ready for the Paris climate conference.

In December, Paris was hosting the United Nations Climate Change Conference, referred to as the Conferences of the Parties 21 (COP 21). This was a big international event and a ripe opportunity to disseminate our country roadmaps to world leaders. The event lasted two weeks, but I could stay for only a few days. I met Marco Krapels there and gave several talks but only one major one. Due to a quirk in scheduling, I was fortunate to be able to give a talk at the Petit Palais, sandwiched in between a talk by the UN Secretary General, Ban Ki-moon, and another one by US Secretary of State, John Kerry. I later quipped to myself that I was the only speaker who did not have a security detail. I had only a few minutes to talk but made the most of it. I laid out for the first time to the world our 100 percent all-sector WWS energy roadmaps for 139 countries, which emitted about 99.5 percent of all world carbon dioxide.

In Paris, many nongovernmental organizations, including those that helped build the 100 percent network, supported the 100 WWS percent goal that The Solutions Project had been disseminating. Paris was filled with the spirit of 100 percent. The Eiffel Tower was lit up with the words "100% Renewable," and a group even formed a human sign stating "100% Renewable." I felt that the 100 percent movement had hit the world stage.

15.15 IMPACTS OF ROADMAPS ON US AND BUSINESS POLICIES

Shortly after completing and publishing the 50-state roadmaps in 2015, we published a companion paper that examined whether it was possible to match power demand among all energy sectors with 100 percent WWS supply continuously among the 48 contiguous US states.[336] The new paper concluded, yes, it was possible. The paper received an award, the Cozzarelli Prize, from the journal it was published in, the *Proceedings of the National Academy of Sciences*.

The combination of the 50-state roadmaps and the grid-reliability study gave confidence to lawmakers and politicians to propose legislation at the national level in the United States on 100 percent clean, renewable energy.

On November 30, 2015, the first of several proposed US federal laws or resolutions calling for 100 percent WWS was introduced into the US Congress. US House Resolution 540 called for the United States to transition to 100 percent clean, renewable energy for all energy sectors. The resolution summary stated, "Expressing the sense of the House of Representatives that the policies of the United States should support a transition to near zero greenhouse gas emissions, 100 percent clean, renewable energy."[419] It was humbling to see the text of the resolution itself acknowledged our work as its scientific basis: "Whereas a Stanford University study concludes that the United States energy supply could be based entirely on renewable energy by the year 2050 using current technologies." US policymakers on the national stage were realizing a solution to the horrendous problems of air pollution, global warming, and energy security was possible. This progress continued. Between 2015 and 2019, seven more resolutions and bills were introduced into the US House of Representatives and Senate proposing that the US go to 100 percent clean, renewable WWS energy in one or more energy sectors (Senate Resolution 632 in 2016, Senate Bill 987 in 2017, House Bill 3314 in 2017, House Bill 3671 in 2017, House Bill 330 in 2019, House Resolution 109 in 2019, and Senate Resolution 59 in 2019). None of these resolutions or bills was ever voted on, but they educated the public and policymakers, fanning the 100 percent movement.

The last two of these resolutions (House Resolution 109 and Senate Resolution 59 in 2019) were the Green-New-Deal proposals for the US to

15.15 IMPACTS OF ROADMAPS ON US AND BUSINESS POLICIES

go to 100 percent clean, renewable, zero-emission energy for all purposes by 2030. Overall, these resolutions contained proposals related to energy, jobs, health care, education, and social justice. In terms of energy, the resolutions called for "100 percent clean, renewable, and zero-emission energy sources" for all energy sectors by 2030 throughout the United States. The 100 percent goal and the 2030 deadline both originated[420,421] from our *Scientific American* paper[331] and New York State energy roadmap paper.[333] Both papers concluded that 100 percent WWS by 2030 is technically and economically feasible, but the 2009 paper cautioned that, for social and political reasons, a complete transition by 2030 was unlikely and may take longer to achieve.

Meanwhile, during 2016, the US was going through a presidential election process. On the Democratic side, the three major candidates were Governor Martin O'Malley, Senator Bernie Sanders, and Senator Hillary Clinton. The 100 percent movement had taken off to such a degree that all three candidates supported it.

Governor O'Malley was the first. On July 2, 2015, he issued a press release, where the number one item stated on his platform was "A complete transition to renewable energy … by 2050."[422]

Next, Senator Sanders formulated a platform calling for a transition "toward a completely nuclear-free clean energy system for electricity, heating, and transportation." To illustrate, he put The Solutions Project 50-state infographic map on his campaign website. Each state infographic showed what a 100 percent WWS system for all purposes would look like in the state in 2050.

Third, although Senator Clinton also supported the temporary use of fossil gas, she stated publicly in a video on October 16, 2015, "We need to be moving as quickly as possible to 100 percent clean, renewable energy. We have a long way to go, but that should be our goal, and we should do nothing that interferes with or undermines our efforts to reach that goal as soon as possible."[423]

The support for 100 percent WWS by the three Democratic presidential candidates culminated in the US National Democratic Party platform mimicking this sentiment by stating, "We believe America must be running entirely on clean energy by mid-century."[424] Thus, the 100 percent movement had grown to a point where the largest US party had adopted the 100 percent goal.

Just two weeks before the presidential election of November 7, 2016, I received a phone call from Senator Sanders. He wanted me to provide him with more details about our 100 percent roadmaps for the 50 states, which were prominent on his campaign website, because he was planning to submit a bill to the Senate calling for the US to go to 100 percent renewables. He subsequently submitted the bill on April 27, 2017. The bill (S.987) was co-sponsored by Senators Merkley, Markey, Booker, and Schatz.

The phone call with Senator Sanders lasted about 45 minutes. I asked him whether he wanted me to suggest policies that might help a 100 percent law to be effective. His reply, in his characteristic voice, was, "No, no, no. That's our job." We further discussed other action that could be taken to educate the public about 100 percent renewables. One such suggestion he had was to write a joint op-ed. Two days after he submitted his bill, S.987, in the Senate, we published a joint op-ed in the *Guardian*, entitled "The American people – not Big Oil – must decide our climate future."[425]

Meanwhile, the 139-country roadmaps were published on September 6, 2017,[284] and reported widely in the press. The Solutions Project then developed infographics for each country summarizing the roadmaps.[413]

Transition highlight
The international roadmaps helped to give confidence to countries that they can each reach 100 percent clean, renewable energy in all energy sectors. As of the end of 2023, seven countries had 100 percent renewable energy policies across all energy sectors: Barbados (by 2030), Comoros (by 2030), Fiji (by 2030), Guyana (by 2025), Nauru (by 2050), the Philippines (by 2050), and Uganda (by 2050).[426] Another 63 countries had net-zero carbon emission targets across all energy sectors by 2025 to 2050.[426] Thus, a total of 69 countries had either 100 percent renewable or net-zero carbon emission targets across all sectors. What is more, 90 countries had some renewable energy target (between 9 and 100 percent) across all sectors by 2025 to 2050. A total of 170 countries had targets in the electricity sector, 43 had targets in the heating and cooling sector, and 49 had targets in the transport sector.[426] In all countries, effective policies are needed to reach the goal.

On November 13, 2017, results from an international public opinion poll about renewable energy were reported.[427] Twenty-six thousand people in 13 countries (Canada, China, Denmark, France, Germany, Japan, Netherlands, Poland, South Korea, Sweden, Taiwan, the United Kingdom, and the United States) were queried. The poll found that 82 percent of all the people who were asked responded that it is "important to create a world fully powered by renewable energy." In the United States, 83 percent supported that statement.

Interestingly, only 66 percent of people in the same poll believed climate change was a significant international problem. The reasons that more

15.15 IMPACTS OF ROADMAPS ON US AND BUSINESS POLICIES

people believed in renewable energy solutions than in the damaging effects of climate change could be extracted as follows: (a) 75 percent of people polled believed their country's renewable technology leadership gave them a sense of pride in their county; (b) 73 percent believed that building and producing renewable energy increased economic growth and jobs; and (c) 69 percent believed that renewables made their country more energy independent. However, only 53 percent believed renewable energy reduced health problems. Ironically, that is their greatest benefit. The results of this poll suggest that people don't need to believe in the problem (climate change), although they should, to believe in the solution (clean, renewable energy). This is one major reason why I like to focus on the solution rather than the problems – more people believe in the solution than the problems.

In fact, because support for renewables is so strong, it crosses political party lines. In the United States, for example, most of the 10 states with the largest amount of wind electricity installed have a population that generally votes for Republicans, whose base has traditionally been the most skeptical about climate change and renewables. Wind farms are installed in those states because of their low cost.

Following the publication of our 139-country roadmap paper, we performed a follow-up study to determine whether matching continuous all-sector power demand with 100 percent WWS supply was possible in 20 world regions encompassing the 139 countries. This study was published in early 2018.[181] It supported the previous grid integration study from 2015 for the US,[336] but it also showed that matching demand with supply and storage on the grid at low cost is possible throughout the world under a variety of conditions beyond those shown in the earlier US study.

Both the 139-country roadmap study and the 20-world-region grid integration study were subsequently updated in 2019 and 2021 with roadmaps and grid studies for 143 countries placed in 24 world regions.[337,303,311] Those studies were updated further in 2022 with roadmaps and grid studies for 145 countries grouped into 24 world regions.[4,65,96] They were updated again in 2024 for 149 countries grouped into 29 world regions.[5,43] A clickable infographic map linking each country to a summary of the 100 percent WWS plan for the country was subsequently developed.[428] All such studies support the potential for a low-cost transition throughout the world.

Similarly, in 2022, the 2015 50-state roadmaps and grid studies for the US were updated with a new study that examined whether the grid could stay stable in all individual US grid regions and in six isolated states.[311] The study found that transitioning the US to 100 percent WWS may more than double electricity use but reduce total end-use energy demand by about

57 percent versus a business-as-usual case. The energy reduction contributed to a 63 percent lower annual private energy cost and 86 percent lower annual social energy cost with WWS than with business-as-usual, averaged over the US. The study also calculated that the costs per unit of energy in California, New York, and Texas may be 11, 21, and 27 percent lower, respectively, when these states are interconnected regionally rather than islanded. Finally, it found that transitioning to WWS for all energy purposes may create about 4.7 million more long-term, full-time jobs than are lost and require only about 0.29 percent of US land for footprint and 0.55 percent of US land for spacing for new WWS electricity and heat generators. This 0.84 percent total land requirement is less than the 1.3 percent occupied by the fossil industry and another 1.24 percent occupied by the agricultural industry for growing corn for ethanol-fueled vehicles.

During 2018, the policy momentum continued toward 100 percent WWS. For example, following the November 2018 US elections, five new governors (in Colorado, Maine, Connecticut, Illinois, and Nevada) were elected whose platforms called for taking their respective states to 100 percent renewable electricity.

Subsequently, on December 18, 2018, Washington, DC, joined two states (Hawaii and California) by passing an enforceable 100 percent renewables law. Hawaii had previously passed a law on June 8, 2015, to go to 100 percent renewable electricity for the entire state by 2045. California had passed SB 100 to do the same on September 10, 2018. Washington, DC's law, though, was to go to 100 percent renewable electricity generation sooner, by 2032.

Then, on March 22, 2019, New Mexico passed a law to go to 80 percent renewable electricity by 2040 and up to 100 percent by 2045.

A few weeks later, on April 11, 2019, the US territory Puerto Rico passed a law to go to 100 percent renewable electricity by 2050.

On April 19, 2019, Nevada passed a law to go to 50 percent renewable electricity by 2030 and up to 100 percent renewable (termed "carbon-free") electricity by 2050.

On May 8, 2019, Washington State enacted a law to go to 100 percent renewables for all new electricity by 2045. The bill states (in its Section 6.a.iii) that to meet the law, electric utilities, when acquiring new resources, must "rely on renewable resources and energy storage."

On May 23, 2019, New Jersey's governor signed an executive order requiring the development of a plan for the state to reach 50 percent renewable electricity by 2030 and "100 percent clean energy by 2050." Clean energy can be renewable electricity or electricity from another technology defined by the state to be "clean."

15.15 IMPACTS OF ROADMAPS ON US AND BUSINESS POLICIES

On June 26, 2019, Maine passed a law to go to 80 percent renewable electricity by 2030 and 100 percent by 2050. An updated law, calling for 100 percent renewables by 2040, was proposed to the state legislature in November 2024.

Not to be outdone, New York passed a law on July 18, 2019, to go to 70 percent renewable electricity by 2030 and effectively to 100 percent by 2045. The law also calls for an 85 percent across-the-board greenhouse-gas-emission reduction in the state by 2050.

On August 16, 2019, the governor of Wisconsin signed an executive order expressing a goal for the state to go 100 percent carbon-free electricity, thus up to 100 percent renewable electricity, by 2050, and ordered state agencies to develop a plan to do this.

On September 3, 2019, the governor of Connecticut signed an executive order for the state to develop a plan to go to 100 percent zero-carbon electricity, thus up to 100 percent renewable electricity, by 2040.

The domino effect continued. On September 17, 2019, the governor of Virginia signed an executive order expressing a goal for the state to reach 30 percent renewable electricity by 2030 and 100 percent carbon-free electricity, thus up to 100 percent renewable electricity, by 2050.

On January 17, 2020, the governor of Rhode Island signed an executive order for the state to go to 100 percent renewable electricity by 2030. This was updated on June 29, 2022, by Governor Dan McKee, who signed into law a bill requiring 100 percent renewable electricity in the state by 2033.

On July 27, 2021, Oregon's governor signed a law for the state to go to 100 percent renewable electricity by 2040, with 80 percent by 2030.

On September 15, 2021, Illinois' governor signed a law for the state to provide 100 percent of its electricity from renewable or zero-carbon sources by 2045, with 40 percent from renewables by 2030.

On October 13, 2021, North Carolina governor Roy Cooper signed a law requiring the state to cut electricity-sector carbon emissions 70 percent by 2030 and reach carbon neutrality by 2050.

On December 9, 2021, Nebraska's largest utility, the Nebraska Public Power District, voted for a nonbinding 100 percent decarbonization goal of the utility by 2050. The only two other utilities in Nebraska, the Omaha Public Power District and the Lincoln Electric System, had already committed to decarbonization by 2050 and 2040, respectively. As such, all three utilities in Nebraska have a zero-carbon goal, which can be met with 100 percent WWS.

On February 7, 2023, Minnesota governor Tim Walz signed a new law requiring 100 percent of the state's electricity generated or procured for use in Minnesota to be from carbon-free sources by 2040.

On November 28, 2023, Michigan Governor Gretchen Whitmer signed into law a bill requiring 100 percent carbon-free electricity generation by the state's utilities by 2040.

On June 19, 2024, the Vermont House and Senate overrode the governor's veto of a 100 percent renewable electricity standard by 2035, passed by both the House and the Senate, making that standard law.

In sum, as of the end of 2024, 19 US states plus Washington, DC, and Puerto Rico had laws or executive orders to go up to 100 percent WWS in their electric-power sector. Nine more states without such laws or orders (South Dakota, Montana, Iowa, Kansas, Wyoming, Oklahoma, North Dakota, Idaho, Texas) produced 33–118 percent of the electricity they consumed from WWS in 2024. This suggests that WWS can grow even without policies supporting it. The reason is its low cost: Eight of the nine no-policy, high-WWS states were among the 17 states with the lowest residential electricity prices in 2024.[257]

Not only have states passed 100 percent WWS laws, but so have over 200 US cities, towns, and counties.[416] Worldwide, by the end of 2020, 617 cities had 100 precent renewable energy targets, and 1,327 cities (with a combined population of over one billion) had some type of renewable energy target or policy.[429]

> **Transition highlight**
> The 100 percent fever caught on not only with state and local governments but also with utilities. Ten US states (California, Illinois, Maryland, Massachusetts, New Hampshire, New Jersey, New York, Ohio, Rhode Island, and Virginia) have laws permitting community-choice-aggregation (CCA) utilities. These utilities take over the electricity-generation portion of a utility bill. Another utility handles the transmission and distribution portion. Starting on March 7, 2010, the California community choice utility, Marin Clean Energy, began offering the option for customers to purchase up to 100 percent of their electricity from clean, renewable WWS sources. Subsequently, the number of community choice utilities offering 100 percent WWS electricity accelerated rapidly across not only California, which itself has 25 CCAs, but also across most other states that allow community choice utilities.

Surprisingly, many traditional utilities also began offering 100 percent WWS options in 2019. For example, in 2019, Xcel Energy, which services Minnesota, Michigan, Wisconsin, North Dakota, South Dakota, Colorado, Texas, and New Mexico, set up an option, called Windsource, that allowed customers to obtain "all of their (electrical) energy from renewables." On August 20, 2019, Appalachian Power even opened a "Wind, Water, and Sunlight (WWS) service" to supply 100 percent renewable power to any customer in

15.15 IMPACTS OF ROADMAPS ON US AND BUSINESS POLICIES

Virginia.[430] On August 21, 2019, Duke Energy obtained approval to allow military, University of North Carolina, and large nonresidential customers to procure 100 percent WWS electricity directly from WWS suppliers.[431]

On another front, during the climate march on September 21, 2014, in New York City, and five years after our *Scientific American* paper on worldwide 100 percent WWS roadmaps, a campaign called RE100 was launched by two nonprofits, Climate Group and CDP. The original purpose of the campaign was to commit 100 companies to 100 percent renewable electricity in the annual average for their global operations by 2020. The first companies to commit were Ikea, Swiss Re, Mars, and H&M. By May 2025, 446 international companies had joined, but with different compliance dates.[432] Among them are 7 out of the 10 largest companies in the world in August 2024 in terms of total value of a company's shares of stocks (Apple, Microsoft, Alphabet, Amazon, Meta, Eli Lilly, and TSMC) as well as 15 of the 25 largest companies (those seven, plus JP Morgan Chase, Walmart, Novo Nordisk, Visa, Mastercard, Procter & Gamble, Johnson & Johnson, and Samsung).

The first company to reach its 100 percent renewables goal for electricity generation was Google, in 2017. This felt rewarding, because Mark Ruffalo, Marco Krapels, and I had presented our first public Solutions Project vision of reaching 100 percent WWS, by combining science, business, and culture, at Google on June 20, 2012.[414] Apple met the 100 percent goal for its global operations during April 2018. Companies have made a difference. Through March 30, 2024, US corporations committing to 100 percent WWS had invested in a cumulative 40 gigawatts of solar PV, or 18 percent of all US PV installations, and 0.57 gigawatts/1,800 gigawatt-hours of battery storage.[433]

Between September 2017 and September 2024, the Climate Group went further, galvanizing 128 international companies to commit to transitioning their company vehicles to 100 percent battery-electric by 2030.[434] On September 19, 2019, for example, Amazon committed to replacing 100,000 delivery trucks with electric trucks. The companies also committed to installing charging stations. Transitioning company vehicles is important, given that up to half of all vehicles on the road in many countries are company-owned vehicles.

Transition highlight
From September through November 2019, the California cities of Berkeley, Menlo Park, and Mountain View banned the use of fossil gas in new residential and/or commercial buildings. Mountain View even required that 100 percent of parking spaces in multiunit housing and

commercial developments have the electrical infrastructure available for battery-electric vehicle charging. During July 2021, Menlo Park went further by proposing an ordinance requiring residents to remove existing fossil-gas appliances (stoves, heaters, dryers) from their homes and offering financing to do this. Banning fossil gas and facilitating vehicle charging are necessary steps toward a fully electrified society. Full electrification is needed for WWS to supply all-purpose energy. By early 2023, over 70 jurisdictions in California and several jurisdictions in six other states (Colorado, Maryland, Massachusetts, New York, Vermont, and Washington State) and Washington, DC, banned fossil gas in new-construction buildings. An appeals court overturned Berkeley's ban in May 2024, prompting other cities in California to pause their bans as well. In May 2023, though, New York became the first state to ban fossil gas and other fossil fuels in most new buildings.

15.16 CLIMATE LAWSUITS

Another tool used to motivate a transition to clean, renewable energy is through the legal system. In 2016, I was asked by a nonprofit legal group, Our Children's Trust, to testify as an expert witness on behalf of 21 youth plaintiffs in a lawsuit, *Juliana v. United States*, against the US federal government. The lawsuit was filed in the District Court for the District of Oregon by Julia Olson, Andrea Rodgers, and Philip Gregory, attorneys representing the youth plaintiffs, and asserted that the US government knowingly violated the youths' due process rights of life, liberty, and property by allowing fossil-fuel mining and combustion. As relief, the youths requested the court to offer "declaratory and injunctive relief" to require the government to protect the atmosphere and prevent the government from taking action to contradict this goal. The purpose of my testimony, articulated in an expert declaration, was to discuss whether fossil fuels were necessary to power the main US energy sectors – electricity, transportation, buildings, and industry – among all 50 United States. Based on the 100 percent WWS energy roadmaps I had developed across all 50 states, I testified that fossil fuels were no longer necessary. Unfortunately, in 2020, the US Ninth Circuit Court of Appeals dismissed the case before trial, finding that the youth plaintiffs lacked standing to sue for an injunction.

Subsequently, however, Our Children's Trust filed lawsuits in state court in Montana, Hawaii, and elsewhere. The Montana case, *Held et al. v. The State of Montana et al.*, went to trial in June 2023. At issue was whether the Montana state government violated the youth plaintiffs' state constitutional rights to a safe and clean environment by promoting fossil fuels. I testified in person on June 15, 2023, in Helena, Montana, that the state could readily transition to 100 percent WWS across all energy sectors, so fossil fuels were no longer needed for energy. The trial court ruled in favor of the youth plaintiffs, resulting in the first court victory worldwide in a climate case based on constitutional grounds. The state appealed, but on December 18, 2024, the Montana Supreme Court upheld the trial court's decision.

In the Hawaii case, *Navahine v. Department of Transportation, State of Hawaii*, the youth plaintiffs similarly argued that the state government violated the youths' constitutional right to a safe and clean environment. In this case, the lawsuit focused on the transportation sector in Hawaii, which produces 72 percent of all greenhouse gas emissions in the state. After submitting an expert report on January 26, 2024, arguing that Hawaii has the ability to transition its transportation to electricity powered by 100 percent clean, renewable sources, and after being deposed by the state's attorneys, I was relieved to hear that the state settled the case with the youth plaintiffs. The settlement called for the state to transition all land, marine, and inter-island air transport in Hawaii to zero emissions (thus, effectively all-electric), in five-year increments, by 2045.

The lesson for me from these cases is that the legal system, while slow and often ineffective at helping to effect a transition to 100 percent WWS, has proven instrumental in some key cases. Additional climate-related cases in the US, Europe, and elsewhere have since been adjudicated, and others are ongoing. Past cases have similarly resulted in a mixture of results. At this point, though, every avenue needs to be explored if we hope to transition the world to clean, renewable energy for all purposes in the time necessary.

15.17 CONCLUSION: WHERE DO WE GO FROM HERE?

Prior to the 2009 *Scientific American* paper, the public discussion focused on whether it was possible to shift from fossil fuels to even 20 percent renewables without collapsing the grid. Our subsequent country, state, city, and town roadmaps and grid studies (summarized in new infographics[428]) – together with the 100 percent WWS movement spawned by The Solutions Project, its founders, members, and partner organizations, all reinforced by additional scientific papers by dozens of independent, international research groups – changed the discussion to *when* and *how fast* 100 percent would be implemented.

I am under no illusion we have finished solving the problem. Although progress has been made in the electricity sector, transitioning other sectors (transport, buildings, industry, agriculture/forestry/fishing, and the military) must be addressed in force as well. However, the goal posts have shifted closer to where they should be, and this was due to cooperation among people with diverse expertise and likeminded passion and goals. The progress made to date is far more than I thought it could have been when I first aspired to help solve air-pollution problems in the late 1970s. A solution to the problems of air pollution, global warming, and energy insecurity requires a large-scale conversion of the world's energy infrastructure. This book has described and analyzed many elements of such a conversion.

The main steps in a transition are to electrify and/or use direct heat for all energy sectors, and to provide the electricity and heat with clean, renewable wind, water, and solar energy and storage. The energy sectors that must be transitioned include the electricity, transportation, buildings, industry, agriculture/forestry/fishing, and military sectors. It is also necessary to eliminate nonenergy emissions.

WWS generators proposed for such a transition include onshore and offshore wind turbines, solar PV, solar heat collectors, CSP plants, geothermal electricity plants, geothermal heat collectors, tidal and ocean-current turbines, wave devices, and hydroelectric facilities. Storage types include electricity, heat, cold, and hydrogen storage. Short- and long-distance transmission is also needed. By combining electricity, heat, cold, and hydrogen generation with storage and demand response, it appears possible to match energy demand with supply on the electric-power grid worldwide, thereby avoiding blackouts with 100 percent WWS.

The resulting 100 percent WWS system worldwide will use an average of about 54 percent less end-use energy than a business-as-usual system.[4] The annual energy cost to consumers will be about 60 percent lower, on average. Because WWS also eliminates health and climate costs of energy, which together are four times the energy cost, the transition to a WWS system will reduce the social (economic) cost that consumers pay by about 90 percent worldwide.

A worldwide transition to WWS will not only reduce up to seven million air-pollution deaths annually and slow, then reverse global warming, but it will also create far more jobs than are lost, use less land than the current energy infrastructure, increase energy security and energy independence, and reduce terrorism risk to electricity grids.

15.17 CONCLUSION: WHERE DO WE GO FROM HERE?

Given the limited time and funding available to solve the world's climate, air-pollution, and energy-security problems, it is essential we focus on known, effective solutions. We should not waste money and allow more damage with inferior options, such as fossil fuels with carbon capture, synthetic direct air capture, blue hydrogen, nonhydrogen electro-fuels, bioenergy with or without carbon capture, small and large nuclear reactors, and geoengineering.

In sum, a concerted international effort can lead to a conversion of the world's energy infrastructure so that by between 2026–2030, the world will no longer build new fossil-fuel electricity-generation power plants or new land- or marine-based vehicles with internal combustion engines. Rather, it will manufacture new wind turbines, solar power plants, and battery-electric and hydrogen-fuel-cell-electric vehicles. Long-distance aviation may take longer to transition. As the WWS infrastructure grows, the remaining existing infrastructure will retire, so that ideally by 2035–2040, and no later than 2050, the world will be powered with 100 percent WWS for all energy purposes.

The main barriers to a conversion to WWS worldwide are not technical, resource based, or economic. Instead, they are social and political. The most difficult places to transition may be countries beset by conflict and countries that have high poverty rates. NIMBYism is also an issue but not a barrier that cannot be overcome. Whereas recycling will be beneficial for keeping costs down, there is no materials limit to a 100 percent WWS transition in all energy sectors.

Because most polluting energy technologies receive government subsidies or tax breaks or are not required to eliminate their emissions, they can run inexpensively for years. Wise policies can promote their rapid replacement with WWS. However, useful policies will be implemented only if policymakers are willing to make changes. Policymakers in democratic countries are elected, whereas those in autocratic countries make decisions dictated by one or a few leaders. Thus, in democratic countries, many people need to be convinced that changes will be beneficial; in autocratic countries, only a few people need to be convinced. In both cases, those who need to be convinced require a social evolution in their thought. Such an evolution comes from a better understanding of the science about, consequences of, and solutions to air pollution, global warming, and energy insecurity.

If the public and policymakers can become confident in understanding the problems and the large-scale solution needed to solve them, they will gravitate toward the solution. Thus, it is important for individuals, advocates, business leaders, scientists, and policymakers to come together to effect a change.

When the 100 percent WWS solution is finally implemented worldwide, the air-pollution and climate problems outlined in this book will be relegated to the annals of history.

References

1. Green, H. and W. Lane, *Particle Clouds*, Van Nostrand, Princeton, NJ, 1969.
2. WHO (World Health Organization), Household air pollution attributable deaths, 2022, https://tinyurl.com/3ht5j9n3 (accessed September 10, 2024).
3. WHO (World Health Organization), Ambient air pollution attributable deaths, 2022, https://tinyurl.com/2zhp4n9y (accessed September 10, 2024).
4. Jacobson, M. Z., A.-K. von Krauland, S. J. Coughlin, et al., Low-cost solutions to global warming, air pollution, and energy insecurity for 145 countries, *Energy & Environmental Sciences*, *15*, 3343–3359, doi:10.1039/d2ee00722c, 2022.
5. Jacobson, M. Z., D. Fu, D. J. Sambor, and A. Mühlbauer, Energy, health, and climate costs of carbon-capture and direct-air-capture versus 100%-wind-water-solar climate policies in 149 countries, *Environmental Science & Technology*, *59*, 3034–3045, doi:10.1021/acs.est.4c10686, 2025.
6. Jacobson, M. Z., Effects of biomass burning on climate, accounting for heat and moisture fluxes, black and brown carbon, and cloud absorption effects, *Journal of Geophysical Research*, *119*, 8980–9002, doi:10.1002/2014JD021861, 2014.
7. Canadell, J. G., P. M. S. Monteiro, M. H. Costa, et al., Global carbon and other biogeochemical cycles and feedbacks. In V. Masson-Delmotte, P. Zhai, A. Pirani, et al., eds., *Climate Change 2021: The Physical Science Basis. Contribution of Working Group I to the Sixth Assessment Report of the Intergovernmental Panel on Climate Change*, Cambridge University Press, Cambridge, UK, 2021.
8. NASA (National Aeronautics and Space Administration) Goddard Institute for Space Studies, Global temperatures, 2024, https://climate.nasa.gov/vital-signs/global-temperature/?intent=121 (accessed September 10, 2024).
9. Copernicus, June 2024 marks 12th month of global temperature reaching 1.5°C above pre-industrial, 2024, https://tinyurl.com/5e7ywdmu (accessed September 10, 2024).
10. Jacobson, M. Z., *100% Clean, Renewable Energy and Storage for Everything*, Cambridge University Press, New York, 2020.
11. Jacobson, M. Z., Strong radiative heating due to the mixing state of black carbon in atmospheric aerosols, *Nature*, *409*, 695–697, 2001.
12. Jacobson, M. Z., Control of fossil-fuel particulate black carbon plus organic matter, possibly the most effective method of slowing global warming, *Journal of Geophysical Research*, *107*(D19), 4410, doi:10.1029/2001JD001376, 2002.

REFERENCES

13. Jacobson, M. Z., Short-term effects of controlling fossil-fuel soot, biofuel soot and gases, and methane on climate, Arctic ice, and air pollution health, *Journal of Geophysical Research*, 115, D14209, doi:10.1029/2009JD013795, 2010.
14. Bond, T. C., S. J. Doherty, D. W. Fahey, et al., Bounding the role of black carbon in the climate system: A scientific assessment, *Journal of Geophysical Research*, 118, 5380–5552, doi:10.1002/jgrd.50171, 2013.
15. Jacobson, M. Z., On the causal link between carbon dioxide and air pollution mortality, *Geophysical Research Letters*, 35, L03809, doi:10.1029/2007GL031101, 2008.
16. Jacobson, M. Z., The enhancement of local air pollution by urban CO_2 domes, *Environmental Science & Technology*, 44, 2497–2502, doi:10.1021/es903018m, 2010.
17. Ritchie, H. and M. Roser, Access to energy, 2024, https://ourworldindata.org/energy-access (accessed September 10, 2024).
18. Vavrin, J. Power and energy considerations at forward operating bases (FOBs). United States Army Corps of Engineers, Engineer Research and Development Center, Construction Engineering Research Laboratory, 2010, https://apps.dtic.mil/sti/tr/pdf/ADA566876.pdf (accessed September 10, 2024).
19. Sambor, D. J., M. Wilber, E. Whitney, and M. Z. Jacobson, Development of a tool for optimizing solar and battery storage for container farming in a remote Arctic microgrid, *Energies*, 13, 5143, doi:10.3390/en13195143, 2020.
20. Krane, J. and R. Idel, More transitions, less risk: How renewable energy reduces risks from mining, trade and political dependence, *Energy Research and Social Science*, 82, 102311, 2021.
21. Memija, A., Mingyang's 20 MW offshore wind turbine stands complete, 2024, https://tinyurl.com/kvmk65fc (accessed September 10, 2024).
22. Durakovic, A., Fixed bottom offshore wind farms 90 metres deep? Offshoretronics says yes, 2021, https://tinyurl.com/47ffk2t8 (accessed September 10, 2024).
23. US DOE (US Department of Energy), Pathways to commercial liftoff: Next-generation geothermal power commercial liftoff, 2024, https://liftoff.energy.gov/next-generation-geothermal-power/ (accessed September 10, 2024).
24. US DOI (US Department of the Interior), Reclamation: Managing water in the west; hydroelectric power, 2005, www.usbr.gov/power/edu/pamphlet.pdf (accessed September 10, 2024).
25. Rahi, O. P. and A. Kumar, Economic analysis for refurbishment and uprating of hydropower plants, *Renewable Energy*, 86, 1197–1204, 2016.
26. Free Dictionary, Installed capacity: Definition, 2019, https://encyclopedia2.thefreedictionary.com/Installed+Capacity (accessed September 10, 2024).
27. Corbley, A., These underwater "kites" are generating tidal electricity as they move, 2021, https://tinyurl.com/jt4t66y4 (accessed September 10, 2024).
28. Bellini, E., Flexible solar panel for vehicle-integrated applications, 2021, https://tinyurl.com/mtj25ncr (accessed September 10, 2024).
29. Lee, K., H.-D. Um, D. Choi, et al., The development of transparent photovoltaics, *Cell Reports*, doi://10.1016/j.xcrp.2020.100143, 2020.

REFERENCES

30. Onyx solar, Website, 2024, https://onyxsolar.com (accessed September 10, 2024).
31. Renovagen, Rollable solar panels, 2021, www.renovagen.com (accessed September 10, 2024).
32. REI (Resilient Energy & Infrastructure), Website, 2024, https://resilientei.com (accessed September 10, 2024).
33. Dragonwings, Website, 2024, www.dragonwings.co (accessed September 10, 2024).
34. Kougias, I., K. Bodis, A. Jager-Waldau, et al., The potential of water infrastructure to accommodate solar PV systems in Mediterranean islands, *Solar Energy*, *136*, 174–182, doi:10.1016/j.solener.2016.07.003, 2016.
35. Hussain, A., A. Batra, and R. Pachauri, An experimental study on effect of dust on power loss in solar photovoltaic module, *Renewables: Wind, Water, and Solar*, *4*, 9, 2017.
36. DOE (US Department of Energy), *Concentrating Solar Power Commercial Application Study: Reducing Water Consumption of Concentrating Solar Power Electricity Generation*, Report to Congress, 2008, www1.eere.energy.gov/solar//pdfs/csp_water_study.pdf (accessed September 10, 2024).
37. Nonbol, E., Load-following capabilities of nuclear power plants, Technical University of Denmark, 2013, http://orbit.dtu.dk/files/64426246/Load_following_capabilities.pdf (accessed September 10, 2024).
38. Utility Dive, Los Angeles considers $3B pumped storage project at Hoover Dam, 2018, https://tinyurl.com/ynb2ea9z (accessed September 10, 2024).
39. Stocks, M., R. Stocks, B. Lu, C. Cheng, and A. Blakers, Global atlas of closed-loop pumped hydro energy storage, *Joule*, *5*, 270–284, 2021.
40. Willuhn, M., Sonnen battery still running after 28,000 full charge cycles, 2021, https://tinyurl.com/y773rh6e (accessed September 10, 2024).
41. Ziegler, M. S. and J. E. Trancik, Re-examining rates of lithium-ion battery improvement and cost decline, *Energy & Environmental Sciences*, *14*, 1635–1651, 2021.
42. McKerracher, C., China's batteries are now cheap enough to power huge shifts, 2024, https://tinyurl.com/yram69nu (accessed September 10, 2024).
43. Jacobson, M. Z., D. J. Sambor, Y. F. Fan, and A. Mühlbauer, Effects of firebricks for industrial process heat on the cost of matching all-sector energy demand with 100% wind-water-solar supply in 149 countries, *PNAS Nexus*, *3*, pgae274, doi:10.1093/pnasnexus/pgae274, 2024.
44. Board, G., Form Energy to begin manufacturing iron air batteries in Weirton to stabilize electrical grid, 2024, https://tinyurl.com/2abtds9y (accessed September 10, 2024).
45. Stiesdal, The GridScale technology explained, 2024, www.stiesdal.com/storage/the-gridscale-technology-explained/ (accessed September 10, 2024).
46. Colthorpe, A., NGK's first sodium-sulfur battery installation in Eastern Europe comes online in Bulgaria, 2023, https://tinyurl.com/324f8c6x (accessed September 10, 2024).

REFERENCES

47. Boukhalf, S. and N. Kaul, 10 disruptive battery technologies trying to compete with lithium-ion, 2019, https://tinyurl.com/mvvbebaj (accessed September 10, 2024).
48. Wang, Y., D. Zhou, V. Palomares, et al., Revitalizing sodium-sulfur batteries for non-high-temperature operation: A crucial review, *Energy & Environmental Science, 13*, 3848–3879, 2020.
49. He, J., A. Bhargav, L. Su, H. Charalambous, and A. Manthiram, Intercalation-type catalyst for non-aqueous room temperature sodium-sulfur batteries, *Nature Communications, 14*, 6568, 2023.
50. Pilkington, B., Introducing an aluminum-ion battery that charges 60 times faster than lithium, 2021, www.azonano.com/article.aspx?ArticleID=5753 (accessed September 10, 2024).
51. Osmanbasic, E., Aluminum-ion batteries get major capacity boost, 2023, www.engineering.com/aluminum-ion-batteries-get-major-capacity-boost/ (accessed September 10, 2024).
52. Poli, N., C. Bonaldo, M. Moretto, and M. Guarnieri, Techno-economic assessment of future vanadium flow batteries based on real device/market parameters, *Applied Energy, 362*, 122954, 2024.
53. Nithyanandam, K. and R. Pitchumani, Cost and performance analysis of concentrating solar power systems with integrated latent thermal energy storage, *Energy, 64*, 793–810, 2014.
54. NREL (US National Renewable Energy Laboratory), Gemasolar thermosolar plant/Solar TRES CSP Project, 2022, https://solarpaces.nrel.gov/project/gemasolar-thermosolar-plant-solar-tres (accessed September 10, 2024).
55. Denholm, P., Y.-H. Wan, M. Hummon, and M. Mehos, The value of CSP with thermal energy storage in the western United States, *Energy Procedia, 49*, 1622–1631, 2014.
56. Torus, Torus flywheel, 2024, www.torus.co/torus-flywheel (accessed September 10, 2024).
57. Crossley, I., Simplifying and lightening offshore turbines with compressed air energy storage, Wind Power, 2018, https://tinyurl.com/4bd36wfb (accessed September 10, 2024).
58. Murray, C., Highview raises 300 million BP to start building 300 MWh liquid air energy storage project in the UK, 2024, https://tinyurl.com/3b8nuspv (accessed September 10, 2024).
59. Allain, R., How much energy can you store in a stack of cement blocks? *Wired*, 2018, www.wired.com/story/battery-built-from-concrete/ (accessed September 10, 2024).
60. Energy Vault, Break through with G-vault, 2024, www.energyvault.com/products/g-vault (accessed September 10, 2024).
61. Gross, B., Efficiency of gravitational mass storage system, 2019, https://twitter.com/Bill_Gross/status/1164617097927806976/photo/1 (accessed September 10, 2024).

REFERENCES

62. Roberts, D., The train goes up, the train goes down: A simple way to store energy, Vox, 2016, www.vox.com/2016/4/28/11524958/energy-storage-rail (accessed September 10, 2024).
63. Hunt, J. D., B. Zakeri, G. Falchetta, et al., Mountain gravity energy storage: A new solution for closing the gap between existing short- and long-term storage technologies, *Energy*, doi:10.1016/j.energy.2019.116419, 2019.
64. Renewell, 132 gigawatt hours of stored potential energy can be activated on the grid on demand, 2024, https://renewellenergy.com/our-impact/ (accessed September 10, 2024).
65. Jacobson, M. Z., Batteries or hydrogen or both for grid electricity storage upon full electrification of 145 countries with wind-water solar? *iScience*, 27, 108988, 2024.
66. Holnicki, P., A. Kaluszko, Z. Nahorski, and M. Tainio, Intra-urban variability of the intake fraction from multiple emission sources, *Atmospheric Pollution Research*, 9, 1184–1193, 2018.
67. Bistak, S. and S. Y. Kim, AC induction motors vs. permanent magnet synchronous motors, 2017, https://tinyurl.com/m7v74x7k (accessed September 10, 2024).
68. Jacobson, M. Z., Should transportation be transitioned to ethanol with carbon capture and pipelines or electricity? A case study, *Environmental Science and Technology*, 57, 16843–16850, 2023.
69. Evarts, E. C., The world's largest EV never has to be recharged, Green Car Reports, 2019, www.greencarreports.com/news/1124478_world-s-largest-ev-never-has-to-be-recharged (accessed September 10, 2024).
70. Wilson, K. A., Worth the watt: A brief history of the electric car, 1830 to present, 2018, https://tinyurl.com/5zcbc4wz (accessed September 10, 2024s).
71. Vinfast, 38% of American cars were electric in 1900 – What about in the future? 2023, https://tinyurl.com/m3hy4mpp (accessed September 10, 2024).
72. Coltura, Electric car range and price comparison, 2024, https://coltura.org/electric-car-battery-range/ (accessed September 10, 2024).
73. IEA (International Energy Agency), Trends in electric cars, 2024, www.iea.org/reports/global-ev-outlook-2024/trends-in-electric-cars (accessed September 10, 2024).
74. Montoya, R., How many electric cars are there in the US?, 2024, www.edmunds.com/electric-car/articles/how-many-electric-cars-in-us.html (accessed September 10, 2024).
75. Tesla, Semi, 2024, www.tesla.com/semi (accessed September 10, 2024).
76. IEA (International Energy Agency), Trends in heavy electric vehicles, 2024, www.iea.org/reports/global-ev-outlook-2024/trends-in-heavy-electric-vehicles (accessed September 10, 2024).
77. Peters, A., India's rail network is nearly 100% electrified. The US is at 1%, 2024, https://tinyurl.com/3ke3zr9s (accessed September 10, 2024).
78. Grasso Macola, I., Electric ships: The world's top five projects by battery capacity, 2020, https://tinyurl.com/mryx2nty (accessed September 10, 2024).

REFERENCES

79. Wilkerson, J. T., M. Z. Jacobson, A. Malwitz, et al., Analysis of emission data from global commercial aviation: 2004 and 2006, *Atmospheric Chemistry and Physics*, *10*, 6391–6408, 2010.
80. Jacobson, M. Z., J. T. Wilkerson, A. D. Naiman, and S. K. Lele, The effects of aircraft on climate and pollution. Part II: 20-year impacts of exhaust from all commercial aircraft worldwide treated individually at the subgrid scale, *Faraday Discussions*, *165*, 369–382, doi:10.1039/C3FD00034F, 2013.
81. Katalenich, S. M. and M. Z. Jacobson, Toward battery electric and hydrogen fuel cell military vehicles for land, air, and sea, *Energy*, 254, 124355, 2022.
82. Eviation, Alice: The all-electric game changer, 2024, www.eviation.com/aircraft/ (accessed September 10, 2024).
83. Harvey, L. D. D., Resource implications of alternative strategies for achieving zero greenhouse gas emissions from light-duty vehicles by 2060, *Applied Energy*, *212*, 663–679.
84. Petitt, J., Inside Redwood Materials, former Tesla CTO's effort to recycle batteries for rare components, 2021, https://tinyurl.com/56v3kjnm (accessed September 10, 2024).
85. Carpenter, S., Salton Sea is key to CA's EV future, contains 1/3 of global lithium supply, 2021, https://tinyurl.com/2ysrap2w (accessed September 10, 2024).
86. Richter, A., Lithium for batteries from the Upper Rhine's Graben's geothermal resources, 2020, https://tinyurl.com/mma4ccea (accessed September 10, 2024).
87. Martin, P., Part 3: Lithium and cobalt-risky materials, 2017, www.linkedin.com/pulse/part-3-lithium-cobalt-risky-materials-paul-martin/ (accessed September 10, 2024).
88. Rai-Roche, S. and L. Stoker, Renewables-plus-storage projects for mining operations in Australia, Madagascar for BHP, Rio Tinto, 2021, https://tinyurl.com/mwucw8ck (accessed September 10, 2024).
89. Parkinson, G., Potash mine to build wind, solar, and battery micro-grid for most of its power needs, 2021, https://tinyurl.com/yyxdhdbv (accessed September 10, 2024).
90. USGS (United States Geological Survey), Lithium, 2024, https://pubs.usgs.gov/periodicals/mcs2024/mcs2024-lithium.pdf (accessed September 10, 2024).
91. Crownhart, C., How new magnets could accelerate climate action, MIT Technological Review, 2024, www.technologyreview.com/2024/01/31/1087413/magnets-climate-action/amp/ (accessed September 10, 2024).
92. Tucker, S., Study: Electric vehicles involved in fewest car fires, 2022, www.kbb.com/car-news/study-electric-vehicles-involved-in-fewest-car-fires/ (accessed November 8, 2024).
93. Barnes, D. H., S. C. Wofsy, B. P. Fehlau, et al., Hydrogen in the atmosphere: Observations above a forest canopy in a polluted environment, *Journal of Geophysical Research*, *108*, 4197, 2003.
94. Bloom Energy, Introducing the world's largest and most efficient solid oxide electrolyzer, 2024, www.bloomenergy.com/bloomelectrolyzer/ (accessed September 10, 2024).

REFERENCES

95. IEA (International Energy Agency), Hydrogen production projects interactive map, 2024, https://tinyurl.com/6ry3tm2s (accessed September 10, 2024).
96. Jacobson, M. Z., A.-K. von Krauland, K. Song, and A. N. Krull, Impacts of green hydrogen for steel, ammonia, and long-distance transport on the cost of meeting electricity, heat, cold, and hydrogen demand in 145 countries running on 100% wind-water-solar, *Smart Energy*, *11*, 100106, 2023.
97. Feng, Z., Stationary high-pressure hydrogen storage, 2018, www.energy.gov/sites/prod/files/2014/03/f10/csd_workshop_7_feng.pdf (accessed September 10, 2024).
98. Hodges, A., A. L. Hoang, G. Tsekouras, et al., A high-performance capillary-fed electrolysis cell promises more cost-competitive renewable hydrogen, *Nature Communications*, *13*, 1304, 2022.
99. Jacobson, M. Z., Effects of wind-powered hydrogen fuel-cell vehicles on stratospheric ozone and global climate, *Geophysical Research Letters*, *35*, L19803, doi:10.1029/2008GL035102, 2008.
100. Sand, M., R. B. Skeie, M. Sandstat, et al., A multi-model assessment of the global warming potential of hydrogen, *Nature Communications*, *4*, 203, 2023.
101. Jacobson, M. Z. and M. A. Delucchi, Providing all global energy with wind, water, and solar power, Part I: Technologies, energy resources, quantities and areas of infrastructure, and materials, *Energy Policy*, *39*, 1154–1169, doi:10.1016/j.enpol.2010.11.040, 2011.
102. Werner, S., International review of district heating, *Energy*, *15*, 617–631, 2017.
103. Lund, H., 4th generation district heating (4GDH): Integrating smart thermal grids into future sustainable energy systems, *Energy*, *68*, 1–11, 2014.
104. Lund, H., P. A. Ostergaard, T. B. Nielsen, et al., Perspectives on fourth and fifth generation district heating, *Energy*, *227*, 120520, 2021.
105. Stagner, J., Stanford University's "fourth-generation" district energy system, District Energy, Fourth Quarter, 2016, https://tinyurl.com/ya5rem6h (accessed September 10, 2024).
106. Stagner, J., Stanford Energy System Innovations, 2024, https://sesi.stanford.edu (accessed September 10, 2024).
107. Cornell University, How lake source cooling works, 2019, https://tinyurl.com/yp5vcfmk (accessed September 10, 2024).
108. IRENA (International Renewable Energy Agency), Thermal energy storage. IEA-ETSAP and IRENA Technology Brief E17, IRENA, Abu Dhabi, 2013.
109. Sibbitt B, D. McClenahan, R. Djebbar, et al., The performance of a high solar fraction seasonal storage district heating system – five years of operation, *Energy Procedia*, *30*, 856–865, 2012.
110. Sorensen, P. A. and T. Schmidt, Design and construction of large scale heat storages for district heating in Denmark, 14th Int. Conf. on Energy Storage, April 25–28, Adana, Türkiye, 2018, https://tinyurl.com/2929a4jp (accessed September 10, 2024).

REFERENCES

111. Ramboll, World's largest thermal heat storage pit in Vojens, 2016, https://tinyurl.com/33yfudm3 (accessed September 10, 2024).
112. Arcon/Sunmark, Large-scale showcase projects, 2017, http://arcon-sunmark.com/uploads/ARCON_References.pdf (accessed September 10, 2024).
113. Damkjaer, L., Gram Fjernvarme 2016, 2016, www.youtube.com/watch?v=PdF8e1t7St8 (accessed September 10, 2024).
114. Ramboll, Pit thermal energy storage: Update from Toftlund, 2020, www.heatstore.eu/documents/20201028_DK-temadag_Rambøll%20PTES%20project.pdf (accessed September 10, 2024).
115. IEA (International Energy Agency), Integrated cost-effective large-scale thermal energy storage for smart district heating and cooling, 2018, https://tinyurl.com/3ym98dv7 (accessed September 10, 2024).
116. Fleuchaus, P., B. Godschalk, I. Stober, and P. Blum, Worldwide application of aquifer thermal energy storage – A review, *Renewable and Sustainable Energy Reviews*, *94*, 861–876, 2018.
117. Butti, K. and J. Perlin, Solar water heaters in California, 1891–1930, California Energy Commission, 6772173, 1980, https://www.osti.gov/biblio/6772173 (accessed April 16, 2025).
118. Dandelion, Geothermal heating and air conditioning is so efficient, it pays for itself, 2018, https://dandelionenergy.com (accessed September 10, 2024).
119. Gibb, D., J. Rosenow, R. Lowes, and N. J. Hewitt, Coming in from the cold: Heat pump efficiency at low temperatures, *Joule*, *7*, 1939–1942, 2023.
120. Fischer, D. and H. Madani, On heat pumps in smart grids: A review, *Renewable and Sustainable Energy Reviews*, *70*, 342–357, 2017.
121. Vzug, AdoraDish V6000 with heat pump, 2024, https://tinyurl.com/5h9hzyp2 (accessed September 10, 2024).
122. Meyers, S., V. Franco, A. Lekov, L. Thompson, and A. Sturges, Do heat pump clothes dryers makes sense for the US market? *ACEEE Summer Study on Energy Efficiency in Buildings*, *9*, 240–251, 2010, https://aceee.org/files/proceedings/2010/data/papers/2224.pdf (accessed September 10, 2024).
123. De Gracia, A. and L. F. Cabeza, Phase change materials and thermal energy storage for buildings, *Energy and Buildings*, *103*, 414–419, 2015.
124. DOE (US Department of Energy), 2021–2022 residential induction cooking tops, 2024, https://tinyurl.com/ysr8ujfv (accessed September 10, 2024).
125. Consumer Reports, 10 best battery lawn mowers of 2024, tested by our experts, 2024, https://tinyurl.com/4ze5yz5c (accessed September 10, 2024).
126. NREL (US National Renewable Energy Laboratory), Microgrids, 2024, www.nrel.gov/grid/microgrids.html (accessed April 16, 2025).
127. Sambor, D. J., H. Penn, and M. Z. Jacobson, Energy optimization of a food-energy-water microgrid living laboratory in Yukon, Canada, *Energy Nexus Journal*, *10*, 100200, 2023.
128. DOE (US Department of Energy), Quadrennial Technology Review, Chapter 6: Innovative clean energy technologies in advanced manufacturing:

REFERENCES

Technology assessment, 2015, www.energy.gov/sites/prod/files/2016/06/f32/QTR2015-6I-Process-Heating.pdf (accessed September 10, 2024).
129. Crippa, M., D. Guizzardi, F. Pagani, et al., GHG emissions of all world countries – JRC/IEA 2023 Report, Publications Office of the European Union, Luxembourg, 2023.
130. IEA (International Energy Agency), Heat, 2023, www.iea.org/reports/renewables-2023/heat# (accessed September 10, 2024).
131. Lin, C. H., C. Wan, Z. Ru, et al., Electrified thermochemical reaction systems with high frequency metamaterial reactors, *Joule*, 8, 1–12, 2024.
132. Butler, T., M. Azih, A. Mitchell, and S. Baltac, Future opportunities for electrification to decarbonize UK industry, A report for Department for Energy Security and Net Zero, 2023, https://tinyurl.com/yc6xtrms (accessed April 16, 2025).
133. Hulls, P. J., Development of the industrial use of dielectric heating in the United Kingdom, *Journal of Microwave Power*, 17, 28–38, 2016.
134. Viking Heat Engines, Heat Booster, 2019, https://heatroadmap.eu/wp-content/uploads/2019/04/Andreas-Muck_Viking-Heat-Engines.pdf (accessed September 10, 2024).
135. McMillan, C., C. Schoeneberger, J. Zhang, et al., Opportunities for solar industrial process heat in the United States. National Renewable Energy Laboratory, Golden, CO, NREL/TP-6A20-77760, 2021, www.nrel.gov/docs/fy21osti/77760.pdf (accessed September 10, 2024).
136. Sunvapor, Website, 2024, www.sunvapor.net (accessed September 10, 2024).
137. Ramaiah, R. and K. S. S. Shekar, Solar thermal energy utilization for medium temperature industrial process heat applications, *IOP Conference Series: Materials Science and Engineering*, 376, 010235, 2018.
138. Forsberg C. W., D. C. Stack, D. Curtis, G. Haratyk, and N. A. Sepulveda, Converting excess low-price electricity into high-temperature stored heat for industry and high-value electricity production, *The Electricity Journal*, 30, 42–52, 2017.
139. Stack, D. C. and C. Forsberg, Performance of firebrick resistance-heated energy storage for industrial heat applications and round-trip electricity storage, *Applied Energy*, 242, 782–796, 2019.
140. Seyitini, L., B. Belgasim, and C. C. Enweremadu, Solid state sensible heat storage technology for industrial applications – A review, *Journal of Energy Storage*, 62, 106919, 2023.
141. Rondo, The Rondo heat battery, 2024, https://rondo.com (accessed September 10, 2024).
142. Antora, Reliable, zero-emissions industrial heat and power, 2024, https://antoraenergy.com (accessed September 10, 2024).
143. Sugita, K., Historical overview of refractory technology in the steel industry, Nippon Steel Technical Report No. 98, Nippon Steel, Tokyo, Japan, 2008, www.nipponsteel.com/en/tech/report/nsc/pdf/n9803.pdf (accessed September 10, 2024).

REFERENCES

144. Forsberg, C. and D. C. Stack, Electrically Conductive Firebrick System, US Patent: 11,877,376 B2, January 16, 2024, https://image-ppubs.uspto.gov/dirsearch-public/print/downloadPdf/11877376 (accessed September 10, 2024).
145. Electrified Thermal Solutions, Website, 2024, https://electrifiedthermal.com (accessed September 10, 2024).
146. Vogl, V., M. Ahman, and L. J. Nilsson, Assessment of hydrogen direct reduction for fossil-free steelmaking, *Journal of Cleaner Production*, 203, 736–745, 2018.
147. Geschwindt, S., Sweden's been stealthily using hydrogen to forge green steel. Now it's ready to industrialize, 2024, https://thenextweb.com/news/sweden-hydrogen-green-steel-vattenfall-ssab-lkab-hybrit (accessed September 10, 2024).
148. Hullinger, J., This startup is making steel with lasers, 2024, https://tinyurl.com/yhwtybd2 (accessed September 10, 2024).
149. Wiencke, J., H. Lavelaine, P.-J. Panteix, C. Petijean, and C. Rapin, Electrolysis of iron in a molten oxide electrolyte, *Journal of Applied Electrochemistry*, 48, 115–126, 2018.
150. Andrew, R. M., Global CO_2 emissions from cement production, 1928–2018, *Earth System Science Data*, 10, doi:10.5194/essd-2019-152, 2019.
151. Choate, W. T., Energy and emission reduction opportunities for the cement industry, 2003, www1.eere.energy.gov/manufacturing/industries_technologies/imf/pdfs/eeroci_dec03a.pdf (accessed September 10, 2024).
152. Brimstone, Cement for our climate future, 2024, www.brimstone.com (accessed September 10, 2024).
153. C-Crete, C-Crete pours world's first basalt-based concrete, a zero-emission product free of Portland cement, 2024, https://ccretetech.com/news/684-2/ (accessed September 10, 2024).
154. Singh, N. B., M. Kumar, and S. Rai, Geopolymer cement and concrete: Properties, *Materials Today Proceedings*, 29, 743–748, 2020.
155. Stone, D., Ferrock basics, 2017, http://ironkast.com/wp-content/uploads/2017/11/Ferrock-basics.pdf (accessed September 10, 2024).
156. Carbon Cure, Carbon Cure, 2018, www.carboncure.com (accessed September 10, 2024).
157. Maldonado, S., The importance of new "sand-to-silicon" processes for the rapid future increase of photovoltaics, *ACS Energy Letters*, 5, 3628–3632, 2020.
158. Dong, Y., T. Slade, M. J. Stolt, et al., Low-temperature molten salt production of silicon nanowires by the electrochemical reduction of CaSiO3, *Angewandte Chemie*, 56, 14453–14457, 2017.
159. Jacobson, M. Z., The short-term cooling but long-term global warming due to biomass burning, *Journal of Climate*, 17 (15), 2909–2926, 2004.
160. Lean, I. J., H. M. Golder, T. M. D. Grant, and P. J. Moate, A meta-analysis of effects of dietary seaweed on beef and dairy cattle performance and methane, *PLoS One*, 16, e0249053, 2021.

REFERENCES

161. Patlolla, S. R., K. Katsu, A. Sharafian, et al., A review of methane pyrolysis technologies for hydrogen production, *Renewable and Sustainable Energy Reviews*, *181*, 11323, 2023.
162. Howarth, R. W. and M. Z. Jacobson, How green is blue hydrogen? *Energy Science and Engineering*, 0, 1–12, doi:10.1002/ese3.956, 2021.
163. Ussiri, D. and R. Lal, Global sources of nitrous oxide. In *Soil Emission of Nitrous Oxide and Its Mitigation*, Springer, Dordrecht, Netherlands, pp. 131–175, 2012.
164. NACAG (Nitric Acid Climate Action Group), Nitrous oxide emissions from nitric acid production, 2014, https://tinyurl.com/bde6zcs4 (accessed September 10, 2024).
165. Boiocchi, R., K. V. Gemaey, and G. Sin, Control of wastewater N_2O emission by balancing the microbial communities using a fuzzy-logic approach, *IFAC-PapersOnLine*, *49*, 1157–1162, 2016.
166. Santin, I., M. Barbu, C. Pedret, and R. Vilanova, Control strategies for nitrous oxide emissions reduction on wastewater treatment plants operation, *Water Research*, *125*, 466–477, 2017.
167. Hong, S., J.-P. Candelone, C. C. Patterson, and C. F. Boutron, Greenland ice evidence of hemispheric lead pollution two millennia ago by Greek and Roman civilizations, *Science*, *265*, 1841–1843, 1994.
168. Nriagu, J. O., A history of global metal pollution, *Science*, *272*, 223–224, 1996.
169. Hong, S., J.-P. Candelone, C. C. Patterson, and C. F. Boutron, History of ancient copper smelting pollution during Roman and Medieval times recorded in Greenland ice, *Science*, *272*, 246–248, 1996.
170. Brimblecombe, P., Air pollution and health history. In S. T. Holgate, J. M. Samet, H. S. Koren, and R. L. Maynard, eds., *Air Pollution and Health*, Academic Press, San Diego, CA, pp. 5–18, 1999.
171. Hughes, J. D., *Pan's Travail: Environmental Problems of the Ancient Greeks and Romans*, The Johns Hopkins University Press, Baltimore, MD, 1994.
172. Brimblecombe, P., *The Big Smoke*, Methuen, London, 1987.
173. McNeill, J. R., *Something New under the Sun*, W. W. Norton & Company, New York, 2000.
174. Rosenberg, N. and L. E. Birdzell, Jr., *How the West Grew Rich*, Basic Books, New York, 1986.
175. Union Gas, Chemical composition of natural gas, 2018, www.uniongas.com/about-us/about-natural-gas/chemical-composition-of-natural-gas (accessed September 10, 2024).
176. De Boer, J. Z., J. R. Hale, and J. Chanton, New evidence of the geological origins of the ancient Delphic oracle (Greece), *Geology*, *29*, 707–710, 2001.
177. IPCC (Intergovernmental Panel on Climate Change), *Special report: Global warming of 1.5º*, 2018, www.ipcc.ch/sr15/ (accessed September 10, 2024).
178. Jacobson, M. Z., Review of solutions to global warming, air pollution, and energy security, *Energy & Environmental Science*, *2*, 148–173, doi:10.1039/b809990c, 2009.

REFERENCES

179. Frangoul, A., Scandinavia's biggest offshore wind farm is officially open, 2019, https://tinyurl.com/y7mhknyn (accessed September 10, 2024).
180. Jacobson, M. Z. and C. L. Archer, Saturation wind power potential and its implications for wind energy, *Proceedings of the National Academy of Sciences, 109,* 15679–15684, doi:10.1073/pnas.1208993109, 2012.
181. Jacobson, M. Z., M. A. Delucchi, M. A. Cameron, and B. V, Mathiesen, Matching demand with supply at low cost among 139 countries within 20 world regions with 100 percent intermittent wind, water, and sunlight (WWS) for all purposes, *Renewable Energy, 123,* 236–248, 2018.
182. IPCC (Intergovernmental Panel on Climate Change), *IPCC special report on carbon dioxide capture and storage.* Prepared by working group III, B. Metz, O. Davidson, H. C. de Coninck, M. Loos, and L. A. Meyer, eds., Cambridge University Press, Cambridge, United Kingdom and New York, NY, USA, 2005, www.ipcc.ch/report/carbon-dioxide-capture-and-storage/ (accessed September 10, 2024).
183. PSR (Physicians for Social Responsibility), Compendium of scientific, medical, and media findings demonstrating risks and harms of fracking and associated gas and oil infrastructure, 2022, https://psr.org/wp-content/uploads/2022/04/compendium-8.pdf (accessed September 10, 2024).
184. Howarth, R. W., R. Santoro, and A. Ingraffea, Methane and the greenhouse gas footprint of natural gas from shale formations, *Climatic Change, 106,* 679–690, 2011.
185. Howarth, R. W., Is shale gas a major driver of recent increase in global atmospheric methane, *Biogeosciences, 16,* 3033–3046, 2019.
186. Alvarez, R. A., D. Zavalao-Araiza, D. R. Lyon et al., Assessment of methane emissions from the US oil and gas supply chain, *Science, 361,* 186–188, 2018.
187. Schneising, O., M. Buchwitz, M. Reuter, et al., Remote sensing of methane leakage from natural gas and petroleum systems revisited, *Atmospheric Chemistry and Physics, 20,* 9169–9182, 2020.
188. MIT (Massachusetts Institute of Technology), *The Future of Natural Gas,* 2011, https://tinyurl.com/529zmd87 (accessed September 10, 2024).
189. US EPA (US Environmental Protection Agency), 2008 US National Emissions Inventory (NEI), 2011, www.epa.gov/air-emissions-inventories/2008-national-emissions-inventory-nei-data (accessed September 10, 2024).
190. Jacobson, M. Z., M. A. Delucchi, G. Bazouin, et al., 100 percent clean and renewable wind, water, sunlight (WWS) all-sector energy roadmaps for the 50 United States, *Energy & Environmental Sciences, 8,* 2093–2117, doi:10.1039/C5EE01283J, 2015.
191. Lazard and R. Berger, Lazard's levelized cost of energy plus, June 2024, www.lazard.com/media/xemfey0k/lazards-lcoeplus-june-2024-_vf.pdf (accessed September 10, 2024).
192. Allred, B. W., W. K. Smith, D. Twidwell, et al., Ecosystem services lost to oil and gas in North America, *Science, 348,* 401–402, 2015.

REFERENCES

193. Reuters, Special Report: Millions of abandoned oil wells are leaking methane, a climate menace, 2020, https://tinyurl.com/yc5huspj (accessed September 10, 2024).
194. Jacobson, M. Z., Short-term impacts of the Aliso Canyon natural gas blowout on weather, climate, air quality, and health in California and Los Angeles, *Environmental Science & Technology*, *53*, 6081–6093, doi:10.1021/acs.est.9b01495, 2019.
195. Global CCS Institute, Global status of CCS 2023, 2024, https://tinyurl.com/6zwvcsye (accessed September 10, 2024).
196. Jaramillo, P., W. M. Griffin, and S. T. McCoy, Life cycle inventory of CO_2 in an enhanced oil recovery system, *Environmental Science and Technology*, *43*, 8027–8032, 2009.
197. Roberts, D., Turns out the world's first "clean coal" plant is a backdoor subsidy to oil producers, 2015, https://tinyurl.com/nsyxy9x8 (accessed September 10, 2024).
198. Schlissel, D. and A. Juhn, Carbon capture and storage, IEEFA, 2023, https://ieefa.org/ccs (accessed September 10, 2024).
199. House, K. Z., C. F. Harvey, M. J. Aziz, and D. P. Schrag, The energy penalty of post-combustion CO2 capture & storage and its implications for retrofitting the US installed base, *Energy & Environmental Science*, *2*, 193–205, 2009.
200. IPCC, 2022: Summary for policymakers. In P. R. Shukla, J. Skea, R. Slade, et al., eds., *Climate Change 2022: Mitigation of Climate Change. Contribution of Working Group III to the Sixth Assessment Report of the Intergovernmental Panel on Climate Change*, Cambridge University Press, Cambridge, UK and New York, NY, doi:10.1017/9781009157926.001, 2022.
201. Jacobson, M. Z., The health and climate impacts of carbon capture and direct air capture, *Energy & Environmental Sciences*, *12*, 3567–3574, doi:10.1039/C9EE02709B, 2019.
202. Ravikumar, D., G. Keoleian, and S. Miller, The environmental opportunity cost of using renewable energy for carbon capture and utilization for methanol production, *Applied Energy*, *279*, 115770, 2020.
203. Schlissel, D. and M. Kalegha, Carbon capture at Boundary Dam 3 is still an underperforming failure, 2024, https://tinyurl.com/ydt6435k (accessed September 10, 2024).
204. Rosenow, J., A meta review of 54 studies on hydrogen heating, *Cell Reports Sustainability*, *1*, 100010, 2024.
205. Chen, W.-H., M.-R. Lin, J.-J. Lu, Y. Chao, and T.-S. Leu, Thermodynamic analysis of hydrogen production from methane via autothermal reforming and partial oxidation followed by water gas shift reaction, *International Journal of Hydrogen Energy*, *35*, 11787–11797, 2010.
206. US Drive, Hydrogen production tech team roadmap, 2017, https://tinyurl.com/3u9c482n (accessed September 10, 2024).

REFERENCES

207. Duan, Y. and D. C. Sorescu, CO_2 capture properties of alkaline earth metal oxides and hydroxides: A combined density functional theory and lattice phonom dynamics study, *Journal of Chemical Physics*, *133*, 074508, 2010.
208. Yamada, R., DAC cost drops to $300 over 30 years, Nikkei GX, May 5, 2024, www.nikkei.com/prime/gx/article/DGXZQOGN210B20R20C24A5000000 (accessed September 10, 2024).
209. Kadiyala, A., R. Kommalapati, and Z. Huque, Evaluation of the lifecycle greenhouse gas emissions from different biomass feedstock electricity generation systems, *Sustainability*, *8*, 1181–1192, 2016.
210. US Department of Energy Alternative Fuels Data Center, E85 (Flex fuel), 2024, https://afdc.energy.gov/fuels/ethanol_e85.html (accessed September 10, 2024).
211. Jacobson, M. Z., Effects of ethanol (E85) versus gasoline vehicles on cancer and mortality in the United States, *Environmental Science & Technology*, *41* (11), 4150–4157, doi:10.1021/es062085v, 2007.
212. Ginnebaugh, D. L., J. Liang, and M. Z. Jacobson, Examining the temperature dependence of ethanol (E85) versus gasoline emissions on air pollution with a largely-explicit chemical mechanism, *Atmospheric Environment*, *44*, 1192–1199, doi:10.1016/j.atmosenv.2009.12.024, 2010.
213. Ginnebaugh, D. L. and M. Z. Jacobson, Examining the impacts of ethanol (E85) versus gasoline photochemical production of smog in a fog using near-explicit gas- and aqueous-chemistry mechanisms, *Environmental Research Letters*, *7*, 045901, doi:10.1088/1748-9326/7/4/045901, 2012.
214. Searchinger, T., R. Heimlich, R. A. Houghton, et al., Use of US cropland for biofuels increases greenhouse gases through emissions from land-use change, *Science*, *319*, 1238–1240, 2008.
215. Delucchi, M., A conceptual framework for estimating the climate impacts of land-use change due to energy crop programs, *Biomass and Bioenergy*, *35*, 2337–2360, 2011.
216. Lark, T. J., N. P. Hendricks, A. Smith, et al., Environmental outcomes of the US renewable fuel standard, *Proceedings of the National Academy of Sciences*, *119*, e2101084119, 2022.
217. Hill, J., A. Goodkind, C. Tessum, et al., Air-quality-related health damages of maize, *Nature Sustainability*, *2*, 397–403, 2019.
218. Hill, J., S. Polasky, E. Nelson, et al., Climate change and health costs of air emissions from biofuels and gasoline, *Proceedings of the National Academy of Sciences*, *106*, 2077–2082, 2009.
219. Salvo, A. and F. M. Geiger, Reduction in local ozone levels in urban Sao Paulo due to a shift from ethanol to gasoline use, *Nature Geoscience*, *7*, 450–458, 2014.
220. Clairotte, M., T. W. Adam, A. A. Zardini, et al., Effects of low temperature on the cold start gaseous emissions from light duty vehicles fuelled by ethanol-blended gasoline, *Applied Energy*, *102*, 44–54, 2013.

REFERENCES

221. Scully, M. J., G. A. Norris, T. M. A. Falconi, and D. L. MacIntosh, Carbon intensity of corn ethanol in the United States: State of the science, *Environmental Research Letters*, *16*, 043001, 2021.
222. Summit (Summit Carbon Solutions LLC), Website, 2024, https://summitcarbonsolutions.com (accessed September 10, 2024).
223. Karam, P. A., How do fast breeder reactors differ from regular nuclear power plants? *Scientific American*, *295* (4), 100, October 2006.
224. IAEA (International Atomic Energy Agency), Trend in electricity supplied, 2024, https://pris.iaea.org/PRIS/WorldStatistics/WorldTrendinElectricalProduction.aspx (accessed September 10, 2024).
225. WNA (World Nuclear Association), World uranium mining production, 2024, https://tinyurl.com/5sutdkyf (accessed September 10, 2024).
226. IAEA (International Atomic Energy Agency), World's uranium resources enough for foreseeable future says NEA and IAEA in new report, 2021, https://tinyurl.com/yc79hjt7 (accessed September 10, 2024).
227. IAEA (International Atomic Energy Agency), Fusion – frequently asked questions, 2024, www.iaea.org/topics/energy/fusion/faqs# (accessed September 10, 2024).
228. Koomey, J. and N. E. Hultman, A reactor-level analysis of busbar costs for US nuclear plants, 1970–2005, *Energy Policy*, *35*, 5630–5642, 2007.
229. Berthelemy, M. and L. E. Rengel, Nuclear reactors' construction costs: The role of lead-time, standardization, and technological progress, *Energy Policy*, *82*, 118–130, 2015.
230. Morris, C., French nuclear power history – the unknown story, 2015, https://energytransition.org/2015/03/french-nuclear-power-history/ (accessed September 10, 2024).
231. Bruckner T., I. A. Bashmakov, Y. Mulugetta, et al., Energy systems. In O. Edenhofer, R. Pichs-Madruga, Y. Sokona, et al., eds., *Climate Change 2014: Mitigation of Climate Change. Contribution of Working Group III to the Fifth Assessment Report of the Intergovernmental Panel on Climate Change*, Cambridge University Press, Cambridge, UK and New York, NY, USA, pp. 511–598, 2014.
232. Garthwaite, J., What should we do with nuclear waste, Stanford Earth, 2018, https://earth.stanford.edu/news/qa-what-should-we-do-nuclear-waste#gs.1sfx0x (accessed September 10, 2024).
233. Denyer, S., Eight years after Fukushima's meltdown, the land is recovering, but public trust is not, *Washington Post*, 2019, https://tinyurl.com/2p9sb6wy (accessed September 10, 2024).
234. Cebulla, F. and M. Z. Jacobson, Alternative renewable energy scenarios for New York, *Journal of Cleaner Production*, *205*, 884–894, 2018.
235. De Coninck, H., A. Revi, M. Babiker, et al., Chapter 4: Strengthening and implementing the global response. In Intergovernmental Panel on Climate Change, Global Warming of 1.5°C report, 2018, https://www.ipcc.ch/site/assets/uploads/sites/2/2019/02/SR15_Chapter4_Low_Res.pdf (accessed April 16, 2025).

REFERENCES

236. Fuhrmann, M., Spreading temptation: Proliferation and peaceful nuclear cooperation agreements (March 9, 2009), *International Security*, *34* (1), Summer 2009, https://ssrn.com/abstract=1356091 (accessed September 10, 2024).
237. IranWatch, Iran's nuclear potential before the implementation of the nuclear agreement, 2015, www.iranwatch.org/our-publications/articles-reports/irans-nuclear-timetable (accessed September 10, 2024).
238. Johnson, G., When radiation isn't the real risk, 2015, www.nytimes.com/2015/09/22/science/when-radiation-isnt-the-real-risk.html (accessed September 10, 2024).
239. BBC News, Japan confirms first Fukushima worker death from radiation, 2018, www.bbc.com/news/world-asia-45423575 (accessed September 10, 2024).
240. Ten Hoeve, J. E. and M. Z. Jacobson, Worldwide health effects of the Fukushima Daiichi nuclear accident, *Energy & Environmental Sciences*, *5*, 8743–8757, 2012.
241. Henshaw, D. L., J. P. Eatough, and R. B. Richardson, Radon as a causative factor in induction of myeloid leukemia and other cancers, *Lancet*, *335*, 1008–1012, 1990.
242. Lagarde, F., G. Pershagen, G. Akerblom, et al., Residential radon and lung cancer in Sweden: Risk analysis accounting for random error in the exposure assessment, *Health Physics*, *72*, 269–276, 1997.
243. Hampson, S. E., J. A. Andres, M. E. Lee, et al., Lay understanding of synergistic risk: The case of radon and cigarette smoking, *Risk Analysis*, *18*, 343–350, 1998.
244. Jacobson, M. Z. and J. E. Ten Hoeve, Effects of urban surfaces and white roofs on global and regional climate, *Journal of Climate*, *25*, 1028–1044, doi:10.1175/JCLI-D-11-00032.1, 2012.
245. King, G., Edison vs. Westinghouse: A shocking rivalry, *Smithsonian Magazine*, 2011, https://tinyurl.com/yyttce77 (accessed September 10, 2024).
246. Sadovskaia, K., D. Bogdanov, S. Honkapuro, and C. Breyer, Power transmission and distribution loses – a model based on available empirical data and future trends for all countries globally, *Electrical Power and Energy Systems*, *107*, 98–109, 2019.
247. IEC (International Electrotechnical Commission), Efficient electrical energy transmission and distribution, 2007, www.iec.ch/basecamp/efficient-electrical-energy-transmission-and-distribution (accessed September 10, 2024).
248. ABB, Review: 60 years of HVDC 2014, https://tinyurl.com/ycya2eht (accessed September 10, 2024).
249. World Bank, Electric power transmission and distribution losses (% of output), 2018, https://data.worldbank.org/indicator/EG.ELC.LOSS.ZS?end=2014&start=2009 (accessed September 10, 2024).
250. Becquerel, A. E., Mémoire sur les effets electriques produits sous l'influence des rayons solaires, *Annalen der Physik und Chemie*, *54*, 35–42, 1841.

REFERENCES

251. Adams, W. G. and R. E. Day, The action of light on selenium, *Proceedings of the Royal Society of London*, *A25*, 113, 1877.
252. Fritts, C. E., On a new form of selenium photocell, *American Journal of Science*, *26*, 465, 1883.
253. Siemens, W., On the electromotive action of illuminated selenium, discovered by Mr. Fritts of New York, *Van Nostrands Engineering Magazine*, *32*, 514–516, 1885.
254. Grondahl, L. O., The copper-cuprous-oxide rectifier and photoelectric cell, *Review of Modern Physics*, *5*, 141, 1933.
255. NREL (US National Renewable Energy Laboratory), Best research-cell efficiency charge, 2024, www.nrel.gov/pv/cell-efficiency.html (accessed September 10, 2024).
256. Bremner, S. P., M. Y. Levy, and C. B. Honsberg, Analysis of tandem solar cell efficiencies under {AM1.5G} spectrum using a rapid flux calculation method, *Progress in Photovoltaics: Research and Applications*, *16*, 225–233, 2008.
257. Jacobson, M. Z., D. J. Sambor, Y. F. Fan, A. Mühlbauer, and M. A. Delucchi, No blackouts or cost increases due to 100% clean, renewable electricity powering California for parts of 98 days, *Renewable Energy*, *240*, 122262, 2025.
258. Dominguez, A., J. Kleissl, and J. C. Luvall, Effects of solar photovoltaic panels on roof heat transfer, *Solar Energy*, *85*, 2244–2255, 2011.
259. Casey, J. P., CGN commissions 400 MW offshore floating solar project in China, 2024, www.pv-tech.org/cgn-commissions-400mw-offshore-floating-solar-project-in-china/ (accessed September 10, 2024).
260. Jacobson, M. Z. and V. Jadhav, World estimates of PV optimal tilt angles and ratios of sunlight incident upon tilted and tracked PV panels relative to horizontal panels, *Solar Energy*, *169*, 55–66, 2018.
261. IEA (International Energy Agency), World Energy Balances, 2024, www.iea.org/data-and-statistics/data-product/world-energy-balances (accessed September 10, 2024).
262. GWEC (Global Wind Energy Council), Global wind report 2025, 2025, www.gwec.net/reports/globalwindreport (accessed May 24, 2025).
263. Jordan, T. G., Evolution of the American windmill: A study in diffusion and modification, *Pioneer America*, *5*, 3–12, 1973.
264. Pavel, C. C., R. Lacal-Arantegui, A. Marmier, et al., Substitution strategies for reducing the use of rare earths in wind turbines, *Resources Policy*, *52*, 349–357, 2017.
265. Pires, O., X. Munduate, O. Ceyhan, M. Jacobs, and H. Snel, Analysis of high Reynolds numbers effects on a wind turbine airfoil using 2D wind tunnel test data, *Journal of Physics: Conference Series*, *753*, 022047, 2016.
266. Schubel, P. J. and R. J. Crossley, Wind turbine blade design, *Energies*, *5*, 3425–3449, 2012.
267. Hu, S-y and J.-h Cheng, Performance evaluation of pairing between sites and wind turbines, *Renewable Energy*, *32*, 1934–1947. 2007.

REFERENCES

268. Sirnivas, S., W. Musial, B. Bailey, and M. Filippelli, Assessment of offshore wind system design, safety, and operation standards, NREL/TP-5000-60573, 2014.
269. Mingyang, Smart wind turbines, 2024, www.myse.com.cn/en/wind-turbine/index.aspx (accessed September 10, 2024).
270. Equinor, Equinor marks 5 years of operations at world's first floating offshore wind farm, 2023, https://tinyurl.com/4rxu3rz9 (accessed September 10, 2024).
271. Wiser, R., D. Millstein, B. Hoen, et al., Land-based wind market report, 2024 Edition, US Department of Energy, 2024, https://emp.lbl.gov/publications/land-based-wind-market-report-2024 (accessed September 10, 2024).
272. Faulstich, S., B. Hahn, and P. J. Tavner, Wind turbine downtime and its importance for offshore deployment, *Wind Energy*, 14, 327–337, 2011.
273. Zhang, J., S. Chowdhury, and J. Zhang, Optimal preventative maintenance time windows for offshore wind farms subject to wake losses, AIAA 2012-5435, 2012.
274. Monitoring Analytics, Quarterly state of the market report for PJM: January through June, 2015, https://tinyurl.com/3zck9cyv (accessed September 10, 2024).
275. Enevoldsen, P. and M. Z. Jacobson, Data investigation of installed and output power densities of onshore and offshore wind turbines worldwide, *Energy for Sustainable Development*, 60, 40–51, 2021.
276. Zhou, L., Y. Tian, S. B. Roy, et al., Impacts of wind farms on land surface temperature, *Nature Climate Change*, 2, 539–543, 2012.
277. Fitch, A. C., Climate impacts of large-scale wind farms as parameterized in a global climate model, *Journal of Climate*, 28, 6160–6180, 2015.
278. Jacobson, M. Z., C. L. Archer, and W. Kempton, Taming hurricanes with arrays of offshore wind turbines, *Nature Climate Change*, 4, 195–200, doi:10.1038/NCLIMATE2120, 2014.
279. American Bird Conservancy, https://abcbirds.org (accessed September 10, 2024).
280. Sovacool, B. K., Contextualizing avian mortality: A preliminary appraisal of bird and bat fatalities from wind, fossil-fuel, and nuclear electricity, *Energy Policy*, 37, 2241–2248, 2009.
281. NWCC (National Wind Coordinating Collaborative), Wind turbine interactions with birds, bats, and their habitats, 2010, www1.eere.energy.gov/wind/pdfs/birds_and_bats_fact_sheet.pdf (accessed September 10, 2024).
282. Smallwood, K. S., Comparing bird and bat fatality rate estimates among North American wind energy projects, *Wildlife Society Bulletin*, 37, 19–33, 2013.
283. May, R., T. Nygard, U. Falkdalen, et al., Paint it black: Efficacy of increased wind turbine rotor blade visibility to reduce avian fatalities, *Ecology and Evolution*, 10, 8927–8935, 2020.
284. Jacobson, M. Z., M. A. Delucchi, Z. A. F. Bauer, et al., 100 percent clean and renewable wind, water, and sunlight (WWS) all-sector energy roadmaps for 139 countries of the world, *Joule*, 1, 108–121, doi:10.1016/j.joule.2017.07.005, 2017.
285. von Krauland, A.-K., F.-H. Permien, P. Enevoldsen, and M. Z. Jacobson, Onshore wind energy atlas for the United States accounting for land use

REFERENCES

restrictions and wind speed thresholds, *Smart Energy*, *3*, 100046, doi:10.1016/j.segy.2021.100046, 2021.

286. Moore, M. A., A. E. Boardman, A. R. Vining, D. L. Weimer, and D. H. Greenberg, Just give me a number! Practical values for the social discount rate, *Journal of Policy Analysis and Management*, *23* (4), 789–812, 2004.

287. OMB (US Office of Management and Budget), Circular A-4, Regulatory Analysis, the White House, Washington, DC, September 17, 2003, https://obamawhitehouse.archives.gov/omb/circulars_a004_a-4/ (accessed September 10, 2024).

288. Drupp, M., M. Freeman, B. Groom, and F. Nesje, Discounting disentangled: An expert survey on the determinants of the long-term social discount rate, *Grantham Research Institute on Climate Change and the Environment Working Paper* No. 172, 2015.

289. IRENA (International Renewable Energy Agency), Renewable power generation costs in 2023, International Renewable Energy Agency, Dubai, 2024, https://tinyurl.com/5ak68syh (accessed September 25, 2024).

290. Batjer, M., S. Berberich, and D. Hochschild, Letter to Governor Gavin Newsom, 2020, www.gov.ca.gov/wp-content/uploads/2020/08/8.17.20-Letter-to-CAISO-PUC-and-CEC.pdf (accessed September 10, 2024).

291. Swenson, A. and A. Lajka, Texas blackouts fuel false claims about renewable energy, 2021, https://tinyurl.com/2bbrrn8u (accessed September 10, 2024).

292. IRENA (International Renewable Energy Agency), IRENASTAT Online data query tool, 2024, https://pxweb.irena.org/pxweb/en/IRENASTAT/ (accessed September 10, 2024).

293. Renewables Now, Chile's Coquimbo region nears 100% renewables share in H1 2019, 2019, https://tinyurl.com/2shztnyt (accessed September 10, 2024).

294. Scottish Energy Statistics Hub, Renewable electricity generation as a percentage of equivalent gross consumption, 2024, https://tinyurl.com/5n7eh473 (accessed September 10, 2024).

295. Mooney, G., South Australia locks in federal funds to become first grid in world to reach 100 percent net wind and solar, 2024, https://tinyurl.com/3ykxzuxx (accessed November 1, 2024).

296. Parkinson, G., South Australia runs on more than 100 pct net renewables in last week of winter, 2024, https://tinyurl.com/y2hcc4hj (accessed September 10, 2024).

297. Linnerud, K., T. Mideksa, and G. S. Eskeland, The impact of climate change on nuclear power supply, *The Energy Journal*, *32*, 149–168, 2011.

298. Ahmad, A., Increase in frequency of nuclear power outages due to climate change, *Nature Energy*, *6*, 755–762, 2021.

299. Elliott, D., M. Schwartz, and G. Scott, Wind resource base, *Encyclopedia of Energy*, *6*, 465–479, 2004.

300. Jacobson, M. Z. and Y. J. Kaufmann, Wind reduction by aerosol particles, *Geophysical Research Letters*, *33*, L24814, 2006.

REFERENCES

301. Pryor, S. C., R. J. Barthelmie, M. S. Bukovsky, L. R. Leung, and K. Sakaguchi, Climate change impacts on wind power generation, *Nature Reviews Earth & Environment*, 1, 627–643, 2020.
302. Beyer, H. G. and B. A. Niclasen, Assessment of the option "wind power to heat for buildings" with respect to meteorological conditions, *Meteorologische Zeitschrift*, 28, 79–85, 2019.
303. Jacobson, M. Z., On the correlation between building heat demand and wind energy supply and how it helps to avoid blackouts, *Smart Energy*, 1, 100009, doi:10.1016/j.segy.2021.100009, 2021.
304. Kahn, E., The reliability of distributed wind generators, *Electric Power Systems*, 2, 1–14, 1979.
305. Archer, C. L. and M. Z. Jacobson, Spatial and temporal distributions of US winds and wind power at 80 m derived from measurements, *Journal of Geophysical Research*, 108(D9), 4289, 2003.
306. Archer, C. L. and M. Z. Jacobson, Supplying baseload power and reducing transmission requirements by interconnecting wind farms, *Journal of Applied Meteorology and Climatology*, 46, 1701–1717, doi:10.1175/2007JAMC1538.1, 2007.
307. Gorman, W., J. Rand, N. Manderlink, et al., Hybrid power plants: Status of operating and proposed plants, 2024 Edition, Lawrence Berkeley National Laboratory, 2024, https://tinyurl.com/ye2m37fc (accessed September 25, 2024).
308. Stoutenburg, E. D., N. Jenkins, and M. Z. Jacobson, Power output variations of co-located offshore wind turbines and wave energy converters in California, *Renewable Energy*, 35, 2781–2791, doi:10.1016/j.renene.2010.04.033, 2010.
309. Stoutenburg, E. K. and M. Z. Jacobson, Reducing offshore transmission requirements by combining offshore wind and wave farms, *IEEE Journal of Oceanic Engineering*, 36, 552–561, doi:10.1109/JOE.2011.2167198, 2011.
310. Jacobson, M. Z., A.-K. von Krauland, S. J. Coughlin, F. C. Palmer, and M. M. Smith, Zero air pollution and zero carbon from all energy at low cost and without blackouts in variable weather throughout the US with 100% wind-water-solar (WWS) and storage, *Renewable Energy*, 184, 430–444, 2022.
311. Jacobson, M. Z., The cost of grid stability with 100% clean, renewable energy for all purposes when countries are islanded versus interconnected, *Renewable Energy*, 179, 1065–1075, 2021.
312. Kempton, W. and J. Tomic, Vehicle-to-grid power fundamentals: Calculating capacity and net revenue, *Journal of Power Sources*, 144, 268–279, 2005.
313. Kempton, W. and J. Tomic, Vehicle-to-grid power implementation: From stabilizing the grid to supporting large-scale renewable energy, *Journal of Power Sources*, 144, 280–294, 2005.
314. Budischak, C., D. Sewell, H. Thompson, et al., Cost-minimized combinations of wind power, solar power, and electrochemical storage, powering the grid up to 99.9% of the time, *Journal of Power Sources*, 225, 60–74, 2013.

REFERENCES

315. Child, M., A. Nordling, and C. Breyer, The impacts of high V2G participation in a 100% renewable Aland energy system, *Energies*, 11, 2206, doi:10.3390/en11092206, 2018.
316. Hart, E. K. and M. Z. Jacobson, A Monte Carlo approach to generator portfolio planning and carbon emissions assessments of systems with large penetrations of variable renewables, *Renewable Energy*, 36, 2278–2286, doi:10.1016/j.renene.2011.01.015, 2011.
317. Kirby, B. J., Frequency regulation basics and trends, ORNL/TM-2004/291, 2004, www.consultkirby.com/files/TM2004-291_Frequency_Regulation_Basics_and_Trends.pdf (accessed September 10, 2024).
318. NRC (US National Research Council), *Real Prospects for Energy Efficiency in the United States*, National Academies Press, Washington, DC, USA, 2010, p. 251, www.nap.edu/read/12621/chapter/6#251, 2010 (accessed September 10, 2024).
319. Erlich, I. and M. Wilch, Primary frequency control by wind turbines, IEEE PES General Meeting, July 25–29, 2010, doi:10.1109/PES.2010.5589911, https://ieeexplore.ieee.org/document/5589911 (accessed September 10, 2024).
320. Roselund, C., Inertia, frequency regulation and the grid, PV Magazine, 2019, https://pv-magazine-usa.com/2019/03/01/inertia-frequency-regulation-and-the-grid/ (accessed September 10, 2024).
321. Sorensen, B., A plan is outlined to which solar and wind energy would supply Denmark's needs by the year 2050, *Science*, 189, 255–260, 1975.
322. Lovins, A. B., Energy strategy: The road not taken, *Foreign Affairs*, 55, 65–96, 1976.
323. Sorensen, B., Scenarios of greenhouse warming mitigation, *Energy Conversion and Management*, 37, 693–698, 1996.
324. Jacobson, M. Z. and G. M. Masters, Exploiting wind versus coal, *Science*, 293, 1438, 2001.
325. Jacobson, M. Z., W. G. Colella, and D. M. Golden, Cleaning the air and improving health with hydrogen fuel-cell vehicles, *Science*, 308, 1901–1905, 2005.
326. Archer, C. L. and M. Z. Jacobson, Evaluation of global wind power, *Journal of Geophysical Research*, 110, D12110, doi:10.1029/2004JD005462, 2005.
327. Czisch, G., Szenarien zur zukünftigen Stromversorgung, kostenoptimierte Variationen zur Versorgung Europas und seiner Nachbarn mit Strom aus erneuerbaren Energien. PhD Dissertation, University of Kassel, 2005, https://kobra.uni-kassel.de/handle/123456789/200604119596 (accessed September 10, 2024).
328. Czisch, G. and G. Giebel, Realisable scenarios for a future electricity supplies based 100% on renewable energies, Riso-R-1608 (EN), 2007, https://tinyurl.com/3ue6wn94 (accessed September 10, 2024).
329. Lund, H., Large-scale integration of optimal combinations of PV, wind, and wave power into the electricity supply, *Renewable Energy*, 31, 503–515, 2006.
330. Hoste, G. R. G., M. J. Dvorak, and M. Z. Jacobson, Matching hourly and peak demand by combining different renewable energy sources, Stanford

University Technical Report, 2009, https://tinyurl.com/55cvrp8r (accessed September 10, 2024).
331. Jacobson, M. Z. and M. A. Delucchi, A path to sustainable energy by 2030, *Scientific American*, November 2009.
332. Delucchi, M. Z. and M. Z. Jacobson, Providing all global energy with wind, water, and solar power, Part II: Reliability, system and transmission costs, and policies, *Energy Policy*, 39, 1170–1190, doi:10.1016/j.enpol.2010.11.045, 2011.
333. Jacobson, M. Z., R. W. Howarth, M. A. Delucchi, et al., Examining the feasibility of converting New York State's all-purpose energy infrastructure to one using wind, water, and sunlight, *Energy Policy*, 57, 585–601, 2013.
334. Jacobson, M. Z., M. A. Delucchi, A. R. Ingraffea, et al., A roadmap for repowering California for all purposes with wind, water, and sunlight, *Energy*, 73, 875–889, doi:10.1016/j.energy.2014.06.099, 2014.
335. Jacobson, M. Z., M. A. Delucchi, G. Bazouin, et al., A 100 percent wind, water, sunlight (WWS) all-sector energy plan for Washington State, *Renewable Energy*, 86, 75–88, 2016.
336. Jacobson, M. Z., M. A. Delucchi, M. A. Cameron, and B. A. Frew, A low-cost solution to the grid reliability problem with 100 percent penetration of intermittent wind, water, and solar for all purposes, *Proceedings of the National Academy of Sciences*, 112 (49), 15060–15065, doi:10.1073/pnas.1510028112, 2015.
337. Jacobson, M. Z., M. A. Delucchi, M. A. Cameron, et al., Impacts of Green-New-Deal energy plans on grid stability, costs, jobs, health, and climate in 143 countries, *One Earth*, 1, 449–463, doi:10.1016/j.oneear.2019.12.003, 2019.
338. Jacobson, M. Z., M. A. Cameron, E. M. Hennessy, et al., 100 percent clean, and renewable wind, water, and sunlight (WWS) all-sector energy roadmaps for 53 towns and cities in North America, *Sustainable Cities and Society*, 42, 22–37, doi:10.1016/j.scs.2018.06.031, 2018.
339. Jacobson, M. Z., A.-K. von Krauland, Z. F. M. Burton, et al., Transitioning all energy in 74 metropolitan areas, including 30 megacities, to 100% clean and renewable wind, water, and sunlight, *Energies*, 13, 4934, doi:10.3390/en13184934, 2020.
340. Lund, H. and B. V. Mathiesen, Energy system analysis of 100% renewable energy systems – The case of Denmark in years 2030 and 2050, *Energy*, 34, 524–531, 2009.
341. Mason, I. G., S. C. Page, and A. G. Williamson, A 100% renewable energy generation system for New Zealand utilizing hydro, wind, geothermal, and biomass resources, *Energy Policy*, 38, 3973–3984, 2010.
342. Connolly, D., H. Lund, B. V. Mathiesen, and M. Leahy, The first step to a 100% renewable energy-system for Ireland, *Applied Energy*, 88, 502–507, 2011.
343. Mathiesen, B. V., H. Lund, and K. Karlsson, 100% renewable energy systems, climate mitigation, and economic growth, *Applied Energy*, 88, 488–501, 2011.

REFERENCES

344. Mathiesen, B. V., H. Lund, D. Connolly, et al., Smart energy systems for coherent 100% renewable energy and transport solutions, *Applied Energy*, *145*, 139–154, 2015.
345. Hart, E. K., E. D. Stoutenburg, and M.Z. Jacobson, The potential of intermittent renewables to meet electric power demand: A review of current analytical techniques, *Proceedings of the IEEE*, 100, 322–334, doi:10.1109/JPROC.2011.2144951, 2012.
346. Hart, E. K. and M. Z. Jacobson, The carbon abatement potential of high penetration intermittent renewables, *Energy & Environmental Science*, *5*, 6592–6601, doi:10.1039/C2EE03490E, 2012.
347. Elliston, B., M. Diesendorf, and I. MacGill, Simulations of scenarios with 100% renewable electricity in the Australian National Electricity Market, *Energy Policy*, *45*, 606–613, 2012.
348. Rasmussen, M. G., G. B. Andresen, and M. Greiner, Storage and balancing synergies in a fully or highly renewable pan-European power system, *Energy Policy*, *51*, 642–651, 2012.
349. Steinke, F., P. Wolfrum, and C. Hoffmann, Grid vs. storage in a 100% renewable Europe, *Renewable Energy*, *50*, 826–832, 2013.
350. Becker, S., B. A. Frew, G. B. Andresen, et al., Features of a fully renewable US electricity-system: Optimized mixes of wind and solar PV and transmission grid extensions, *Energy*, *72*, 443–458, 2014.
351. Frew, B. A., S. Becker, M. J. Dvorak, G. B. Andresen, and M. Z. Jacobson, Flexibility mechanisms and pathways to a highly renewable US electricity future, *Energy*, *101*, 65–78, 2016.
352. Bogdanov, D. and C. Breyer, North-east Asian super grid for 100% renewable energy supply: Optimal mix of energy technologies for electricity, gas, and heat supply options, *Energy Conversion and Management*, *112*, 176–190, 2016.
353. Bogdanov, D., J. Farfan, K. Sadovskaia, et al., Radical transformation pathway towards sustainable electricity via evolutionary steps, *Nature Communications*, *10*, 1077, doi:10.1038/s41467-019-08855-1, 2019.
354. Child, M. and C. Breyer, Vision and initial feasibility analysis of a decarbonized Finnish energy system for 2050, *Renewable and Sustainable Energy Reviews*, *66*, 517–536, 2016.
355. Aghahosseini, A., D. Bogdanov, L. S. N. S. Barbosa, and C. Breyer, Analyzing the feasibility of powering the Americas with renewable energy and inter-regional grid interconnections by 2030, *Renewable and Sustainable Energy Reviews*, *105*, 187–205, 2019.
356. Blakers, A., B. Lu, and M. Socks, 100% renewable electricity in Australia, *Energy*, *133*, 471–482, 2017.
357. Barbosa, L. S. N. S., D. Bogdanov, P. Vainikka, and C. Breyer, Hydro, wind, and solar power as a base for a 100% renewable energy supply for South and Central America, *PLoS One*, doi:10.1371/journal.pone.0173820, 2017.

REFERENCES

358. Lu, B., A. Blakers, and M. Stocks, 90–100% renewable electricity for the South West Interconnected System of Western Australia, *Energy, 122*, 663–674, 2017.
359. Gulagi, A., D. Bogdanov, and C. Breyer, A cost optimized fully sustainable power system for Southeast Asia and the Pacific Rim, *Energies, 10*, 583, doi:10.3390/en10050583, 2017.
360. Esteban, M., J. Portugal-Pereira, B. C. Mclellan, et al., 100% renewable energy system in Japan: Smoothening and ancillary services, *Applied Energy, 224*, 698–707, 2018.
361. Zapata, S., M. Casteneda, M. Jiminez, et al., Long-term effects of 100% renewable generation on the Colombian power market, *Sustainable Energy Technologies and Assessments, 30*, 183–191, 2018.
362. Sadiqa, A., A. Gulagi, and C. Breyer, Energy transition roadmap towards 100% renewable energy and role of storage technologies for Pakistan by 2050, *Energy, 147*, 518–533, 2018.
363. Barasa, M., D. Bogdanov, A. S. Oyewo, and C. Breyer, A cost optimal resolution for sub-Saharan Africa powered by 100% renewables in 2030, *Renewable and Sustainable Energy Reviews, 92*, 440–457, 2018.
364. Caldera, U. and C. Breyer, Role that battery and water storage play in Saudi Arabia's transition to an integrated 100% renewable energy power system, *Journal of Energy Storage, 17*, 299–310, 2018.
365. Liu, H., G. B. Andresen, and M. Greiner, Cost-optimal design of a simplified highly renewable Chinese network, *Energy, 147*, 534–546, 2018.
366. Teske, S., D. Giurco, T. Morris, et al., Achieving the Paris Climate Agreement Goals, 2019, https://tinyurl.com/34uk5uam (accessed September 10, 2024).
367. Hansen, K., B. Mathiesen, and I. R. Skov, Full energy system transition towards 100% renewable energy in Germany in 2050, *Renewable and Sustainable Energy Reviews, 102*, 1–13, 2019.
368. Oyewo, A. S., A. Aghahosseini, M. Ram, and C. Breyer, Transition towards decarbonized power systems and its socio-economic impacts in West Africa, *Renewable Energy, 154*, 1092–1112, 2020.
369. Marczinkowski, H. M. and L. Barros, Technical approaches and institutional alignment to 100% renewable energy system transition of Madeira Island – Electrification, smart energy and the required flexible market conditions, *Energies, 13*, 4434, 2020.
370. Li, M., M. Lenzen, D. Wang, and K. Nansai, GIS-based modelling of electric-vehicle-grid integration in a 100% renewable electricity grid, *Applied Energy, 262*, 114577, 2020.
371. Alves, M., R. Segurado, and M. Costa, On the road to 100% renewable energy systems in isolated islands, *Energy, 198*, 117321, 2020.
372. Kiwan, S. and E. Al-Gharibeh, Jordan toward a 100% renewable electricity system, *Renewable Energy, 147*, 423–436, 2020.

REFERENCES

373. Zozmann, E., L. Goke, M. Kendziorski, et al., 100% renewable energy scenarios for North America – Spatial distribution and network constraints, *Energies*, *14*, 658, 2021.
374. Cole, W. J., D. Greer, P. Denholm, et al., Quantifying the challenge of reaching a 100% renewable energy power system for the United States, *Joule*, *5*, 1732–1748, 2021.
375. Pombo, D. V., J. Martinez-Rico, and H. M. Marczinkowski, Towards 100% renewable islands in 2040 via generation expansion planning: The case of Sao Vicent, Cape Verde, *Applied Energy*, *315*, 118869, 2022.
376. Marocco, P., R. Novo, A. Lanzini, G. Mattiazzo, and M. Santarelli, Towards 100% renewable energy systems: The role of hydrogen and batteries, *Journal of Energy Storage*, *57*, 106306, 2023.
377. Oyewo, A. S., S. Sterl, S. Khalili, and C. Breyer, Highly renewable energy systems in Africa: Rationale, research, and recommendations, *Joule*, *7*, 1437–1470, 2023.
378. Icaza, D., D. Vallejo-Ramirez, C. G. Granda, and E. Marin, Challenges, roadmaps, and smart energy transition toward 100% renewable energy markets in American islands: A review, *Energies*, *17*, 1059, 2024.
379. Kuriyama, A., X. Liu, K. Naito, A. Tsukui, and Y. Tanaka, Importance of long-term flexibility in a 100% renewable energy scenario for Japan, *Sustainability Science*, *19*, 165–187, 2024.
380. Shaikh, R. A., D. J. Vowles, A. Dinovitser, A. Allison, and D. Abbott, Robust capital cost optimization of generation and multitimescale storage requirements for a 100% renewable Australian electricity grid, *PNAS Nexus*, *3*, pgae127, 2024.
381. Breyer, C., S. Khalili, D. Bogdanov, et al., On the history and future of 100% renewable energy systems research, *IEEE Access*, *10*, 78176–78218, 2022.
382. Brown, T. W., T. Bischof-Niemz, K. Blok, et al., Response to "Burden of proof: A comprehensive review of the feasibility of 100% renewable electricity systems," *Renewable and Sustainable Energy Reviews*, *92*, 834–847, 2018.
383. Diesendorf, M. and B. Elliston, The feasibility of 100% renewable electricity systems: A response to critics, *Renewable and Sustainable Energy Reviews*, *93*, 318–330, 2018.
384. Edelstein, S., Report: EV battery costs hit a new low in 2021, but they might rise in 2022, Green Car Reports, 2021, www.greencarreports.com/news/1134307_report-ev-battery-costs-might-rise-in-2022 (accessed September 10, 2024).
385. Levine, S., The electric: How the China playbook has changed the rules in EVs and batteries, June 2024, https://tinyurl.com/2tv5fa75 (accessed September 10, 2024).
386. Abbott, C., USDA says land near solar and wind farms tends to remain in agriculture, 2024, https://tinyurl.com/25pp5yux (accessed September 23, 2024).
387. Macdonald-Smith, A., South Australia's big battery slashes $40m from grid control costs in first year, 2018, https://tinyurl.com/4ecu7abf (accessed September 10, 2024).

388. Spector, J., "Cheaper than a peaker": NextEra inks massive wind+solar+storage deal in Oklahoma, 2019, https://tinyurl.com/23yn69sc (accessed September 10, 2024).
389. McCarthy, D., Chart: Almost all new US power plants are carbon free, 2024, https://tinyurl.com/3vpcktbb (accessed September 10, 2024).
390. O'Malley, I., China built out record amount of wind and solar power in 2024, 2025, https://tinyurl.com/3wepy6kp (accessed May 25, 2025).
391. Miller, C., M. Plucinski, A. Sullivan, et al., Electrically caused wildfires in Victoria, Australia are over-represented when fire danger is elevated, *Landscape and Urban Planning, 167,* 267–274, 2017.
392. Larsen, P. H., A method to estimate the costs and benefits of undergrounding electricity transmission and distribution lines, *Energy Economics, 60,* 47–61, 2016.
393. Thomas, H., Richard Branson's disappointing space jaunt, 2021, https://tinyurl.com/bdcryde5 (accessed September 10, 2024).
394. Metz, M., J. London, and P. Rosler, Gasoline superusers, 2021, https://tinyurl.com/392yckxv (accessed September 10, 2024).
395. Nilsen, E. and R. Wilson, Americans have saved billions with a law they know next to nothing about, 2024, www.cnn.com/2024/08/16/climate/ira-tax-credit-savings/index.html (accessed September 10, 2024).
396. Jacobson, M. Z., Developing, coupling, and applying a gas, aerosol, transport, and radiation model to study urban and regional air pollution. PhD Dissertation, Department of Atmospheric Sciences, University of California, Los Angeles, 1994.
397. Jacobson, M. Z., R. Lu, R. P. Turco, and O. B. Toon, Development and application of a new air pollution modeling system. Part I: Gas-phase simulations, *Atmospheric Environment, 30B,* 1939–1963, 1996.
398. Jacobson, M. Z., Development and application of a new air pollution modeling system. Part II: Aerosol module structure and design, *Atmospheric Environment, 31A,* 131–144, 1997.
399. Jacobson, M. Z., Development and application of a new air pollution modeling system. Part III: Aerosol-phase simulations, *Atmospheric Environment, 31A,* 587–608, 1997.
400. Jacobson, M. Z., Simulations of the rates of regeneration of the global ozone layer upon reduction or removal of ozone-destroying compounds. *EOS Supplement,* F119, Fall 1995.
401. Jacobson, M. Z., Global direct radiative forcing due to multicomponent anthropogenic and natural aerosols, *Journal of Geophysical Research, 106,* 1551–1568, 2001.
402. Jacobson, M. Z., GATOR-GCMM: A global through urban scale air pollution and weather forecast model. 1. Model design and treatment of subgrid soil, vegetation, roads, rooftops, water, sea ice, and snow, *Journal of Geophysical Research, 106,* 5385–5401, 2001.

REFERENCES

403. Jacobson, M. Z., GATOR-GCMM: 2. A study of day- and nighttime ozone layers aloft, ozone in national parks, and weather during the SARMAP field campaign, *Journal of Geophysical Research, 106*, 5403–5420, 2001.
404. Jacobson, M. Z., Y. J. Kaufmann, and Y. Rudich, Examining feedbacks of aerosols to urban climate with a model that treats 3-D clouds with aerosol inclusions, *Journal of Geophysical Research, 112*, D24205, doi:10.1029/2007JD008922, 2007.
405. Jacobson, M. Z., Studying the effects of aerosols on vertical photolysis rate coefficient and temperature profiles over an urban airshed, *Journal of Geophysical Research, 103*, 10593–10604, 1998.
406. Jacobson, M. Z., Isolating nitrated and aromatic aerosols and nitrated aromatic gases as sources of ultraviolet light absorption, *Journal of Geophysical Research, 104*, 3527–3542, 1999.
407. New York Times, Text of President Bush's remarks on global climate, 2001, https://tinyurl.com/nr8t6up8 (accessed September 10, 2024).
408. Dvorak, M., C. L. Archer, and M. Z. Jacobson, California offshore wind energy potential, *Renewable Energy, 35*, 1244–1254, doi:10.1016/j.renene.2009.11.022, 2010.
409. Dvorak, M. J., E. D. Stoutenburg, C. L. Archer, W. Kempton, and M. Z. Jacobson, Where is the ideal location for a US East Coast offshore grid? *Geophysical Research Letters, 39*, L06804, doi:10.1029/2011GL050659, 2012.
410. Dvorak, M. J., B. A. Corcoran, J. E. Ten Hoeve, N. G. McIntyre, and M. Z. Jacobson, US East Coast offshore wind energy resources and their relationship to peak-time electricity demand, *Wind Energy, 16*, 977–997, doi:10.1002/we.1524, 2013.
411. Jacobson, M. Z., J. H. Seinfeld, G. R. Carmichael, and D. G. Streets, The effect on photochemical smog of converting the US fleet of gasoline vehicles to modern diesel vehicles, *Geophysical Research Letters, 31*, L02116, doi:10.1029/2003GL018448, 2004.
412. TED (Technology, Education, Development), Does the world need nuclear energy? Public debate, Mark Z. Jacobson and Stewart Brand, Long Beach, California, February 11, 2010, www.ted.com/talks/debate_does_the_world_need_nuclear_energy?language=en (accessed September 10, 2024).
413. Solutions Project, Our 100% Clean Energy Vision, 2019, www.thesolutionsproject.org/why-clean-energy/ (accessed September 10, 2024).
414. Ruffalo, M. A., M. Krapels, and M. Z. Jacobson, A plan to power the world with wind, water, and sunlight, Talks at Google, Google, Inc., Mountain View, California, June 20, 2012, www.youtube.com/watch?v=N_sLt5gNAQs (accessed September 10, 2024).
415. Ruffalo, M. Z., M. Z. Jacobson, M. Krapels, and video from J. Fox, Powering the world, US, and New York with wind, water, and sunlight, The Nantucket Project, Nantucket, Massachusetts, October 6, 2012, http://vimeo.com/52038463 (accessed September 10, 2024).

REFERENCES

416. Sierra Club, Saying farewell to Ready For 100, 2022, www.sierraclub.org/articles/2022/04/saying-farewell-ready-for-100 (accessed September 10, 2024).
417. Letterman, D., *The Late Show with David Letterman*, New York City, October 9, 2013, https://vimeo.com/83279421 (accessed September 10, 2024).
418. California Senate, SB 100 FAQs, 2018, https://focus.senate.ca.gov/sb100/faqs (accessed September 10, 2024).
419. House (US House of Representatives), H.Res.540, 2015, www.congress.gov/bill/114th-congress/house-resolution/540/text (accessed September 10, 2024).
420. Shepherd, M., The climate science behind the green new deal – A layperson's explanation, 2019, https://tinyurl.com/68ujfkax (accessed September 10, 2024).
421. Green Party US, Green New Deal – Full Language, 2018, www.gp.org/gnd_full (accessed September 10, 2024).
422. O'Malley, M., A jobs agenda for our renewable energy future, 2015, www.p2016.org/omalley/omalley070215climate.html (accessed September 10, 2024).
423. Clinton, H., Hillary is ready for 100, October 16, 2015, www.c-span.org/clip/campaign-2016/user-clip-hillary-is-readyfor100/4557641 (accessed May 25, 2025).
424. DNC (Democratic National Committee), 2016 Democratic Party Platform, July 9, 2016, www.presidency.ucsb.edu/documents/2016-democratic-party-platform (accessed May 25, 2025).
425. Sanders, B. and M. Jacobson, The American people, not Big Oil, must decide our climate future, *The Guardian*, April 29, 2017, https://tinyurl.com/bdd2rktx (accessed September 10, 2024).
426. REN21 (Renewable Energy Policy Network for the 21st Century), Renewables 2024 global status report, 2024, www.ren21.net/gsr-2024/modules/global_overview/02_policy/ (accessed September 10, 2024).
427. Lillian, B., Orsted survey: Eight out of 10 support 100% global renewable energy, November 13, 2017, https://tinyurl.com/ph8f5zuc (accessed September 10, 2024).
428. Stanford Solutions Project, Infographic roadmaps to transition cities, states, and countries to 100% wind-water-solar (WWS) for all energy purposes, 2024, https://sites.google.com/stanford.edu/wws-roadmaps/home (accessed September 10, 2024).
429. REN21 (Renewable Energy Policy Network for the 21st Century), Renewables in cities: 2021 global status report, 2019, www.ren21.net/wp-content/uploads/2019/05/REC_2021_full-report_en.pdf (accessed September 10, 2024).
430. Boyer, G., Appalachian power rolls out 100% renewable option for customers, WFXR News, 2019, https://tinyurl.com/ynd5zubk (accessed September 10, 2024).

REFERENCES

431. Duke Energy, More renewable energy options available under Duke Energy's Green Source Advantage, 2019, https://tinyurl.com/46sfu87p (accessed September 10, 2024).
432. RE100, The world's most influential companies committed to 100% renewable power, 2025, http://there100.org (accessed May 25, 2025).
433. SEIA (Solar Energy Industries Association), Solar means business 2024, 2024, https://seia.org/research-resources/solar-means-business/ (accessed November 20, 2024).
434. Climate Group, EV100 members, 2019, www.theclimategroup.org/ev100-members (accessed September 10, 2024).

Index

100 percent movement, 351, 373,
376–377, 380–381
100 percent network, 376, 379
100 percent WWS
Cost. *See* Energy cost
Roadmaps. *See* Roadmaps, 100 percent WWS
Studies on, 319
Technologies. *See* WWS (wind, water, solar), Technologies
Transition timeline. *See* Transition timeline
100.org, 376

Absorption of light, 242
Acid rain, 2, 128, 203, 271, 352–353
Adams, William, 226
Adipic acid, 137
Aerosol, 1–2, 6, 9–10, 12–13, 199, 202, 242, 288, 357–358
Defined, 1
Aerosol particle, 1–3, 6, 9, 12–13, 154, 199, 202–203, 242, 288, 357–358
Defined, 1
Afghanistan, 16, 247, 359
Africa, 265, 288, 320, 323
Agricultural fires, 132
Agricultural machines, 276, 331
Agricultural sector, xvi, 132, 137, 174–175, 263, 275–276, 291, 326, 335, 390
Agricultural waste, 9, 133, 175
Agriculture industry, 138
Air-conditioning, 19, 84, 87–89, 99, 108, 110, 112, 203, 289, 300–301, 322, 341
Air heating, 84, 100, 110, 144, 149, 277, 290, 341
Air pollution, xiv–xvi, 1–3, 8–9, 12–14, 18–19, 32–33, 53, 70, 106, 109, 122, 125, 130, 132, 138–140, 142, 145, 151, 153, 155–156, 158, 160–165, 167–168, 170, 173–174, 176, 178–179, 181, 183, 185, 189–190, 194, 202–203, 219, 243, 271, 288–290, 300, 320, 329–330, 335, 339–340, 342, 344, 346, 348, 351–353, 355–359, 361–362, 364–366, 374–375, 380, 390–392
Deaths, 2–3, 155, 189, 271, 288, 337, 375, 390
Illnesses, 288
Impacts on wind speed, 300
Airbus, 82
Aircraft. *See* Vehicles, Aircraft
Long-haul flights, 338
Transition timeline, 338
Aircraft, electric. *See* Vehicles, Battery-electric, Aircraft
Airfoil, 253–254
Alabama, 47, 65
Albedo, 138
Alberta, 90
Alex, Ken, 376
Algae, xvi
Aliso Canyon fossil-gas disaster, 288
All-of-the-above policy, 138, 339
Alphabet, 387
Altamont Pass wind farm. *See* Wind farm, Altamont Pass
Alternating current. *See* Electric current, Alternating current (AC)
Frequency, 207, 256, 317
Alternator, 212, 219
Alumina, 121
Aluminum, 43, 69, 105, 117, 126–127, 130, 209, 213, 252, 316

INDEX

Amazon, 387
American Motor Company, 57
Ammonia, 51, 131, 155, 165, 177, 179
   Manufacturing, 20–21, 131
Ampère, André-Marie, 210
Ancillary services, 315
   Frequency regulation. *See* Frequency regulation
   Load following. *See* Load following
   Regulation. *See* Regulation
   Replacement reserves. *See* Replacement reserves
   Spinning reserves. *See* Spinning reserves
   Supplemental reserves. *See* Supplemental reserves
   Voltage control. *See* Voltage control
Anderson, Chris, 363
Anderson, Robert, 55
Andrews, Wallace C., 85
Anemometer, 251–252
Angle of attack, 254–255
Antarctic ozone depletion, 2
Anthropogenic emissions, xvi, 2, 6–9, 11, 14, 19, 126, 136–137, 145, 147
Anthropogenic heat flux. *See* Heat flux, Anthropogenic
Anthropogenic water vapor. *See* Water vapor, Anthropogenic
Apollo, Temple of, 142
Apple, 387
Appliances, efficient. *See* Efficiency, Energy, Appliances
Aquifer-thermal-energy storage. *See* Storage, Aquifer-thermal-energy storage
Arc furnace. *See* Furnaces, Arc
Arc lamp, carbon, 116
Archer, Cristina L., 319, 361
Arctic sea ice, 154
Arkansas, 70
Array losses. *See* Wind farm, Array losses
Arsenic, 226
Aspdin, Joseph, 126
Asthma, 2
Australia, 70, 128, 131, 164, 201, 227–228, 265, 334
   South, 295

Autothermal reforming of methane, 72, 167–170
Azerbaijan, 17

Bacteria, 2, 9, 72, 134, 136
Bailey, William, 97
Band-gap energy, 229–231, 234–235
Band-gap wavelength, 230–231, 235
Barth, Jannette, 370
Basalt, 127
Baseload power, 26, 29–30, 33, 36, 292, 303, 320–321, 361
Basic oxygen furnace, 123–125
Bastnasite, 253
Batchelor, Charles, 216
Bats, wind turbine impacts on, 246, 270–272
Batteries, 15–16, 20–21, 24, 32–33, 36, 38, 40–44, 48, 51–52, 55–56, 61, 64–66, 68–71, 75, 77, 81–82, 89, 106–108, 110–112, 155, 190, 209, 213, 248, 257, 277, 292, 295–296, 301, 305, 307, 309–310, 313–314, 316–318, 322, 327, 334–336, 341–343, 349
   Aluminum ion, 43
   Basalt-stone, 42
   Cell, 38, 40
   Colocated with PV, 236
   Cost. *See* Energy cost, Battery cost
   Cycles, 37, 40, 43, 188, 292
   Efficiency, 40, 48–49
   Fires, 71
   Iron-air, 42
   Lithium-ion, 39, 43, 69–70, 309
   Lithium-iron-phosphate, 40, 61
   Long-duration storage, 309
   Pack, 38, 41, 64, 66–67, 79, 81, 324
   Rechargeable, 55
   Recycling, 69
   Salt-water, 43
   Sodium-sulfur, 42–43
   Swapping, 56
   Vanadium flow, 44
Battery-electric vehicles. *See* Vehicles, Battery-electric
BECCS. *See* Bioenergy, With carbon capture and storage (BECCS)
Becquerel, Edmond, 226

INDEX

Before the Flood, 379
Beinecke, Frances, 370
Belgium, 94, 143
Bell Labs, 227
Betz, Albert, 258
Betz limit, 258
Betz's law, 258
Biden, Joe, 378
Biking, 344, 349
Biodiesel, xvi, 2, 51, 176, 179–181
  Algae, 181
  Soy, 181
Bioenergy, xv–xvi, 2, 7, 9, 11, 149, 157, 164, 175, 178–179, 202, 275, 277, 280–282, 290–291, 320, 327, 332, 347, 350, 352
  With carbon capture and storage (BECCS), 175, 178
Biofuel burning, 6, 123
Biofuels, xvi, 9, 51, 107, 175, 179, 181–182, 184, 275, 278, 281, 319–320, 339
Biogas, 9, 143, 175, 177, 319–320
Biogenic gas, xvi
Biomass, xv–xvi, 2, 7, 9, 11, 86, 114–115, 117, 120, 123, 132–134, 137, 139–140, 148, 151, 170, 174–179, 274, 288, 319–321, 337, 346
  Agriculture residues, 175
  Contaminated wastes, 176
  Energy crops, 176–177
  Forestry residues, 176
  Industry residues, 176–177
  Park and garden wastes, 176
Biomass burning, 6–7, 9, 11, 132–134, 137, 170, 174, 288
Birds, wind turbine impacts on, 246, 266–267, 270–272
  Raptors, 271–272
Black, Joseph, 143, 171
Black carbon, 9–11, 67, 80, 145, 157, 161, 203, 242, 359–360, 376
Black-lung disease, 18
Blackouts, 155, 291–293, 297, 318, 324, 390
  California, 293
  Texas, 293
Blast furnace, 123–126

Blathy, Otto, 218, 220
Block Island wind farm, 249. *See also* Wind farm, Block Island
Blyth, James, 248
Boats, electric, 64
Boiler. *See* Resistance boiler
Boilers, 84, 86, 88, 90, 92, 95, 101, 279–280, 341
Bollinger semi-truck, 60
Borehole-thermal-energy storage. *See* Storage, Borehole
Boron, 231–232, 253
Boundary Dam Project, 162
Boyle, Robert, 140
Bradley, C. S., 220
Braedstrup. *See* Storage, Borehole
Brand, Stewart, 363
Branson, Richard, 344
Brazil, 28, 180, 223, 293, 300
Breeder reactor. *See* Nuclear reactor, Breeder
Bridge fuel, 138, 151, 153, 155
Bronchitis, 2
Bronze Age, 121
Brown, Harold, 221
Brown carbon, 6, 9–11, 80, 133, 161, 203–204, 242
Brown Jr., Edmund G., 375
Brune, Michael, 372
Brush, Charles F., 248
Buckiewicz, Mike, 373
Building materials
  Sustainable, 342
Bulgaria, 43
Bunker tanker, electric, 66
Bunsen, Robert, 144
Bunsen burner, 144
Bush, George W., 360
Business-as-usual, 101, 274–275, 277, 280–282, 285–286, 288, 290, 323–325, 348, 384, 390
Butanol, 180–181
BYD (Build Your Dreams), 61

Cable car, 62–63
Calcium carbonate, 123, 126–129, 171
Calicchia, Marsha, 366

INDEX

California, 13, 26, 30, 58, 62, 70, 80, 82, 95–97, 111, 156, 228, 248, 287, 293, 296, 299, 320–321, 326, 336, 351, 356, 361–364, 368, 371, 373, 375–377, 384, 386
   Days with 100 percent WWS, 296
   Reasons for high electricity prices, 288
California Air Resources Board, 57
Californium
   252, 185
Canada, 5, 90, 162, 188, 201, 207, 265, 293, 299, 302, 323, 359, 377, 382
Cap and trade, 347
Capacitor, 213
Capacity factor, 45, 264, 284, 359
   Wind farm, 264
   Wind turbine, 259–260
      US-average, 260
Capital cost. *See* Energy cost, Capital cost
Carbon capture, 15–16, 18, 125, 147–148, 150–151, 154, 157–166, 169, 173–174, 178, 194, 201, 320, 339, 350, 362
   Efficiency, 159, 165, 168–169
   Energy penalty, 159
   Projects, 161, 166–167, 169
   And storage (CCS), 147, 157–158, 164–165, 179, 196
   And use (CCU), 158, 162, 164
Carbon dioxide, 4–7, 10, 27, 67, 80, 115, 122–133, 135, 142–143, 145, 150, 153–154, 157–174, 176–179, 193–195, 199, 202, 282, 290, 322, 332–333, 340, 344, 346, 349, 359
   Compression, 159, 163
   Emissions, 153, 157, 171, 179, 332
      Combustion, 115
      Process, 115
   Process emissions, 123, 127, 130
   Time needed to recover, 332
Carbon footprint, 344
Carbon monoxide, 80, 106, 123, 133, 155, 161, 167–170, 340
Carbon tax, 346
Carbothermic reduction of silica, 130
Cargo ship, electric, 66
Catalytic converter, 83
Catamaran, 66
Cavendish, Henry, 72

CCU. *See* Carbon capture, And use (CCU)
Cement manufacturing, 7, 114, 126–127, 164, 332
Centrifugal diffusion, 197
Cesium, 137, 200
Chad, 3
Chain saw, electric, 107, 302
Chapin, Daryl, 227
Charcoal, 16, 140, 143
Chargers, electric vehicle, 16, 349
Chernobyl nuclear meltdown, 199
Chicago World's Fair, 222
Chile, 295
Chillers, 86, 88, 95, 341
China, 3, 28, 38, 41, 60, 63, 65, 80, 86, 94, 139, 143, 148, 192–193, 200, 223, 241, 253, 300, 323–324, 334, 382
Chromia, 121
Climate march, New York, 378
Clinton, Hillary, 381
CNG. *See* Fossil gas, Compressed
CO_2-equivalent emissions, 145–147, 149–151, 153, 155, 163–164, 172–173, 176–178, 181–182, 185, 193–194, 199, 289–290, 344, 362
   Total, 151
Coal, xv–xvi, 2, 7, 9, 15–18, 37–38, 45, 61–65, 70, 85–86, 105, 114, 123, 128–129, 138–143, 145, 148, 151, 153–158, 161–166, 169–170, 172–173, 176–179, 186, 194, 214–215, 219, 227, 255, 261, 271, 274–275, 280–281, 292–293, 298, 303, 317, 319, 327, 335, 337, 346, 359, 361–362, 373
   Sea, 140
Coal gasification, 72, 166, 169–170, 319
Coastal Virginia wind farm. *See* Wind farm, Coastal Virgina
Cobalt, 40
Coefficient of performance, 100–101, 279–281
Cold
   Demand, 291, 301, 304, 306, 312, 322
      Flexible, 305, 312
      Inflexible, 305, 312
   Production, 262–263, 313, 390
   Storage. *See* Storage, Cold
   Supply, 74, 291

425

INDEX

Colorado, 37, 220, 384, 386
Combined-cycle gas turbine, 37, 152–154, 173, 261
Command-and-control policy, 346
Commercial sector, 275
Commitments to 100 percent WWS
  Companies. *See* Companies committed to 100 percent WWS
  States. *See* States committed to 100 percent WWS
Community-choice-aggregation (CCA) utility, 111, 343, 386
Companies committed to 100 percent WWS, 387
Compressed natural gas. *See* Fossil gas, Compressed
Compressed-air energy storage. *See* Storage, Compressed air
Compressor, 49
Computer modeling, 58, 80, 96, 140, 268–269, 282–283, 302, 314–315, 319–320, 322, 332, 355–358, 361, 364
  Air pollution model, 356
  GATOR-GCMOM, 358, 361
  GATOR-MMTD, 357
  UCLA General Circulation Model, 357
  Weather prediction model, 356
Concentrated solar power (CSP), xvi, 19–21, 33–36, 44–45, 149–150, 155, 225, 235, 276, 283, 299, 302, 310, 316, 320, 322, 326–327, 362, 390
  Central tower, 34, 44–45, 120
  Cooling, 34, 299
  Parabolic dish, 34, 120
  Parabolic trough, 34, 44
Concentrator PV cells. *See* PV cells, Concentrator
Concrete, 12, 48, 74, 86, 88, 102–103, 114–115, 127–129, 150–151, 158, 167, 252, 342, 349
  Geopolymer, 127–128
  Industry, 126
  Manufacturing, 126
  Recycling, 129
  Trapping carbon dioxide in, 129, 158
Conduction, 25, 102
Conduction band, 229–231, 233

Conductivity
  Electrical, 225
  Thermal, 102
Conductor, electrical, 209
Connecticut, 247, 384–385
Construction machines, 276, 331
  Electric, 64
Container ship, electric, 66
Contrails, 67, 80, 175
COPD (chronic obstruction pulmonary disease), 2
Copper, 70, 97, 105, 117, 139, 209, 215, 217, 227–228, 234, 252, 257, 316
Countries
  In conflict, 336
  In poverty, 337
Cozzarelli Prize, 380
Croatia, 215
Crop residues, 16
Crotched Mountain wind farm. *See* Wind farm, Crotched Mountain
Cruise ship, electric, 67
Cuba, 323–324
Cuomo, Andrew M., 365, 367, 370–371
Cuprous oxide, 226–228
Curacao, 249
Curtailment, 32, 88, 260, 262, 311, 313
Czisch, Gregor, 320

Daimler, 78, 80–81
Daimler semi-truck, 60
Darrieus, Georges Jean Marie, 251
Dart leader, 206
Data
  Local, 282
  Remotely sensed, 282
  Satellite, 282–283
Data center, 16, 26, 86, 108, 113, 189
Davidson, Robert, 61, 63
Day, Richard, 226
DC electricity. *See* Electric current, Direct current (DC)
de Cervantes, Miguel, 247
de Ferranti, Sebastian Ziani, 217
Deaths
  From air pollution. *See* Air Pollution, Deaths
  From ethanol vehicles, 183

INDEX

From Fukushima meltdowns, 200
From heat, xiv
From underground uranium mining, 201
Delucchi, Mark A., 320, 329, 363–364, 366, 370
Demand response, 20, 112, 273, 291–292, 301, 304–308, 310, 312, 314–316, 319, 321–323, 330, 345, 348–349, 390
Democratic Party platform, 381
Denaturant, 180
Dengue fever, 289
Denmark, 65, 86, 91–94, 148, 248–249, 295, 319–321, 382
Depletion region, 232–233
Deri, Miksa, 218
DiCaprio, Leonardo, 367, 373, 378–379
Dielectric heater, 20, 115, 119, 277, 280, 301
  Microwave heating, 119
  Radio-frequency heating, 119
Diesel fuel, xv, 2, 9, 11, 16, 51–52, 67, 70, 72, 159, 164, 166–167, 174–175, 179–182, 274, 278, 281, 349
Diet, animal, 134
Digester, 177
  Methane, 134
Direct air capture
  How it works, 171
  Natural, 170
  Problems, 172
  Synthetic, 170, 174, 320, 339, 350
Direct current. *See* Electric current, Direct current (DC)
Direct-drive wind turbine. *See* Wind turbine, direct drive
Direct heating, 114
Disappearing wind syndrome, 300
Discount rate, 285–286
  Private, 285–286
  Social, 285–286
Distributed energy. *See* Power plants, Decentralized
Distributed PV, 31
  Advantages, 236
  Behind-the-meter, 31, 236, 297
  In-front-of-the-meter, 31, 236
Distribution, 274

Distribution, electricity, 19–20, 31, 80, 85, 90, 96, 111–112, 222–224, 250, 257–258, 271, 274–275, 284–285, 291, 302, 323, 330, 336, 386
  Line losses, 260, 262
District cooling, 85–86
District heating, 42, 84–89, 91–92, 95–96, 323, 330, 341, 343, 348
  Efficiency, 96
  Fifth-generation, 87
  First-generation, 85
  Fourth-generation, 86
  Second-generation, 85
  Stanford SESI project, 86, 95
  Third-generation, 86
Dominican Republic, 324
Don Quixote, 247
Drake Landing Solar Community, 90–92
Drake, Edwin, 144
Dredger, electric, 66
Dronninglund. *See* Storage, Pit-thermal-energy
Drought, 13, 298–299
Dung, xvi, 2, 16, 105, 175
Dvorak, Michael J., 361, 366
Dynamo, 215–217, 220–221, 255

Earth-sun distance, 241
Eberhard, Martin, 58
Edison, Thomas, 52–53, 142, 144, 214–217, 219–221
Efficiency
  Electrolyzer. *See* Electrolyzer, Efficiency
  Energy, 19, 107–109, 274–275, 277, 281–282, 291, 302, 316, 341, 345–348, 369–370, 376
    Appliances, 21, 108, 302
    Hydrogen compression, 76
    Hydrogen-fuel-cell, 77
      Electric vehicle, 76–77
      For electricity generation, 75
    Plug-to-wheel, 53–54, 76–77, 278–279
    Tank-to-wheel, 53, 76, 278–279
Eiffel Tower, 380
Eindhoven. *See* Storage, Aquifer-thermal-energy storage
Einstein, Albert, 229

INDEX

El Salvador, 295
Electric bicycle. *See* Vehicles, Battery-electric, E-bikes
Electric chair, 221
Electric cracker, 20, 118, 277
Electric current, 41, 105, 115–117, 119, 207, 209–210, 225–226
   Alternating current (AC), 19, 31, 58, 207–208, 211, 218, 220–221, 240, 256, 261
   Direct current (DC), 19, 31, 117, 207, 211, 215, 218, 240, 261
Electric field, 207–208, 213–214, 225, 231, 233
Electric Launch Company, 64
Electric load. *See* Electricity, Demand
Electric motorcycle. *See* Vehicles, Battery-electric, E-motorcycles
Electric scooters. *See* Vehicles, Battery-electric, E-scooters
Electric vehicles. *See* Vehicles, Battery-electric
Electrical circuit, 207–209
Electricity, xvi, 11, 16, 19–34, 36–38, 40–49, 51–55, 58, 61–66, 70, 72–77, 84, 86, 88–90, 95–97, 99–101, 105–107, 110–112, 114–118, 120, 124–131, 139, 142, 144, 146–156, 162–165, 167–168, 171, 173, 175–179, 185–187, 193–196, 198–199, 201, 203–228, 230, 234, 240, 246, 248, 250, 252, 255–257, 260–263, 267, 270–271, 273–282, 284–285, 291–295, 299–303, 305–324, 327, 330, 334, 337, 340, 342–343, 345–346, 348–349, 361–362, 364, 371, 376–377, 381, 383–387, 390–391
   Alternating current (AC), 221
   Demand, 209, 291, 301, 304–306, 308, 312, 316, 322
      Day, 297
      Flexible, 29, 305–308, 310–311, 323
      Inflexible, 305–308, 310–311, 313, 315
      Night, 297
   Direct current (DC), 221
   Lightning. *See* Lightning
   Single-phase, 212
   Spot price. *See* Spot price of electricity
   Static, 205
   Storage, 21, 24, 33, 36, 38, 42–43, 48, 89–90, 196, 291, 307, 309–313, 316, 323–324, 327, 348, 390
   Supply, 74, 291, 304, 312–313
   Three-phase, 212
   Wired, 206
   WWS, 18–19
Electricity storage. *See* Electricity, Storage
Electro-fuels, 159, 174, 350
   Carbon-based, 175
Electrolysis, 21, 72, 76, 124–126, 166, 285, 316, 319
Electrolyzer, 20, 49, 66, 72–73, 75–77, 279, 311
   Efficiency, 76
   First, 73
   Polymer electrolyte membrane, 73
   Solid oxide, 73
Electromagnetism, 210–211
Electron-beam heaters, 20, 115, 118–119, 301, 331
Electrons, 39–40, 116, 118–119, 205–208, 225–233
Eli Lilly, 387
Emissions
   Carbon dioxide. *See* Carbon dioxide, Emissions
   Carbon dioxide-equivalent. *See* CO2-equivalent emissions
   Covering land, 150
   Lifecycle, 145–148, 154, 177, 188, 193–194
   Negative, 178
   Opportunity cost, 146–148, 193
   Vehicle, 52
Emphysema, 2
End-use energy. *See* Energy, End-use
Energy
   End-use, 262, 267, 270, 274–275, 277–278, 281–282, 286, 291, 319, 338, 383, 390
   End-use demand, 101, 240, 253, 266, 275, 277, 282, 320
   Primary, 274–275, 291
   Sectors, xvi–xvii, 104, 205, 266–267, 273–275, 319, 321–322, 326, 329, 340, 344, 371, 376, 380–382, 390–391
Energy, clean, renewable, xvii, 1, 3, 15–16, 18, 96, 107, 109, 117, 125, 145, 273,

INDEX

290–291, 329, 333–334, 337, 345, 347, 351, 365, 371, 375, 379–383
Energy consumption. *See* Energy, End-use
Energy cost
  Annual cost, 324
  Battery cost, 40, 50, 309, 324
  Capital cost, 112–113, 195, 285, 324–325, 338
  Climate cost, 325
  Health cost, 325
  Levelized cost. *See* Levelized cost of electricity
  Social cost. *See* Social cost of energy
  Storage cost, 324
  Transmission cost, 323
Energy efficiency. *See* Efficiency, Energy
Energy insecurity, xv, 14–15, 17–19, 138–139, 145, 202–203, 320, 339–340, 390
Energy security, 1, 18, 109, 139, 151, 185, 199, 320, 329–330, 339, 361–362, 364, 366, 380, 390
Energy-efficient appliances. *See* Efficiency, Energy, Appliances
England, 140, 228. *See also* United Kingdom
Enhanced geothermal systems, 26–27, 190, 283, 298
Enhanced oil recovery, 158, 161–164, 169, 179
Ethanol, xvi, 2, 51, 161, 174–176, 179–181, 183, 361
  Cellulosic, 180
  Corn, 179, 182
  E85, 180, 182–183
  Refinery, 179, 182, 185
  Sugarcane, 180
  Water consumption, 183
  With carbon capture, 184
Ethene, 142
Europe, 5, 78, 86, 101, 140, 200, 207, 215, 222, 247–248, 256, 265, 298, 300, 302, 317, 320, 323–324
European Union, 17, 86, 223, 359
EV1, 57
Evelyn, John, 140
Eviation, 68
Externality cost, 285

Facebook, 369
Faraday, Michael, 210

Feed-in tariff, 345
Fermentation, 184
Ferrock, 127–129
Ferry, electric, 64–65
Fertilizer, 9, 131–132, 137
Finland, 191
Firebricks, 20–21, 74, 120–121, 280, 305, 310–312, 322
  Cost, 310
  Direct resistance heating, 121
  Metallic resistance heating, 121
Fireplace, electric, 106, 302
Flexible demand. *See* Electricity, Demand, Flexible
Floating platform. *See* Wind turbine, Floating
Florida, 30, 97, 234
Fly ash, 128, 161
Flywheel, 20–21, 36, 46, 155, 302, 305, 311, 316, 318
Footprint area, 326, 362
  Agriculture, 326
  Solar PV and CSP, 326
  Wind farm, 263, 326
Forbidden band, 229
Ford *F-150*, 54, 60, 185
Fossil fuels, xiv–xv, 1–2, 11, 14, 16, 18, 51, 73, 107, 109, 117, 120, 124, 149, 151, 156–157, 159–160, 164–165, 172–173, 179, 181, 193, 195, 203, 267, 277–278, 280–282, 288, 290–291, 320, 326, 332, 339, 347, 361, 366, 369, 389
Fossil gas, xv–xvi, 2, 7, 9, 15, 17, 29, 37–38, 45, 86, 88, 95, 97, 101, 105–106, 109–112, 114, 129, 131, 136, 138–139, 142–145, 148, 151–157, 161, 163–168, 173, 176, 178, 194, 261, 271, 274–276, 278–281, 292–293, 298, 319, 321, 327, 334–337, 339–343, 347, 365–366, 370, 373, 381, 388
  Abandoned wells, 156, 167
  Active wells, 156
  Bans, 387
  Compressed, 152
  Demand, 297
  Distribution, 152
  Electricity, 151–152
  Electricity production, 144

429

INDEX

Fossil gas (cont.)
  Heat, 151
  Heating, 144
  Land requirements, 156
  Leaks, 134, 167
  Liquefied, 51, 152, 162, 164
  Processing, 152
Fossil-fuel industry, 138
Foundations. *See* Wind turbine, Foundations
Fox, Josh, 365–366, 370, 373
Fracking, 152, 157, 271, 365–366, 370–371, 373
France, 116, 134, 192–193, 199, 216–218, 226, 228, 292, 298, 382
Fredonia Gas Light Company, 144
Frequency. *See* Alternating current (AC), Frequency
Frequency regulation, 317
Fritts, Charles, 226–227
Fuel cells, 75
  Hydrogen, xvii, 16, 19, 49, 74–81, 135, 168, 177, 277–279, 301, 330–331, 338, 347, 361
Fuel oils, 114
Fuel rod, 187, 197–198
Fuel-cell stack, 75, 79, 83
Fukushima nuclear meltdowns, 195, 199–200
Fuller, Calvin, 227
Funicular, 61–62
Furnaces, 20
  Arc, 20, 115–117, 209, 277, 280, 301, 327, 331
  Induction, 20, 117, 277, 280, 301, 327, 331
  Efficiency, 117
  Resistance, 20, 115, 117, 124, 277, 280, 301, 327, 331

Gallium, 226, 234–235
Galvanometer, 210
Gamma rays, 186
Garbage patch, Pacific Ocean, 133
Gas, defined, 1, 143
Gasland, 365, 373
Gasoline, xv, 2, 9, 51–53, 56–58, 64–65, 72, 74, 106–107, 111–112, 149, 159, 164, 167, 174–175, 177, 179–180, 182–183, 274, 278–279, 281, 349, 361
Gaulard, Lucien, 218

Gaza, 336
Gearbox, 22, 79, 251–252, 256, 261
Geared wind turbine. *See* Wind turbine, Geared
Gearless wind turbine. *See* Wind turbine, Direct drive
General Motors, 57
Generator, 21–22, 24–28, 33–34, 37, 44–49, 52, 150, 153, 186, 212, 215, 218–219, 223, 250–252, 255–257, 262, 284–285, 299, 310, 316, 318
  AC asynchronous, 255, 257
  AC synchronous, 255
  Induction, 256
  Nameplate capacity, 259
  Permanent-magnet, 256
  Rotor, 256
  Three-phase AC, 220
  Wound-field, 256
Geoengineering, 174, 201–204, 391
  Carbon capture, 202
  Problems, 202
  Solar radiation management, 201–202
  White roofs, 202–204
Geopolymer cement, 127
Geopolymer concrete. *See* Concrete, Geopolymer
Georgia, US, 192
Geothermal
  Reservoir, 26–27
Geothermal, flash steam plant, 26
Geothermal energy, 25
  Electricity, 20, 26, 292, 321, 323, 362
  Enhanced. *See* Enhanced geothermal systems
  Heat, xvi, 19–20, 25–26, 87–88, 301, 303, 306, 322–323, 390
Geothermal plants, 21, 26
  Binary, 26–27, 70, 150
  Dry steam, 26–27
  Flash steam, 26–27
Germanium, 226, 235
Germany, 47, 63, 70, 80–81, 140, 144, 216, 249, 295, 298, 382
Geysers Resort Hotel, 26
Gillibrand, Kirsten, 369
Global ozone reduction, 2

INDEX

Global warming, 2–3, 5–14, 19, 67, 109,
    133–134, 136, 138–139, 145, 150,
    153–154, 170, 174, 176, 185, 189–190,
    194, 199, 202–203, 267–268, 289, 292,
    298–300, 320, 329, 333, 339–340, 352,
    359–362, 366, 375, 380, 390
  Historical, 332
  Impacts on
    Air pollution, xiv
    Droughts and floods, xiii
    Fossil and nuclear power, 298
    Heat-related deaths, xiv
    Hurricanes, xiii
    Hydropower, 299
    Sea levels, xiii
    Solar power, 299
    Wildfires, xiv
    Wind power, 300
Global warming potential, 136
Google, 368–369, 387
Gorgon Project, 164
Gram. *See* Storage, Pit-thermal-energy
Graphite, 130
Gravitational storage. *See* Storage,
    Gravitational
Greece, 139, 142, 249
Green bank, 370
Green building standards, 108, 348
Green hydrogen, 19, 49, 72–73, 125, 170,
    198, 276–278, 330
Green New Deal, 290, 324, 365, 380
Greenhouse effect, natural, 3–5
Greenhouse gases, xvi, 4–10, 12–14, 70, 134,
    136, 150, 154, 160, 179, 202, 266–267,
    289, 339, 352, 359, 371, 380, 385
Greenland, 139
Gregory, Philip, 388
Grid stability, 291, 304, 310
  Ancillary services. *See* Ancillary services
  Loss of load expectation, 312
  Measures to ensure, 312
  Oversizing
    Nameplate capacity, 307–308, 313
    Storage, 313
    Transmission, 314
  Steps to ensure, 311
  Vehicle-to-grid. *See* Vehicle-to-grid

Weather forecasting, 314
With fossil fuels, 293

Haber–Bosch process, 131
Haiti, 223, 324
Halladay, Daniel, 247, 250
Hallidie, Andrew Smith, 62–63
Hallwachs, Wilhelm, 227–228
Halogens, 6–7, 132, 136, 203
Harbour Air, 68
Hart, Elaine K., 321, 364
Hart, William, 144
Hawaii, 287, 384
Heart disease, 2
Heat, 4, 9, 12, 14, 104
  Demand, 291, 301, 304, 306, 308,
    311–312, 322
    Flexible, 305, 311–312
    Inflexible, 305, 311–312
  From radioactive decay, 4, 25
  Primordial, 4
  Production, 73, 87, 126–127, 140, 149,
    262–263, 313, 315, 390
  Storage. *See* Storage, Heat
  Supply, 74, 291, 304
Heat, waste, 21, 53, 75–76, 87, 89, 96,
    148–149, 153, 274, 278, 280, 282
Heat capacity, 102
Heat exchanger, 89
Heat flux, anthropogenic, 6, 11–12, 14, 145,
    147–149
Heat-island effect, urban. *See* Urban heat-
    island effect
Heat pumps, 19–20, 25, 42, 45, 74, 76,
    84, 86–90, 93–101, 108, 110, 112,
    119–120, 277, 279–282, 291, 301, 303,
    305, 311, 318, 323, 327, 331, 337,
    341–343, 347
  Air-source, 87, 100
  Clothes dryer, 20, 101, 323
  Coefficient of performance. *See*
    Coefficient of performance
  Dishwasher, 101, 323
  Ductless minisplit, 98, 110
  Efficiency, 99
  Ground-source, 87, 99
  Hot tub, 101

INDEX

Heat pumps (cont.)
  Washing machine, 323
  Waste-source, 87
  Water heater, 100
  Water-source, 87
Heat storage. *See* Storage, Heat
Heat stroke, 13, 289
Heat wave, 296
Heron of Alexandria, 246
Heroult, Paul, 116
Hindenburg, 72
Hino Motors, 80
Hochschild, David, 376
Hole, 91, 229, 232, 364
Holly, Birdsill, 85
Holly Steam Combination Company, 85
Home Retrofits, 342
Home, all-electric, 109, 112, 341
Honda, 57, 80
  Insight, 57
Hope, Sarah, 373
Horace, 139
Hoste, Graeme, 362
Howarth, Robert, 366, 370
Hub height. *See* Wind turbine, hub height
Hummer EV, GMCC, 60
Hurlbut, Brandon, 378
Hurricanes, xiii, 13, 15, 246, 259, 266, 268–270, 289, 298
  Category-5, 259
  Eye wall, 269–270
  Katrina, 269
  Maria, 15
  Sandy, 269
  Storm surge, 13, 266, 268–270
  Surface pressure, 270
  Wind speed, 13, 268, 270
  Wind turbine impacts on. *See* Wind turbine, Impacts on hurricanes
HVAC. *See* Transmission, HVAC
HVDC. *See* Transmission, HVDC
Hydraulic fracturing. *See* Fracking
Hydroelectricity. *See* Hydropower
Hydrofoil powerboat, electric, 66
Hydrogen, xvii, 7, 11, 16, 19, 27, 34, 51, 66, 72–83, 124–125, 129, 135–136, 149, 161, 165–170, 177–178, 181, 200, 234, 248, 275–276, 278–279, 285, 291, 301, 305–307, 311, 316, 319, 322–323, 327, 330–331, 338–339, 347–348, 361–362
  Blue, 166, 169, 174
  Brown, 166, 174
  Combustion, 51, 165
  Compressed, 74
  Cryogenic, 74
  Demand, 291, 301, 304, 306, 312, 322
    Flexible, 305
    Inflexible, 305–306, 312
  Density, 74
  Electrolytic. *See* Green hydrogen
  Gray, 131, 166, 169, 174
  Green, 21, 124, 131, 174, 273. *See also* Green hydrogen
  Leaks, 77–78
  Load, 312
  Production, 72, 78, 262–263, 313, 390
  Removal from air, 72
  Sources to air, 72
  Supply, 74, 291
  Turquoise, 166
Hydrogen direct reduction, 124–125
Hydrogen fuel cells. *See* Fuel cells, Hydrogen
Hydrogen storage. *See* Storage, Hydrogen
Hydrogen-fuel-cell-electric vehicles. *See* Vehicles, Hydrogen-fuel-cell-electric
Hydropower, 28–29, 36–38, 200, 274, 276, 284, 292–293, 295, 299, 310–311, 314, 316, 320–323, 377
  Conventional, 21, 28, 36, 38, 100–101, 152, 280, 282, 284, 322, 346
  Reservoir. *See* Storage, Hydropower reservoir
  Run-of-the-river, 21, 28
Hyundai, 81

Ice storage. *See* Storage, ice
Iceland, 86, 172, 223, 293, 323–324
Idaho, 187
Illinois, 384–386
India, 3, 32, 63, 65, 198, 223, 246, 293
Indirect heating, 114
Indoor air pollution, 2
Induction, electromagnetic, 117, 256

INDEX

Induction cookers, electric, 105, 110, 302, 337, 341, 343
Induction cooktop, 20, 323
Induction furnace. *See* Furnaces, Induction
Inductor, 210, 214, 218, 318
Industrial heat, 20, 114
   From electricity, 115
   From fuels, 114
   From steam, 115
Industrial processes, 20, 26, 45, 137, 139, 144, 276
Industrial Revolution, 3, 5, 8, 139–140, 142, 332, 334, 352
Industrial sector, 114, 275, 281, 329
Inflation Reduction Act, U.S., 350
Inflexible demand. *See* Electricity, Demand, Inflexible
Ingraffea, Tony, 366, 370
Installed capacity, 29
Insulation, 21, 42, 74, 86, 91–92, 103, 108, 145, 302, 341–342
Insulator, electrical, 209
Intake fraction, 52
Interconnecting
   Countries, 314
   Wave resources, 303
   Wind and solar resources, 284, 302, 304, 313
   Wind and wave resources, 303
   Wind resources, 302–303, 313
Intergovernmental Panel on Climate Change, 194, 196, 199
Intermittency, 24, 33, 46–47, 153, 155, 292, 317, 320, 323, 361
Inverter, 31, 54, 76–77, 79, 279, 341, 343
Investment subsidies, 346
Iodine-131, 200
Iowa, 55, 296, 369
   WWS electricity percentage, 287
Iran, 198, 246
Iraq, 198
Ireland, 249
Iron, 41, 72, 105–106, 115, 117–118, 123–130, 141, 210, 214, 242, 252–253, 257, 275
Iron Age, 121
Iron nitride, 71, 253, 257
Ironmaking, 114, 123–124, 126

Israel, 17, 323
Itaipu Dam, 28
Italy, 143

Jacobson, Mark Z., 319–321, 329
Jamaica, 324
Japan, 17, 63, 66, 80, 101, 199–200, 240, 382
Jet fuel, 174
Jobs, 61, 107, 327–328, 332, 334, 360, 377, 379, 381, 383–384, 390
   Changes in number due to WWS, 328
   Construction, 327–328
   Direct, 327
   Gains, 327
   Indirect, 327
   Induced, 328
   Losses, 327
   Operation, 327–328
   Permanent, 328
Joby Aviation, 82
Johnson & Johnson, 387
JP Morgan Chase, 387

Kansas, 296
Kazakhstan, 17, 188, 201
Kemp, Clarence, 96
Kenya, 295
Kerosene, xv, 2, 9, 274, 278
Kerry, John, 369, 379
Kiln, 121
Ki-moon, Ban, 379
Kinetic energy, 11, 22, 30, 46, 54, 118, 149, 186, 246, 251, 254, 258, 262–263, 265–268, 270
Kirchhoff's voltage law, 208
Krapels, Marco, 365–366, 368–370, 375–376, 378–379, 387
Kyoto Protocol, 319, 358–360

la Cour, Paul, 248
Land requirements
   Behind-the-meter PV, 237
   Corn ethanol, 183
      In the US, 327
   Fossil gas, 156
   Solar PV, 183
   US fossil-fuel industry, 152

INDEX

Landfill, 7, 134–135, 175, 177
Landfill gas, 135, 176
Late Show with David Letterman, 373
Latent heat storage materials, 103
Lavoisier, Antoine Laurent, 143
Lawn mowers, 20, 327
   Electric, 106–107, 302
Lawsuits, climate, 388
   Held v. Montana, 388
   Juliana v. United States, 388
   Navahine v. Hawaii, 389
Leaf blowers, 20, 38, 106, 277, 302
   Electric, 106, 305
Leaks
   Carbon dioxide. *See* Carbon dioxide, Leaks
   Fossil gas. *See* Fossil gas, Leaks
   Hydrogen. *See* Hydrogen, Leakage
LeBron, Philippe, 143
LED lights, 19–20, 108, 110, 302, 341, 343, 345
Letterman, David, 373–374
Levelized cost of electricity, 195, 285–286
Libya, 223
Lifecycle emissions. *See* Emissions, Lifecycle
Lift force, 253–254
Light bulb, 208–209, 213, 215
Lightning, 2, 8, 205–206, 298
Limestone, 123, 127. *See also* Calcium carbonate
Liquefied gases, 114
Liquefied natural gas. *See* Fossil gas, Liquefied
Liquid, defined, 1
Lithium, 39–40, 43
Lithuania, 249, 295
Liu, Lucy, 374
LNG. *See* Fossil gas, Liquefied
Load following, 29, 33, 36, 38, 155, 315–316
Loan guarantees, 346
London, 61, 63, 82, 95, 140, 143, 217
Lopez, Andres, 378
Los Angeles, 80, 97, 351, 355–357
Lovins, Amory, 319
Lu, Rong, 356
Lund, Henrik, 320–321
Lung cancer, 2, 151, 189–190, 200–201
Luxembourg, 295
Lyme disease, 289

Magellan, xvii
Magnesium carbonate, 171
Magnesium oxide, 126, 171
Maine, 287, 296, 384–385
Maintenance, 22, 24, 30, 46, 61, 106–107, 112, 160, 169, 252, 261, 285, 292, 312, 323, 327
Malaria, 289
Mali, 70
Mandatory emission limits, 346
Manufactured gas, 143
Manure, 7, 134, 175
Marstal. *See* Storage, Pit-thermal-energy
Maryland, 96, 215, 386
Massachusetts, 63, 215, 220, 369, 386
Mastercard, 387
Masters, Gilbert M., 319, 352–353, 359
Mathiesen, Brian V., 321
Matthews, Chris, 369
Mauritius, 323
Menlo Park, California, 388
Mercury, 139, 161
Messmer Plan, 193
Meta, 387
Metakaolin, 128
Metal, 28, 39, 82, 89, 105, 114–117, 119, 123–124, 171, 187, 205–206, 213, 225–226, 228, 233
Methane, xvi, 4, 6–7, 72, 77, 132–136, 142–143, 145, 152, 154–156, 159, 163, 166–170, 173–175, 177, 366
Methane pyrolysis, 72, 135, 169, 177
Methanogenic bacteria, xvi, 134
Methanol, 159, 174
Mexico, 30, 336, 386
Michigan, 386
Microgrid, 16–17, 21, 32, 38, 70, 73, 75, 113, 165, 273, 301, 337–338
   Mobile, 113
Microsoft, 387
Middle East, 17, 247, 320
Military sector, 276
Minckeleers, Jan Pieter, 143
Mining, 18, 52, 70, 73, 125, 130, 146, 151, 154–155, 157, 159–163, 165–168, 170, 172–174, 189–190, 196, 201, 203, 253, 271, 275, 281–282, 326–327, 347

INDEX

Minnesota, 385–386
Miracle technologies, xvii
Molten oxide electrolysis, 126
Molten salt, 33–34, 44–45, 130
Monazite, 253
Montana, 296
    WWS electricity percentage, 287
Morris, Henry G., 56
Morrison, William, 55
Motor, 21, 37, 46–49, 51–55, 57, 61, 63, 69, 71, 73, 75–77, 79, 81–82, 107, 212, 217, 220, 252, 256, 279, 318
    Permanent-magnet, 53, 278
    Three-phase AC, 220
Mouchet, Augustin, 226
Mountain gravity storage. *See* Storage, Mountain gravity
Mountaintop removal, 271
Multijunction PV cells. *See* PV cells, Multijunction (tandem)
Municipal financing, 346–347
Munter, Leilani, 367, 375, 378
Musk, Elon, 58, 367
Myanmar, 336

Nacelle, 251–252, 259
Nameplate capacity, 21–22, 26, 28–30, 37, 44–45, 70, 75, 110, 248, 253, 255, 259–260, 264, 266, 307–308, 312–313
    Growth in, 297
    Oversizing, 308, 313
Namibia, 188
National Renewable Energy Laboratory, 321
National Resources Defense Council, 368
Natural gas. *See* Fossil gas
NDACCS. *See* Direct air capture, Natural
Needham, Rick, 369
Neodymium, 69, 71, 252–253, 257
Net metering, 349
Netherlands, 94, 249
Nevada, 200, 384
New Jersey, 56, 384, 386
New Mexico, 384
New York, 65, 85, 116, 144, 195, 220–222, 269, 365–368, 370–371, 375, 378, 381, 384–386

New York City, 56, 85, 142, 214, 217, 220, 269, 370, 372–374, 387
New York Steam Company, 85
New Zealand, 66, 69, 265, 321, 323
Newcomen, Thomas, 141
Newfoundland, 30
Niagara Falls, 62, 65, 222
Niger, 201
Nigeria, 3
Night ventilation, 104
Nikola, 58, 60, 81
Nikola semi-truck, 60
NIMBYism, 335, 391
    Transmission lines, 335
    Wind farms, 335
Nitrate particles, 6, 12
Nitrogen gas, 4, 137
Nitrogen oxides, 80, 154–155, 161
Nitrous oxide, 4, 6–7, 9, 132–133, 136–137, 174, 179
Nonenergy emissions, xv, xvii, 14, 132, 332–333, 390
Nonmetal, 225
Norquist, Grover, 369
North Carolina, 385
North Dakota, 101, 369, 386
North Korea, 3, 198
Norway, 65–66, 249, 293
Novo Nordisk, 387
NRDC. *See* National Resources Defence Council
Nuclear fission, 185, 187
Nuclear fusion, 190
Nuclear industry, 138
Nuclear power, 15–18, 37, 151, 185–186, 188, 190–192, 194, 196, 199–200, 271, 292, 320, 347, 362–364
    Contribution to air pollution, 194
    Contribution to global warming, 194
    Cost, 194
    Lifecycle emissions, 193
    Meltdowns, 151, 189–190, 199. *See also* Nuclear reactor, Meltdown
    Mining risk, 201
    Radioactive waste. *See* Radioactive waste
    Weapons proliferation, 147, 151, 188–190, 194, 196, 198–199

INDEX

Nuclear reactor, 185–187, 191, 193, 195, 197–198, 298
  Barakah 1 to 4, 192
  Boiling water, 186
  Breeder, 187–188
  Construction time, 191
  Darlington, 193
  Diablo Canyon, 288
  Fast, 188
  Fessenheim, 193
  Flamanville 3, 192
  Fukushima Dai-ichi, 200
  Haiyang 1 and 2, 192
  Hinkley Point C, 191
  Light-water, 187, 197
  Meltdown, 195, 200
  Olkiluoto 3, 191
  Once-through, 187–188, 198, 200
  Planning-to-operation time. *See* Planning-to-operation time
  Pressurized water, 186–187
  Refurbishment time, 193
  Ringhals, 193
  Shidao Bay, 192
  Small modular, 138, 188–189
  Taishan 1 and 2, 192
  Vogtle 3 and 4, 192
Nuclear Regulatory Commission, 191
Nuclear subsidies, 195
Nuclear weapons. *See* Nuclear power, Weapons proliferation

O'Malley, Martin, 381
Obama, Barak H., 369
Ocean-current energy devices, xvi, 21, 30, 224, 276
Offshore wind farm. *See* Wind farm, Offshore
Ohio, 248, 386
Ohl, Russel, 227
Oil, xv, 2, 9, 15, 17, 33–34, 44, 52, 66, 70, 85, 107, 138–139, 144–145, 148, 152, 156, 158, 161–163, 165, 167, 169, 172, 176, 180–181, 186, 274, 276, 280–281, 317, 335–337, 373, 376, 382
Oil embargo, Arab, 57, 193
Oklahoma, 296
Olds, Ransom, 56

Olson, Julia, 388
Onshore wind. *See* Wind farm, Onshore
Open-cycle gas turbine, 152–153, 292
Opinion poll on renewable energy, 382
Opportunity cost emissions. *See* Emissions, Opportunity cost
Optimal tilt angle, 243–244
Oracle of Delphi, 142
Oregon, 217, 296, 385
Organic PV cells. *See* PV cells, Organic
Orsted, Hans Christian, 210
Our Children's Trust, 388
Output subsidies, 345
Oversizing nameplate capacity. *See* Grid stability, Oversizing nameplate capacity
Oxygen gas, 4, 5, 7–8, 11, 41, 72–73, 75, 116, 123–124, 126, 134–135, 137, 142, 149, 167–169, 176
Ozone, 2–8, 133, 180, 203, 242, 288, 359–361
  Layer, 5, 8

Pacific Gas & Electric, 336
Pakistan, 3, 198
Papin, Denis, 140
Paracelsus, 72
Paraguay, 28
Paris, 127, 215–216, 226, 379
  Climate conference, 379
Parker, Thomas, 55
Partial-peak electricity price, 112
Particulate matter, 106, 155
Passive heating and cooling, 96, 101, 108, 302
Payback time of investment in WWS, 113, 325, 338
Peaking power plants. *See* Power plants, Peaking
Pearson, Gerald, 227
Pennsylvania, 144, 199, 215, 219, 365
Penstock, 28–29, 37
Permanent magnet, 46, 71, 252–253, 256–257
  Iron nitride. *See* Iron nitride
  Neodymium. *See* Neodymium
Permitting, 5, 348–349, 386
Perovskite, 235
Petra Nova Project, 163
Phase-change materials, 103

INDEX

Philippines, 3
Phosphorus, 69, 116, 123, 231–232
Photoelectric effect, 228–229
Photoelectrochemical water splitting, 72–73, 166
Photosynthesis, 5, 7, 133, 151, 170, 177, 183
Photovoltaic effect, 226–229
Photovoltaics, xvi, 15–16, 21, 31–33, 36, 65–66, 70, 73, 75–76, 95–97, 106, 110–112, 130, 149, 177, 195, 203, 225, 227–228, 231, 233, 235, 240–241, 243–244, 257, 274, 276, 283, 295, 299, 302, 316, 318, 320, 322–323, 326–327, 334–336, 341, 362
  Floating, 224, 240–241
  Tilting. *See* Tilting, solar panel
  Utility scale, 177, 244, 283, 326
    Hybrid farm with wind, 189
Phytoplankton, xvi
Piero Conti, 26
Pipelines, 77, 96, 151–152, 156, 275–276, 291, 327, 335–337, 340
  Carbon dioxide, 158, 170, 184–185
  Oil-and-gas, 144, 165, 172, 239
Pitch, 247, 252, 259, 318
  Control, 255, 259, 261
  System. *See* Wind turbine, Pitch system
Pit-thermal-energy storage. *See* Storage, Pit-thermal-energy
Planning-to-operation time, 146, 148, 191–195
Planté, Gaston, 55
Platinum, 82–83, 226
Plutarch, 142
Plutonium, 185–189, 196–198
p-n junction, 227, 232–233
Pneumonia, 2–3
Policies
  Cap and trade. *See* Cap and trade
  Carbon tax. *See* carbon tax
  Command and control. *See* Command and control
  Feed-in-tariff. *See* Feed-in-tariff
  Green bank. *See* Green bank
  Incentives for energy efficiency, 345
  Investment subsidies. *See* Investment subsidies
  Laws requiring demand response, 345
  Loan guarantees. *See* Loan guarantees
  Mandatory emission limits. *See* Mandatory emission limits
  Municipal financing. *See* Municipal financing
  Net metering. *See* Net metering
  Output subsidies. *See* Output subsidies
  Pollution tax. *See* Pollution tax
  Production tax credit. *See* Production tax credit
  Purchase incentives and rebates. *See* Purchase incentives and rebates
  Renewable portfolio standards. *See* Renewable portfolio standards
  Revenue-neutral carbon tax. *See* Revenue-neutral carbon tax
  By sector, 347
Political will, 329
Pollution tax, 346
Polonium, 18
Portland cement, 126–129
Portugal, 62, 295
Potassium carbonate, 171
Power, 209
  Reactive. *See* Reactive power
  Real. *See* Real power
Power plants
  Biomass, 298, 330
  Centralized, 15–16, 19, 84–87, 142, 172, 261–262, 341–342
  Coal, 15, 139–140, 148, 153–154, 157, 162–164, 173, 215, 261, 298, 330, 358, 372
  CSP. *See* Concentrated solar power
  Decentralized, 15–16
  Distributed, 261
  Fossil fuel, 16
  Fossil gas, 15, 29, 38, 45, 155, 298, 330, 335
  Geothermal, 70, 86, 150, 326–327. *See also* Geothermal plants
  Hydropower. *See* Hydropower
  Maintenance. *See* Maintenance
  Nuclear, 15–18, 37, 255, 261, 271, 298, 330, 348

INDEX

Power plants (cont.)
  Oil, 298, 330
  Peaking, 29, 33, 36, 38, 46, 155, 323
  Photovoltaic (PV). *See* Photovoltaics
  Wind. *See* Wind farms, Wind turbines
Powerhouse, 28–29
Pressure cooker, 140
Primary energy. *See* Energy, Primary
Prius, Toyota, 57
Private cost of energy, 284–286, 290, 323–325, 384
Process emissions, 127
Procter & Gamble, 387
Production tax credit, 348
Public transport, 20, 107–108, 344, 349
Puerto Rico, 15, 384
Pumped hydropower. *See* Storage, Pumped hydro
Purchase incentives and rebates, 346
PV array, 236, 369
PV cells, 73, 75, 225–228, 230–236
  Amorphous silicon, 233–234
  Cadmium-telluride, 234
  Concentrator, 235
  Copper-indium-gallium-diselenide, 234
  Gallium-arsenide, 234
  Microcrystalline silicon, 234
  Multijunction (tandem), 234
  Organic, 234
  Perovskite-silicon, 235
  Polycrystalline silicon, 233
  Single-crystal silicon, 233
  Thin-film, 233
PV panel, 16, 32, 97, 110, 177, 203–204, 225, 231, 236, 240, 243, 257, 299, 322, 326–327, 341
  Bendable, 32
  Foldable, 32
  Horizontal, 242, 244
  Mobile, 32
  Tilted. *See* Tilting, solar panel
  Tracking, 243–244
  Transportable, 32
Pyrolysis. *See* Methane pyrolysis

R1T, Rivian, 60
Radiant heating, 89

Radiation
  Infrared. *See* Heat
  Solar. *See* Sunlight
Radiators, 89
Radioactive waste, 18, 187, 189–190, 195–196, 200, 362
Ramp rate
  Combined-cycle gas turbine, 153
  Open-cycle gas turbine, 153
  Power plant, 37–38, 45, 292, 316–317
Rare-earth metals, 253
Rated power. *See* Nameplate capacity
Rayleigh distribution, 257–258
RE100, 387
Reactive power, 318
Readman, James B., 116
Real power, 318
Rechtschaffen, Cliff, 375–376
Rectifier, 49
Reducing energy use, 107, 109, 277, 281, 345, 376
Refrigerant, 98–100, 110
Refrigeration, 84–85, 88, 306, 308, 322, 334
Regenerative braking, 52, 54
Regulation, 315–316
Renewable portfolio standards, 345, 348
Rental units, WWS solutions for, 343
Replacement reserves, 318
Residential sector, 275, 280
Resistance, electrical, 209
Resistance boiler, 20
Resistance furnace. *See* Furnaces, Resistance
Resistance heating, electric, 20, 115, 117–118
  Direct, 118
  Firebricks, 277, 280, 310–311, 331
  Indirect, 118
Resistance kiln, 20
Return stroke, 206
Revenue-neutral carbon tax, 346
Revolution Wind farm. *See* Wind farm, Revolution
Rhode Island, 385–386
Rice paddies, 7, 134–135
Rice-husk ash, 128
Riverlight Project. *See* Storage, Aquifer-thermal-energy storage

438

INDEX

Roadmaps, 100 percent WWS, 200, 273, 320–321, 329, 351, 366, 370–373, 377–380, 382, 389
  139 countries, 379, 382
  143 countries, 383
  145 countries, 383
  50 states, 378, 380–381, 383
  California, 371, 376
  Cities, 373
  New York, 367
  Washington State, 373
Rockets, 72, 74–75
Rodgers, Andrea, 388
Roman Empire, 139
Rome, 139
Rooftop
  Area, 283
  Solar PV. *See* Solar, Rooftop
Ruffalo, Mark A., 365, 368–370, 372, 374, 378, 387
Russia, 3, 5, 17, 86, 187, 199, 265, 293, 302, 323

Saffir–Simpson scale, 268
Saint-Laurent nuclear accident, 199
Samsung, 387
San Bruno fossil-gas disaster, 156, 288
San Francisco, 62–63, 81, 364, 373, 375
San Gorgonio Pass wind farm. *See* Wind farm, San Gorgonio
Sanders, Bernie, 381–382
Sanderson Brothers Steel Company, 116
Saudi Arabia, 15
Savery, Thomas, 140
SB 100, California law, 377, 384
SB 1383, California law, 376
SB 350, California law, 376–377
Scattering of light, 242
Schneider, Stephen, 374
Scientific American, 362–365, 381, 387, 389
Scobies, Stan, 366, 370
Scotland, 55, 116, 249, 260
SDACCS. *See* Direct air capture, Synthetic
SDACCU. *See* Direct air capture, Synthetic
Selenium, 226–227, 234
Semiconductor materials, 225–227, 229–230

Sensible heat storage materials, 102
Severy, Melvin, 227
Shale, 126, 152, 366
Shallenberger, Oliver B., 220
Ships. *See* Vehicles, Marine
  Timeline to transition, 338
Shockley–Queisser limit, 231
Siemens, Carl W., 116
Siemens, Werner, 227
Sierra Club, 371–373, 376
  100 percent WWS cities campaign, 373
Signorelli, Gianluca, 366
Silica, 121, 126, 128, 130, 233
Silicon, 7, 58, 114–116, 126–128, 130–131, 226–236, 349
  Amorphous, 233–234
  Multicrystalline, 233
  Polycrystalline, 233
  Single-crystal, 233
Silicon dioxide. *See* Silica
Silicon manufacturing, 114, 130–131, 349
Slag, 123, 128
Sluice gate, 28, 37
Small modular reactor. *See* Nuclear reactor, Small modular
Smart grid, 330
Smith, Willoughby, 62, 226
Social cost analysis, 286
Social cost of carbon, 325
Social cost of energy, 284–285, 325, 331, 338, 384
  Ending carbon dioxide emissions, 325
Solar
  Resources, 239
  Rooftop, 15, 21, 96–97, 110, 147–148, 203, 245, 276, 322, 326, 341, 343, 348–349
Solar Challenger, 228
Solar eclipse, 297
Solar heat, xvi, 84, 90–91, 291, 301, 307, 311, 390
Solar Impulse, 68
Solar photovoltaics. *See* Photovoltaics
Solar water heaters, 96
  Active direct circulation, 97
  Active indirect circulation, 97
  Passive, 97

INDEX

Solar-thermal heaters, 119–120, 277
  Central-tower, 120
  Flat-plate, 120
  Linear-Fresnel, 120
  Parabolic dish, 120
  Parabolic trough, 120
Solid, defined, 1
Solstice, summer, 241
Solutions Project, The, 364, 367–373, 375–376, 378, 380–382, 387
  Infographics, 368, 373, 378, 382
Somalia, 3
Sorensen, Bent, 319
South America, 30, 265
South Australia. *See* Australia, South
South Carolina, 195
South Dakota
  WWS electricity percentage, 287
South Fork Wind Farm, 249. *See also* Wind farm, South Fork
South Korea, 207, 323, 382
South pole, 245
Southwick, Alfred P., 220–221
Space shuttle, 72, 74–75
Space tourism, 344
Spacing area, wind farm, 263–265, 326, 335, 362, 384
Spain, 45, 247, 249
Spinning reserves, 318
Spot price of electricity, 287
Sprague, Frank, 52, 63
Stall control
  Active, 255, 259
  Passive, 255, 259
Standard test conditions, 231
Stanford Energy System Innovations project. *See* District heating, Stanford SESI project
Stanford University, 86, 95–96, 352–353, 357, 368, 370–371, 374, 379–380
Stanley, Jr., William, 219–220
States committed to 100 percent WWS, 386
Stator, 256
Steam engine, 26, 140–142, 215
Steam reforming of methane, 72, 131, 166–170, 177–178, 319

Steel, 7, 35, 42, 46, 62, 70, 73–74, 88, 95, 102–103, 105–106, 109, 114–117, 119, 122–125, 128, 141, 158, 165, 177, 252, 275, 342, 349
  Green (fossil-free), 73, 125
  Manufacturing, 20–21, 122
Steelmaking, 123
  Basic oxygen method, 123
  Primary, 123
  Secondary, 123
Stockholm airport. *See* Storage, Aquifer-thermal-energy storage
Storage
  Aquifer, 87–88, 93–94, 310
  Aquifer-thermal-energy storage, 93
  Battery. *See* Batteries
  Battery-electric, 350
  Borehole, 20, 26, 88, 90–91, 135, 310
  Efficiency, 92
  Cold, 19–21, 88–89, 291, 307, 311, 313, 322–323, 327, 348, 390
  Community, 348
  Compressed air, 20–21, 36, 47–48, 155, 302, 310–311, 316, 318, 327
  CSP, 20, 44, 309, 316, 327
  Energy, 336
  Flywheel. *See* Flywheel
  Gravitational, 20–21, 36, 48, 155, 302, 310, 316, 318, 327
  Grid-hydrogen, 20, 75
  Heat, 19–21, 90, 92, 94, 102–103, 291, 307, 311, 313, 322–323, 327, 390
  Home-battery, 348
  Hydrogen, 20–21, 73–75, 79, 291, 307–308, 311–313, 322, 327, 390
  For grid electricity, 49
  Hydrogen-fuel-cell-electric
  For electricity storage, 350
  Hydropower reservoir, 20, 28–29, 36–37, 150, 155, 299
  Ice, 89–90, 322, 327
  Mountain gravity, 49
  Pit-thermal-energy, 92–93, 327
  Pumped hydro, 21, 37–38, 48, 155, 292, 302, 305, 309–310, 318, 322
  Reservoirs, 37
  Tank, 88

INDEX

Undergound thermal energy, 90, 94, 322
Water-pit, 20, 88, 306, 310
Water-tank, 20, 87–88, 91, 95–96, 301, 306, 311–312, 322
Storage capacity, maximum, 22
Storage discharge rate, maximum, 22
Storm surge. *See* Hurricanes, Storm surge
Stoutenburg, Eric, 364
Stranded assets, 331
Stratosphere, 8
Straubel, J. B., 58
Streamer, 206
Streetcars, electric, 53, 62–63
Stroke, 2
Subways, electric, 63
Sudan, 336, 359
Sulfate particles, 6, 12, 80, 133
Sulfur dioxide, 27, 154–155, 161
Sulfur oxides, 80
Summers, Larry, 369
Sunlight, xvi, 4–5, 8–10, 12, 19, 25, 31, 33–35, 45, 73, 102, 104, 108, 119, 149, 154, 177, 201–204, 225–227, 229–231, 233–235, 240–245, 281–284, 288, 291, 299–300, 304, 312, 314, 341, 356–357, 370
  Diffuse, 242
  Direct, 242
  Ultraviolet, 5, 230
Supercapacitors, 52, 213–214
Supplemental reserves, 318
Sustainable aviation fuel, 174
Sweden, 65, 69, 73, 86, 94, 125, 193, 222, 248, 293, 298, 382
Switzerland, 54, 62, 64, 75, 144, 298
Synthetic fuels, 179
Syria, 198

Taiwan, 323, 382
Tajikistan, 299
Tarpenning, Marc, 58
TED debate, 363
Tehachapi Pass wind farm. *See* Wind farm, Tehachapi Pass
Telecommuting, 107, 344
Temperature, global average, 5

Tesla, Inc., 58
  Cars, 58
  Roadster, 58
  Semi-truck, 60
  Trucks, 60
  Vehicles, 53
Tesla, Nikola, 58, 207, 215–216
Texas, 47, 70, 163, 293, 384, 386
Thermal conductivity. *See* Conductivity, thermal
Thermal-mass materials, 20, 102, 104, 108
Thermogenic gas. *See* Fossil gas
Thin-film PV cells. *See* PV cells, Thin-film
Thompson, William, 217
Three Gorges Dam, 28
Three-Mile Island nuclear accident, 199
Thunder, 206
Tibet, 246
Tidal turbines, xvi, 21, 30–31, 36, 224, 276, 285, 292, 300, 302, 326–327, 362, 390
Tides, 30, 268, 314
Tilting, solar panel, 32, 110, 225, 243–244
Time-of-use pricing, 112
Toftlund. *See* Storage, Pit-thermal-energy
Tokelau, 295
Toon, Owen B., 356
Toyota, 57, 80–81
Tragedy of the commons, 346
Trains, electric, 63, 349
Transformer, 48, 130, 217–220, 222–223, 318
  Step-down, 218, 220, 222–223
  Step-up, 218, 220, 222–223
  ZBD, 218
Transition timeline, 329–330
Transmission, 15, 19, 21, 28, 31, 111, 146, 152, 205, 212, 214–215, 217–220, 222–224, 240, 261, 263, 274–275, 284–285, 291–292, 301–303, 307, 312–315, 318, 320, 322–323, 325, 327–328, 330, 335–337, 348, 386, 390
  Downed lines, 15
  HVAC, 19–20, 219, 222–223, 261
  HVDC, 19–20, 222–223, 261, 314, 320
  Interconnecting. *See* Interconnecting
  Line losses, 215, 222–224, 260, 262, 314
  Overhead, 61, 63, 336
  Underground, 336

INDEX

Transportation, xvi, 19, 42, 51, 53, 55, 58, 61–62, 65, 72, 139, 144, 175, 179–180, 273–279, 281–282, 301, 319, 334, 340, 344, 349, 362, 381, 390
    Sector, 275, 329
Tropic of Cancer, 243
Tropic of Capricorn, 243
Troposphere, 8
TSMC, 387
Tugboat, electric, 66–67
Turco, Richard P., 355–356
Turkey, 66
Turkmenistan, 167

US House Bill
    330, 380
    3314, 380
    3671, 380
US House Resolution
    109, 380
    540, 380
US Senate Bill 987, 380
US Senate Resolution
    59, 380
    632, 380
UCLA, 355–358
Ukraine, 336
Ultraviolet. *See* Sunlight, Ultraviolet
Underground thermal-energy storage. *See* Storage, Underground thermal energy
United Arab Emirates, 192
United Kingdom, 48, 55, 126, 191, 295
United Nations Climate Change Conference, 379
United Nations General Assembly, 378
United States, 29, 47, 59, 62, 64, 80, 113, 143, 152, 156, 167, 180–181, 191, 195, 200, 207, 221, 223, 240, 247–248, 265, 286, 299, 302–303, 314, 319, 323, 334, 355, 359, 361, 380–383
Uprating, 29
Uranium, 4, 9, 14, 17–18, 151, 185–189, 194, 196–198, 201, 203, 271, 274, 277, 281–282, 291
    235, 185–188, 197–198
    238, 186–188, 197–198
    Enrichment, 197
    Highly enriched, 197
    Low-enriched, 197
Urban heat-island effect, 6, 12
Utility PV. *See* Photovoltaics, Utility-scale

Vacuum furnace, 119
Valence band, 229–233
Valence electrons, 228
Valence shell, 228, 231
Value of statistical life, 289
Van Helmont, John Baptist, 143
Van Horn, Jodie, 371–372
Variable resource, 24, 33, 292, 313
Vehicle charging, 112, 341, 343
Vehicles
    Aircraft, xvii, 51, 67–69, 74–75, 79–82, 228, 276, 279, 330–331, 338
        Long-haul flights, 277
        Short-haul flights, 277
    Battery-electric, 19, 51–53, 55–59, 69, 77, 182–183, 276, 278, 305, 314, 338, 343, 349, 362
        Aircraft, 55, 67, 80, 276, 331
        Buses, 55, 59–60, 331
        Cable cars, 61
        Container ships, 66
        Dump trucks, 64
        E-bikes, 59, 71
        E-motorcycles, 59, 71
        E-scooters, 59, 71
        Ferries, 65
        Marine, 64
        Passenger cars, 55, 59, 111, 276, 331
        Semi-trucks, 60
        Ships, 276
        Streetcars, 61
        Tractors, 64
        Trains, 61, 276, 331
        Trams, 61
        Trucks, 55, 59, 331
        Tugboats, 66
    Ground, 55, 80
    Hydrogen fuel cell, 181–182
    Hydrogen-fuel-cell-electric, 19, 51, 76–80, 83, 276, 278–279, 291, 319, 338, 362

INDEX

Aircraft, 79, 81, 331
  Cargo airship, 82
  Ferries, 81
  Military, 79
  Trains, 80
  Trucks, 78, 81, 331
  Marine, xvii, 51, 55, 77, 79, 276, 279, 330, 338
  Ships. *See* Vehicles, Marine
Vehicle-to-grid, 314
Venezuela, 28
Ventilated façade, 103, 108
Vermont, 386
Vested interests, 333
Vietnam, 66
Vineyard Wind 1. *See* Wind farm, Vineyard Wind 1
Virginia, 30, 53, 63, 385, 387
Visa, 387
Vojens. *See* Storage, Pit-thermal-energy
Volcanoes, 5, 25
Volta, Alessandro, 143, 226
Voltage, 43, 208, 216, 222, 236, 315, 318
Voltage control, 318
Volvo semi-truck, 60

Walmart, 387
Walz, Tim, 385
Wank, Jon, 368–370, 378
Warming particles, 9, 12–14, 133
Washington, DC, 269, 369, 384
Washington State, 30, 69, 296, 321, 373, 377, 384
Waste burning, 132–133
Wastewater treatment, 177
Water bricks, 103, 113
Water heating, 19–20, 84, 87, 91–92, 97, 101, 110, 112, 144, 225, 277, 308, 322, 341, 343
Water vapor, anthropogenic, 6–7, 9, 14, 145, 147
Watt, James, 141
Wave power, 24
  Devices, 21, 25, 224, 322, 326–327, 390
Weapons proliferation. *See* Nuclear power, Weapons proliferation
Weather forecasting, 315–316

Weatherization, 20, 108–109, 282, 327, 342–343
Weeknd, 374
Weibull distribution, 258
Weihl, Bill, 369
West Nile virus, 289
West Virginia, 42
Westinghouse, George, 219–221
Wetlands, 134
White House visit, 378
White roofs. *See* Geoengineering, White roofs
Whitmer, Gretchen, 386
Wildfires, xiv, 13, 21, 132, 289, 298, 335–336
  Effects on electricity prices, 288
William I, Emperor, 216
Wind
  Energy penetration, 296
  Offshore, xvi, 15, 21, 23–25, 146, 148, 224, 241, 248–249, 259–263, 265–266, 268–270, 276, 300, 302, 322, 335, 361–362, 371, 390
  Onshore, xvi, 15, 21, 146–147, 177, 189, 194–195, 246, 249, 261–263, 265, 276, 283, 302, 322–323, 326, 335, 362, 371, 390
Wind farm
  Altamont Pass, 248
  Array losses, 262–263, 265
  Block Island, 249
  Coastal Virginia, 249
  Crotched Mountain, 248
  Footprint area. *See* Footprint area, Wind farm
  Hybrid with PV, 189
  Installed power density, 264–265
  Offshore, 23, 249, 261, 326
  Output power density, 264–265
  Revolution, 249
  San Gorgonio Pass, 248
  South Fork, 249
  Spacing area. *See* Spacing area, Wind farm
  Tehachapi Pass, 248
  Vineyard Wind 1, 249
Wind power potential, saturation, 266
  Land, 266
  World, 266

INDEX

Wind resources, 265, 320
Wind speed, 23–24, 246, 250–252, 254–255, 257–260, 267, 269, 282, 320, 356, 359
  Correlation with heat demand, 284, 302
  Cut-in, 258
  Cut-out, 259
  Destruction, 259, 268, 270
  Impacts of air pollution on, 300
  Impacts of global warming on, 300
  Probability distribution, 257
  Rated, 259–260
Wind turbine, 15–16, 21–24, 29–30, 47, 71, 77, 108, 111, 148–150, 165, 172, 174, 182–183, 185, 224, 246, 248–272, 276, 279, 281, 283, 293, 300, 318–319, 322, 326–327, 335, 341, 353, 359, 390–391
  Backyard, 341, 343
  Blades, 251
    Diameter, 260
  Braking system, 251
  Capacity factor. *See* Capacity factor, Wind turbine
  Controller, 251
  Darrieus, 251
  Destruction wind speed. *See* Wind speed, Destruction
  Direct drive, 22
  Direct drive (gearless), 22, 252
  Downtime losses, 261–262
  Floating, 24, 30
  Foundations, 23, 126
  Gearbox. *See* Gearbox
  Geared, 22, 252
  Generator. *See* Generator
  Heat exchanger, 252
  High-altitude, 24
  Horizontal-axis, 246, 250–251
  Hub height, 22–23, 47, 250, 258, 260, 264, 266, 320
  Impacts on
    Bats. *See* Bats, wind turbine impacts on
    Birds. *See* Birds, wind turbine impacts on
    Climate, 267, 300
    Hurricanes, 268–270
  Nacelle. *See* Nacelle
  Nameplate capacity, 260
    Worldwide, 249
  Offshore
    Fixed-bottom, 248–249, 259
    Floating, 249
  Painted, 272
  Parts, 251
  Pitch system, 251–252, 254–255
  Power curve, 258
  Power output, 257–260, 262
  Rotor, 251, 256
  Shaft, 251, 255
  Three-bladed, 250
  Tip height, 264
  Tower, 251
  Two-bladed, 250
  Typhoon-class, 259, 268
  Vertical-axis, 246, 251
  Yaw drive, 251–252, 261
  Yaw motor, 251
Wind vane, 247, 251–252
Windmill, 246–247, 250
  Daniel Halladay's, 247
  Heron of Alexandria's, 246
Window awning, 104, 108
Window blind, 104, 108
Window film, 104
Windows
  Sealing cracks, 342
  Smart glass, 341
  Triple-glazed, 108–109, 341
Wisconsin, 385–386
Wright, Ian, 58
Wright Electric, 69
WWS (wind, water, solar), 17
  All-electric home. *See* Home, all-electric
  Avoided climate cost, 290
  Avoided health cost, 289
  Cost, 286
  Generator mix, 307
  Impacts of climate change on, 298
  Policies, 331. *See also* Policies
  Reduced energy needs due to, 290
  Reduced private cost due to, 290
  Reduced social cost due to, 290
  Reduced solar cost due to, 331

Resources, 24, 139, 284, 292, 308, 313, 320, 323, 337
Roadmaps. *See* Roadmaps, 100 percent WWS
Solutions for renters. *See* Rentals, WWS solutions for
Technologies, xv, 14, 19, 362, 390

Transition in impoverished countries, 338
Transmission timeline. *See* Transition timeline
Wyoming, 296

ZeroAvia, 82
Zipernowsky, Karoly, 218

For EU product safety concerns, contact us at Calle de José Abascal, 56–1°, 28003 Madrid, Spain or eugpsr@cambridge.org.

www.ingramcontent.com/pod-product-compliance
Ingram Content Group UK Ltd.
Pitfield, Milton Keynes, MK11 3LW, UK
UKHW020025241225
466378UK00007B/42